Fire Serv[ice]
Rapid Interven[tion]
Crews

Principles and Practice

Joe Nedder

JONES & BARTLETT
LEARNING

Jones & Bartlett Learning
World Headquarters
5 Wall Street
Burlington, MA 01803
978-443-5000
info@jblearning.com
www.jblearning.com

Jones & Bartlett Learning books and products are available through most bookstores and online booksellers. To contact Jones & Bartlett Learning directly, call 800-832-0034, fax 978-443-8000, or visit our website, www.jblearning.com.

Production Credits
Chief Executive Officer: Ty Field
President: James Homer
Chief Product Officer: Eduardo Moura
Executive Publisher: Kimberly Brophy
Vice President of Sales, Public Safety Group: Matthew Maniscalco
Director of Sales, Public Safety Group: Patricia Einstein
Executive Acquisitions Editor: William Larkin
Editor: Alison Lozeau

Production Editor: Cindie Bryan
Senior Marketing Manager: Brian Rooney
VP, Manufacturing and Inventory Control: Therese Connell
Composition: diacriTech
Cover Design: Kristin E. Parker
Director of Photo Research and Permissions: Amy Wrynn
Cover Image: © Joe Nedder/Jones & Bartlett Learning.
    Flames: Tommy Maenhout/Dreamstime.com
Printing and Binding: Courier Companies
Cover Printing: Courier Companies

Library of Congress Cataloging-in-Publication Data
Nedder, Joe.
Fire service rapid intervention crews: principles and practice / Joe Nedder.
    pages cm
Includes bibliographical references and index.
    ISBN 978-1-4496-0976-4 (paperback)—ISBN 1-4496-0976-7 (paperback)
1. Fire extinction—Safety measures. 2. Fire extinction—Accidents. 3. Fire fighters—Health and hygiene. 4. Rescue work.
5. Emergency management. 6. Command and control at fires. I. Title.
    TH9182.N43 2014
    628.9'2—dc23
                                    2014002814
6048

Printed in the United States of America
18 17 16 15 14   10 9 8 7 6 5 4 3 2 1

Brief Contents

Contents

Skill Drills

© Photos.com

Instructor, Student, and Technology Resources

Instructor Resources

■ Instructor's ToolKit CD

Preparing for class is easy with the resources on this CD, including:

- **PowerPoint Presentations** Provides you with a powerful way to make presentations that are educational and engaging to your students. These slides can be modified and edited to meet your needs.
- **Lesson Plans** Provides you with complete, ready-to-use lesson plans that include all of the topics covered in the text. Offered in Word documents, the lesson plans can be modified and customized to fit your course.
- **Test Bank** Contains multiple-choice questions, and allows you to create tailor-made classroom tests and quizzes quickly and easily by selecting, editing, organizing, and printing a test along with an answer key, including page references to the text.
- **Image and Table Bank** Provides you with a selection of the most important images and tables found in the textbook. You can use them to incorporate more images into the PowerPoint presentations, make handouts, or enlarge a specific image for further discussion.

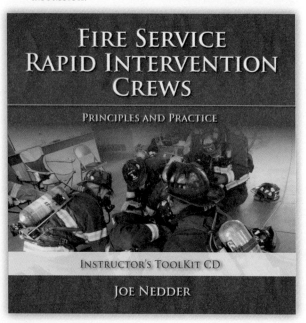

Technology Resources

■ Navigate Course Manager

Combining our robust teaching and learning materials with an intuitive and customizable learning platform, Navigate Course Manager gives you the tools to build a solid, knowledgeable foundation with world-class content. With Navigate Course Manager, learning is no longer confined to the four walls of the classroom. Now you can learn anytime and anywhere, when it is ideal for you.

World-class content joins instructionally sound design in a user-friendly online interface to give students a truly interactive, engaging learning experience. Course Management Tools simplify the management and delivery of curriculum and assessments to students enabling anytime, anywhere access to learning. Instructors can track real-time progress, manage assignments, and view results in the grade book.

Acknowledgments

As with any project of this size, there are so many people to thank. I would like to acknowledge some who helped me through this endeavor these past 4 years. Fire fighter Jamie Desautels, Northborough Massachusetts Fire Department, assisted me in writing all of the chapter opening Case Studies and Chapter Wrap-Ups, and helped me to move forward whenever I hit a block while writing—his help was invaluable. Assistant Chief Jim Peltier, Berlin Massachusetts Fire Department, assisted me with many of the skill drills by helping me take something complex and convert it into words that were easy to understand. He also assisted me with his technical expertise during all of the photography. Fire fighter Ken Franks, Southborough Massachusetts Fire Department, assisted me with all matters regarding technical rope skills, and reviewed everything I wrote regarding ropes and rescue, to make sure it was accurate. He also assisted with his technical expertise for much of the photography. Lieutenant Will Trezek, Chicago Illinois Fire Department was invaluable in assisting me with the section on RASP. Will was instrumental in the development of the RASP system for the Chicago Fire Department and his expertise in helping with this book is greatly appreciated. Thanks to Fire fighter Tom Hogan, Southborough Massachusetts Fire Department, for his help creating the construction drawings of the Multi-Purpose Prop found in Appendix A. Thanks to Deputy Chief John Sullivan, Worcester Massachusetts Fire Department, for his knowledge, expertise, and thoughts when I would turn to him for help. Thanks to Chief David Durgan, Northborough Massachusetts Fire Department, Chief Bruce Ricard, Berlin Massachusetts Fire Department, Chief Joe Mauro, Southborough Massachusetts Fire Department, and Battalion Chief Jim Vieira, Bristol Rhode Island Fire Department, for the use of their facilities for numerous photography sessions. Without their help we would have had a difficult time producing the photos needed for this text. Thanks also to the manufacturers who were a great help to me with equipment: David Flynn from MSA Safety; Brandon Millan from Scott Safety Products; Sam Morton and Matt Hunt from Sterling Rope; John Weaver from Lions; Scott Gohl from Dragon Fire Gloves; and Chris Moran from Lonestar Axe LLC. These are people who really do care about us as fire fighters and not just as customers. In addition, I would like to thank from Jones & Bartlett Learning, my editor, Alison Lozeau, who put up with all my craziness and helped to make this book what it is, and my production editor, Cindie Bryan, for her assistance along with Alison at two full days of photo shoots and for all she does to produce books such as this. A special acknowledgment for my wife Janet, who has endured this ordeal for almost 4 years, listening to me express my frustrations throughout—thank you for your support, encouragement, and most of all patience!

Joe Nedder
March 2014

Reviewers and Contributors

Jones & Bartlett Learning and the author would like to thank the reviewers and contributors of *Fire Service Rapid Intervention Crews: Principles and Practice*.

Christopher F. Best
NC Office of State Fire Marshal (*Retired*)
Rougemont, North Carolina

Robert M. Cembor
Massachusetts Fire Academy
Warwick Fire Department
Warwick, Rhode Island

Randall J. Childs
Jacksonville Fire Department
Jacksonville, Alabama

Captain Aaron J. Collette
Burlington Fire Department
Burlington, Vermont

Lieutenant Christopher Corbin
South Burlington Fire Department
South Burlington, Vermont

Chief Jeff Degeal
Ashland Fire Department
Ashland, Illinois

Jamie Desautels
Northborough Fire Department
Northborough, Massachusetts

Cliff Dyson
Alabama Fire College
Tuscaloosa, Alabama

Kenneth W. Franks III
Southborough Fire Department
Southborough, Massachusetts

Deputy Fire Chief Neil Fulton
Norwich Fire Department
Norwich, Vermont

Chief Kevin Gallagher
Acushnet Fire Department
Acushnet, Massachusetts

Alex Gerarve
Hammond Fire Department
Hammond, Louisiana

Todd A. Homan
Pennsylvania State Fire Academy
Lewistown, Pennsylvania

Justin Howard
New Albany Fire Department
New Albany, Mississippi

Jacob P. Johnsrud
Manitowoc Fire Department
Manitowoc, Wisconsin

Captain Ken LeBelle
NRI Fire Photos
Lincoln Fire Department
Lincoln, Rhode Island

Chad Landis
LSU Fire and Emergency Training
Baton Rouge, Louisiana

Captain James D. Lea
Training Officer, Spartanburg Fire
Department
Spartanburg, South Carolina

Lieutenant Darrick Lundeen
Fairfield Fire Department
Fairfield, Connecticut

Captain Daniel Manning, MS
Professor Pima Community College
Tucson, Arizona
Fort Huachuca Fire Department
Fort Huachuca, Arizona

Captain Toby Martin
Roanoke County Fire Rescue
Roanoke, Virginia

Jim Nelson
Nicolet Area Tech College
Rhinelander, Wisconsin

Kevin O'Brien
Burlington Fire Department
Burlington, Vermont

Assistant Chief James Peltier
Berlin Fire Department
Berlin, Massachusetts

Michael D. Penders
Guilford Fire Department
Guilford, Connecticut

Jeff Pricher
Scappoose Rural Fire District
Scappoose, Oregon

Bill Sampson
Connecticut Fire Academy
Wilton Fire Department
Wilton, Connecticut

Jeff Seaton
South Bay Regional Public Safety Training
Consortium
San Jose, California

Captain Joshua J. Smith
Fire Service Training Coordinator, Mitchell
Community College
Statesville Fire Department
Statesville, North Carolina

Nathan Smitherman
Alabaster Fire Department
Alabaster, Alabama

Deputy Fire Chief William P. Sullivan
Massachusetts Fire Academy
Malden Fire Department
Malden, Massachusetts

Glenn Sutphin
Bluefield Fire Department
Bluefield, West Virginia

John Tommaney
Southborough Fire Department
Southborough, Massachusetts

Lieutenant Will Trezek
Chicago Fire Department
Chicago, Illinois

Lieutenant William Tuttle
Fairfield Fire Training Center
Fairfield, Connecticut

Assistant Fire Chief Robert S. Wilson
Sioux City Fire Rescue
Sioux City, Iowa

Photo Shoot Acknowledgments

We would like to thank the following people and organizations for their collaboration on the photo shoots for this project. Their assistance was greatly appreciated.

Ashland Fire Department, Massachusetts
Doug Dow
Lieutenant Lyn Moraghan

Bellingham Fire Department, Massachusetts
Jason Bangma
Brad Kwatcher
Mark Lister
Gregory Prew
Robert Provost III

Berlin Fire Department, Massachusetts
Brendon Gilchrist
Brandon MacNeill
Travis Pacific
John Pauline

Assistant Chief Jim Peltier
Tom Steele

Cheshire Fire Department, Massachusetts
1st Assistant Chief Robert F. Lamb

Cross St. Associates, Uxbridge, Massachusetts
Catherine Nedder

Great Barrington Fire Department, Massachusetts
Ryan Brown

Northborough Fire Department, Massachusetts
Jamie Desautels
James Foley
Douglas Pulsifer
Michael Serapligia

Northbridge Fire Department, Massachusetts
Lieutenant Peter Cavalieri
Richard Latour, Jr

Southborough Fire Department, Massachusetts
Kenneth W. Franks III
Thomas Hogan
Danny Martins
Kenneth Strong
John Tommaney
David Wills

Uxbridge Fire Department, Massachusetts
Zachary Holzman
Jason Marchand
Lieutenant Richard Nedder

Dedication

This book is dedicated to Assistant Chief Jack Peltier, *Retired* (1941 to 2012), and Lieutenant Pat Lynch, *Retired*, both of whom were instrumental in the writing of this book.

Jack Peltier was my teacher, mentor, confidant, fire service Dad, and very close friend. As my teacher, he inspired me and lit a burning passion in me, and so many other fire fighters, to be the best fire fighter I could be and to share my knowledge with others willing to learn. He taught me to thirst for knowledge, never be complacent, and to accept change as a constant occurrence in my fire service life. As my mentor, he made sure I realized the importance of ongoing training and to never stop learning and training. It was Jack who first told me that I needed to share my passion and knowledge for rapid intervention with the fire service by writing a book, and that even as a "small town" fire fighter, I had a lot to contribute and could make a difference.

Jack's career began in Marlborough, Massachusetts, when he joined the fire department in 1955 as an on-call fire fighter and was appointed career in 1965. He retired from the Marlborough department as Assistant Chief in 1990 and went on to join the Berlin Massachusetts Fire Department as a "call man," fulfilling the role of Incident Safety Officer, continuing to serve until his unexpected death. He became an Instructor for the Massachusetts Firefighting Academy in 1974 and was still teaching when he passed away in December of 2013. He was a member of the International Society of Fire Service Instructors (ISFSI) and a past President. Jack was also an active figure in the Congressional Fire Service Caucus, where he was an advocate for fire fighter safety training. He attended, taught, and worked for the Fire Department Instructors Conference (FDIC) for 40 years, first attending in 1972. He was well-known nationally as an advocate for fire fighter safety and training, and was also well-known and respected by equipment and apparatus manufacturers throughout the U.S. One of the biggest disappointments in my life is that, in trying to keep this dedication a surprise, I never got to tell him about it face-to-face, but I know he knows, and continues to watch over all of us fire fighters as he always did. We will never forget you Jack!

I first met Lieutenant Pat Lynch of the Chicago Illinois Fire Department in 2000, while attending a Rapid Intervention training class being conducted in Massachusetts. From the beginning, I saw that Pat was dedicated to fire fighter safety and survival, and he taught me to be an RIC Instructor that could make a difference. Pat had two statements he always liked to make. The first was to always be proactive, which is sound wisdom and a good training point that is repeated often in this text. The second statement was that he was a "student of the game," in reference to athletes who are always training and improving their skills. Pat believed that fire fighters need to regard themselves as students, because there is always something new to learn or improve upon, which is what makes great fire fighters. Pat taught me many skills and techniques and how to convey my message clearly during RIC training. From Pat I learned that regardless of the size job you are on, fire is fire, and the dangers are still the same. As fire fighters, we have to dedicate ourselves to train hard and to always be looking to improve.

Pat joined the Chicago Fire Department in 1975 and retired in 2012. During his tenure, he served on Snorkel Squad 1, Squad 5, was promoted to Lieutenant and assigned to Squad 1, finishing his career on Squad 2. Even with these credentials, Pat always emphasized the importance of executing the basics to the fullest, whether rapid intervention or Fire Fighter I/II, because they are our foundation—true words that can and do make a difference every day in the fire service.

Preface

When asked why I am so passionate about rapid intervention and survival skills training, I can't think of a better reason than the photo enclosed. Our families, loved ones, and those we care about want us to return safe and sound from every call. The young men in this photo are Steven and Shawn Peltier, and although they not blood relatives, I care about them very much. As so many of us once dreamed when we were young, these two brothers want to be fire fighters when they grow up. They idolize fire fighters, attend training sessions with their Dad, and actually read the fire service trade publications. We all know that our youth look to us for direction and guidance, and what better way to provide direction and guidance than to show, by our actions, the importance of rapid intervention and why it is a skill set that needs constant refreshing in order to always be prepared to rescue one of our own so they can return to their family.

Rapid intervention has been an important part of my fire fighter skill set since April 2000, when I first met Rick Kolomay from the Shaumburg Illinois Fire Department and Bob Hoff from the Chicago Illinois Fire Department. They had been hired by the Commonwealth of Massachusetts Firefighting Academy to come and run numerous training programs on Rapid Intervention and Fire Fighter Survival Skills. With Rick and Bob were two additional instructors, Lieutenants Will Trezek and Pat Lynch, both of the Chicago Illinois Fire Department. After all these years, we have maintained our friendships and our sincere dedication to rapid intervention and saving

fellow fire fighters. The lessons, skills, and techniques taught were eye opening, and when I was given the opportunity as a State Instructor to teach the Massachusetts Firefighting Academy's new Rapid Intervention Crew Program, I jumped at it! Since then, this has been my personal passion and goal, and in the course of these 14 years, I can proudly state that I have helped thousands of fire fighters learn rapid intervention and survival skills. With this book, it is my hope that the message and skills spread, and that all fire fighters and fire service leaders understand and support the need to have a Rapid Intervention Crew *on every fireground.*

About the Author

Joe Nedder joined the fire service as a volunteer in 1977. He was an active member as both a volunteer and later as an on-call fire fighter with different fire departments for over 36 years, retiring from the Uxbridge Massachusetts Fire Department in October of 2013. During his career, he held different positions including Lieutenant, Captain, and Training Officer. Joe started assisting with training in the mid-1980s and, in 1991, became a Certified Fire Service Instructor. He worked as an Instructor for the Massachusetts Firefighting Academy of over 16 years, where he was an active member of the Rapid Intervention and Fire Fighter Survival training team. In 2003, he founded his own fire service training

company, Cross St. Associates, in answer to the frequent calls he was receiving to help train local fire departments. Since then, his company has grown, and he has conducted training throughout the Northeast. Joe has contributed articles to *Fire Engineering Magazine* and has taught several classes at the Fire Department Instructor Conference since 2010. Although retired from active fire fighter duty, he continues to be active in training. He is known for his passion when training and for his commitment to all fire fighters—regardless of the size of their departments—to share, teach, and do all he can to provide them with knowledge and skills.

Introduction

From the beginning, the purpose of this book has been to bring awareness of the need for rapid intervention, and therefore, this book is devoted to basic rapid intervention skills. I have always believed that if taught the basics, fire fighters will continue to seek out more advanced training as they move forward; but they have to start with the basics. In the process of writing of this book, numerous people have suggested it include more advanced skills (e.g., dealing with collapse), however I have chosen not to for two reasons. The first is that the focus for training should be on basic RIC skills, as they are fundamental to more advanced skills and need to be taught and learned before they can be built upon. The second reason is that many of the more advanced skills are very technical in nature and will most likely not involve all fire fighters. Fire fighters should, therefore, focus on the basics, practice to retain those skills, and then look for the next horizon of skill and training.

RIC training should be three things: (1) pertinent, (2) as realistic as possible, and (3) safe. All the skills described here have an explanation of their purpose and pertinence. To replicate realistic scenarios, note that all the fire fighters executing the skills in this book are fully geared up and "on air," which is an important point for training. RIC skills are hard and demanding, and need to be taught and practiced in the manner they would be executed (i.e., "on air"). For safety in learning, you will also see that belay lines are used whenever a rescue fire fighter or "downed fire fighter" was in an elevated emergency scenario. Obviously in a real-life rescue situation, safety belay lines would not be used, but in training, they should always be used for the safety of all.

Whether you are on a large urban department or a small rural volunteer department, the information and skills in this book apply. Whereas larger urban departments have the ability to have many experienced fire fighters on scene from the first due Company and the ability to get additional manpower on scene quickly with an RIC activation, smaller rural departments often lack the manpower for the duration of the incident. It is these types of situations where rapid intervention needs to be looked at in a broader way. Because it is important that all firegrounds have a qualified RIC standing by, one way to accomplish this is to promote the idea that all fire fighters should regionally learn the same RIC skills. This promotes effective mutual aid, since RICs from different departments will then all have the same skill set.

Finally, don't be fooled! It has been suggested that all that is needed on a fireground is a two-member initial rapid intervention crew (IRIC). However, if you are in trouble—lost, injured, or in a medical emergency—would you want to have only half of a team standing by for your rescue or have the IC use extra minutes to gather additional RIC members before beginning the RIC activation? We must accept the fact that, as members of the fire service, we have a moral and ethical obligation to each other to be well-trained and ready at a moment's notice to commit to an RIC activation and make every effort to save one of our own. I hope this book provides you with the inspiration, motivation, and skills to do just that.

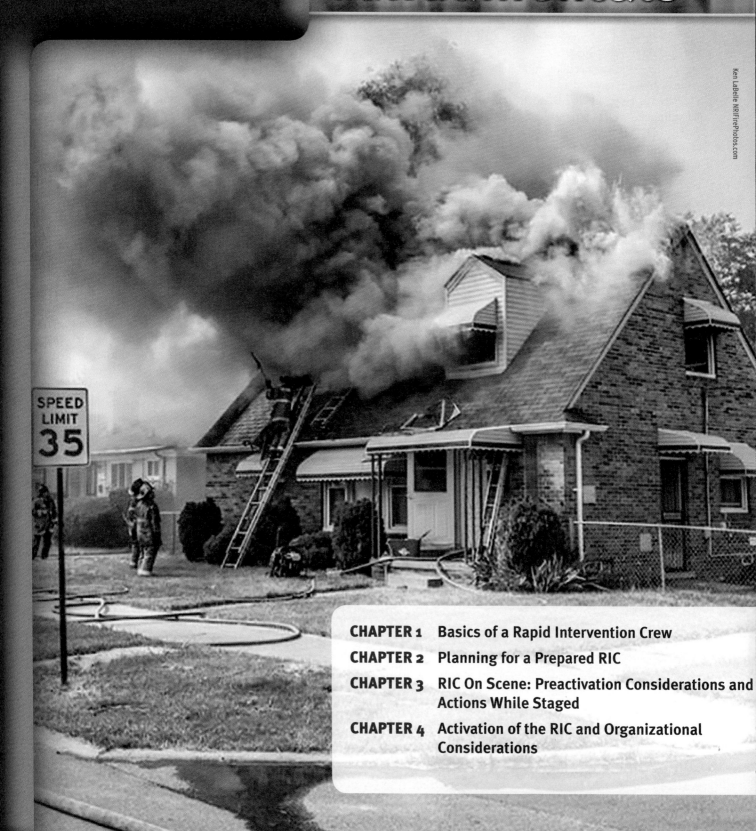

SECTION

I

Fundamentals

Basics of a Rapid Intervention Crew

Knowledge Objectives

After studying this chapter you will be able to:

- Define the term rapid intervention and explain what it is. (NFPA 1407, 4.1.1, p 4)
- Define the term rapid intervention crew (RIC) and explain criteria that must be met in order for RICs to be effective. (NFPA 1407, 4.1.1, 4.2.4, pp 4, 7, 8)
- List reasons why fire fighters may resist the concept of rapid intervention. (p 5)
- List the consequences of a fire fighter's death. (p 6)
- Describe obstacles within the fire service culture that need to be overcome in order to protect fire fighter personnel. (p 7)

Skills Objectives

- There are no skill objectives for Rapid Intervention Crew candidates. NFPA 1407 contains no Rapid Intervention Crew Job Performance Requirements for this chapter.

Additional NFPA Standards

- NFPA 1001, *Standard for Fire Fighter Professional Qualifications*
- NFPA 1021, *Standard for Fire Officer Professional Qualifications*
- NFPA 1500, *Standard on Fire Department Occupational Safety and Health Program*

You Are the Rapid Intervention Crew Member

Your department has responded to a working structure fire with heavy fire showing on arrival. You are on the company assigned as the rapid intervention crew (RIC). All companies are working aggressively and things seem to be going well when over the radio the words "MAYDAY, MAYDAY, MAYDAY" are heard. A fire fighter is lost and running low on air. His MAYDAY is not only a cry for help, but also a cry of fear that he really is in trouble. The Incident Commander activates the RIC to rescue the downed fire fighter.

1. Why is the RIC so important to the life and safety of all fire fighters?
2. Is your crew capable, ready, and equipped to activate and attempt the rescue of the downed fire fighter?
3. How confident are you of your RIC skills training? Do you have a solid foundation in basic fire fighter skills to back up your RIC skills?

Introduction

The term rapid intervention, which first appeared in the 1997 edition of National Fire Protection Association (NFPA) 1500, *Standard on Fire Department Occupational Safety and Health Program*, has been referred to by different acronyms, the most common being rapid intervention team (RIT) and fire fighter assist and search team (FAST). In many ways, this began a national effort to bring forward the skill set of saving distressed fire fighters. Rapid intervention crew (RIC) was included in the 2010 edition NFPA 1407, *Standard for Training Fire Service Rapid Intervention Crews*. The word "crew" is defined as personnel at the scene of an incident who are assembled to perform a specific task, and its use is more in keeping with the Incident Command system as defined by the National Incident Management System (NIMS). Regardless of what your local department might call this important function, we must ensure that it is always properly staffed, in place, and ready to go. Here, the term RIC will be used when talking about the company. The term RIT will also be used in the discussion of tools and equipment in the field, because many equipment terms and labels are already well-established in the fire service. There is no need to rename equipment or skills if the intent is understood (an example of this is the RIC rescue air unit or RIT pack).

Many fire departments have embraced the concept of the RIC, making every effort to commit and train fire fighters to save their own. These departments have taken a proactive approach to fire fighter survival and rescue, shown by automatic assignment of RIC engines and ladders and automatic mutual aid for an RIC in areas where adequate staffing is a concern. However, other departments have not adopted the concept. To properly implement the concept of rapid intervention, RICs must be created and trained in the skills needed. This text is intended to help fire departments of all sizes understand the need for rapid intervention, promote the attitude that will lead to effective rapid intervention, train their members on the basic skills of rapid intervention, understand how rapid intervention crews work, and, most importantly, be prepared if a MAYDAY is called for a downed fire fighter.

What Is a Rapid Intervention Crew?

At a fire scene, fire fighters' responsibilities include rescuing and protecting civilians (life safety), extinguishing the fire (incident stabilization), and preserving as much property as possible (property conservation). But who is there to protect and save fire fighters if a situation goes wrong?

The term rapid intervention is used to describe a team of fire fighters who are trained, prepared, and standing by at the scene of a fire to rescue a fellow fire fighter who is in peril. A fire fighter in peril is one who is:

- Lost, trapped, or injured
- Low or out of air, or who has a self-contained breathing apparatus (SCBA) emergency while in an immediate danger to life and health (IDLH) environment
- Experiencing a medical emergency while operating inside a structure fire (heart attack, loss of consciousness, respiratory difficulties, etc.)

The RIC is a "rescue team" for fire fighters. Their job is to stand by and activate quickly if a MAYDAY is called for a downed fire fighter **FIGURE 1-1**. The team should not be involved in other activities unless those activities are in support of fire fighter safety and possible rescue—examples of this would be assisting in laddering and "softening" the building by breaking locks and other hardened or secured points of emergency entry and egress. All this should be done under the orders of the Incident Commander. Whatever support activity the RIC might be involved in, the activity must be one that does not require them to go on air, and they must be able to stop and deploy immediately. Most importantly, the RIC must be equipped and ready to go at all times. If RICs are not present for whatever reason, there is no designated resource available to immediately protect fellow fire fighters who are in peril.

FIGURE 1-1 An RIC needs to be equipped and on active standby at all times.

© Joe Nedder/Jones & Bartlett Learning

Crew Notes

TRAINING = SKILLS
SKILLS = COMPETENCE
COMPETENCE = LIFE SAFETY
LIFE SAFETY MEANS EVERYONE GOES HOME!

Safety

MAYDAY is an emergency procedure word used internationally as a distress signal in voice procedure radio communications. It derives from the French *Venez m'aider*, meaning "come help me."

Common Reasons Given for Not Staffing a Proper RIC

Fire fighters by their very nature want to be in the middle of the action! To take four fire fighters and tell them to "Stand over there—you're the RIC" goes against everything they have felt and believed in the past. Fire fighters do not want to be assigned to the RIC and will do everything they can to avoid it. This is a lethal attitude because it is crucial to have an RIC in place, ready to be called to immediate action when a fire fighter is down.

Other reasons given for not staffing a proper RIC include:

- We did not need it in the past, so why now?
- We do not have enough staffing.
- If something goes wrong, I can assemble a team quickly.
- My members do not believe in it.

These reasons suggest that it is not important to have a resource dedicated to the immediate rescue of fellow fire fighters. They are excuses! In today's day and age it is common knowledge that fire fighters are exposed to greater dangers than in the past and that they are expected to do their job with significantly fewer personnel. Less staff and greater danger equate to greater risk. Claiming that an RIC is not needed promotes recklessness and a disregard to life and leads to unsafe practices, endangering others. Let's look at the excuses offered and examine what can be done to overcome them.

Crew Notes

Just because something did not exist in the past does not mean it was not needed, or that it is not needed now.

■ We Did Not Need It in the Past, So Why Now?

In the past, we did not need a lot of things! Full personal protective equipment (PPE), SCBAs, motorized fire apparatus, and aluminum ladders were among the items considered unnecessary. These changes represented progress that allowed fire fighters to more effectively practice their profession. Like all change, these concepts may not have been immediately embraced by all parties, but eventually became the standard as people realized the need for safer and improved practices.

Like most people, fire fighters have been traditionally afraid of change. When SCBAs were first introduced, they were met with resistance. It required changes in procedure and learning to do things a different way. It was thought they were heavy, cumbersome, and really not needed in typical situations. It took many years before the use of SCBA became automatic and commonplace. Today's fire fighters are equipped with better turnout gear and SCBAs, which allow them to go deeper into the building. Compared to the past, today's fire fighter works in greater danger, in higher temperatures, and surrounded by more toxic gas and smoke. A fire fighter who is lost, trapped, or in medical distress in this type of environment needs help. The RIC is equipped and trained to provide the assistance needed to save one of our own.

■ We Do Not Have Enough Staffing

Staffing has become a major issue in today's fire service. Career departments are faced with budget cuts, layoffs, and reduction in staff. Volunteer departments experience a reduced number of people willing or able to give of their time freely. Lives are complicated, and volunteering as a fire fighter, which is a major time commitment, is not as possible as once before. To overcome this issue, fire departments need to change their run cards. In larger communities, an option is to add an additional company to the initial alarm. In smaller communities, if you do not have the staff, an option is to set up prearranged mutual aid agreements with neighboring communities to send an RIC automatically so that help is there if you call.

■ If Something Goes Wrong, I Can Assemble a Team Quickly

This excuse usually ties back into "not enough staff" or "we did not need it before." When a MAYDAY is transmitted, the downed fire fighter does not have the time for you to assemble a company. Compared to having an RIC immediately ready, how long will it take to assemble a crew? Whether it is 3, 4, or 5 minutes, someone's life depends on a crew responding immediately and quickly. Even if a crew can be assembled rapidly, tools will need to be assembled as well. Forming a team on an as-needed basis can and will lead to failure.

My Members Do Not Believe in It

To not believe in rapid intervention is to believe that fire fighters do not get into trouble, that fire fighters do not call for help, and that it is OK for fire fighters to die on the job even if they could have been saved by more immediate and rapid action. Believing that calling a MAYDAY is the sign of a weak, unworthy fire fighter is another lethal attitude.

Part of the rapid intervention skill set is to have the right attitude—namely, when fire fighters get into trouble, they need help. This attitude says life is sacred and that fire fighters are not expendable. We need to take this positive attitude and couple it with the skills and knowledge that will make us effective members of an RIC.

To create this positive attitude, we need to present the RIC in a way that fire fighters will relate to. We must show:

- Why we really need it
- What it means to us
- Department commitment
- How we will train, implement, and continue to train for rapid intervention

How Do We Present Rapid Intervention to Our Departments?

Whenever a fire fighter dies in the line of duty, the National Institute for Occupational Safety and Health (NIOSH) will investigate the death and issue its findings via a report. The reports are published for all to see and are an opportunity to learn from the past. By reviewing them, we will understand what happened and will learn the recommendations given to help prevent a similar tragedy from happening again. Some example reports to review are:

- NIOSH Report #2005-05: Career Captain Dies After Running Out of Air at a Residential Structure Fire—Michigan
- NIOSH Report #2008-34: Volunteer Fire Fighter Dies While Lost in Residential Structure Fire—Alabama
- NIOSH Report #2010-10: One Career Fire Fighter/Paramedic Dies and a Part-Time Fire Fighter/Paramedic Is Injured When Caught in a Residential Structure Flashover—Illinois
- NIOSH Report #2007-28: A Career Captain and an Engineer Die While Conducting a Primary Search at a Residential Structure Fire—California

These NIOSH reports are just four out of hundreds that are available for review. It is important for us as fire fighters to study

what has happened and look for and implement practices that will help prevent such deaths in the future. When you review a NIOSH line-of-duty death (LODD) report, you will often see the following two items called out as recommendations:

"NIOSH investigators concluded that, to minimize the risk of similar occurrences, fire departments should: (1) ensure that a Rapid Intervention Team is in place before conditions become unsafe; and (2) train fire fighters on actions to take while waiting to be rescued if they become lost or trapped inside a structure."

LODD reports are accurate and to the point, but they are also by their nature "clinical" and must be so. The death of a fire fighter is a tragic and horrible thing. One of our own is gone forever. Let's consider the many effects of the death of a fire fighter.

Death of a Fire Fighter

Sometimes fire fighters do not want to accept that their job is extremely dangerous. We hear a lot about "everyone goes home;" conversely, according to NIOSH, 81 fire fighters died in 2012 in the line of duty. A death in the line of duty has long-lasting effects on those left behind, with the survivors bearing the full impact. Spouses, children, grandchildren, parents, and extended family suffer the loss mentally and financially. Colleagues will also suffer, wondering what went wrong and why they could not save the person lost.

To protect our families, our colleagues, and ourselves we must use rapid intervention, which can have a significant impact in whether a fire fighter lives or dies. Without a qualified and capable RIC available, there is no chance of impact! We, as fire fighters, must expect and demand that an RIC team be present on all firegrounds. Some of the techniques can be transferred to civilian rescue, but the core skills are specific to rescuing a downed fire fighter. We must further accept the fact that in order to staff and activate a capable RIC, we need to have the proper training, skills, and equipment to save fellow fire fighters. The skills and techniques of rapid intervention are designed to save us. Remember: If you are the one calling the MAYDAY, your life depends on the RIC. Rapid intervention might be the difference as to whether you live or die!

Crew Notes

To make rapid intervention successful, total commitment is needed from those involved. Note that buy-in and commitment are two different things. When someone "buys in," they are saying "OK, let's give it a try." While this may be a first step, it is not commitment. Commitment means training, training, and more training. It means having the right attitude, and it means supporting the issue. Commitment on your part will make a difference if a fire fighter is in trouble.

Culture Change

To make the commitment to rapid intervention a reality, we need a culture change within the fire service. We need to accept that:

- Our lives are not expendable.
- We can and do die.
- Basic skills are important to learn and maintain.
- When in trouble, calling a MAYDAY is the right thing to do.
- Rapid intervention is a key position that must be filled at all structure fires.

It is hard to understand how in the 21st century many fire fighters still struggle with these concepts. When we talk about life safety, we need to remember that our lives come first and that a culture change in this direction is necessary to retain fire fighter personnel. With this culture change comes rapid intervention and fire fighter survival skills.

Safety

To make the commitment to rapid intervention a reality, we need a culture change within the fire service. A staffed and capable RIC must be present at all structure fires.

Minimum Basic RIC Requirements

Evaluation of Your Physical Abilities

Members of any RIC must be physically able to perform the assigned duties and functions of the crew. Evaluate yourself— are you in good physical shape? Can you work under stressful and adverse conditions? Do you truly understand how your health and fitness affect air supply? Are you physically able to perform any job that might be required of the RIC? The answers to these questions are important, as your inability to perform might have an adverse effect on the rescue of the downed fire fighter. The job of a fire fighter is dangerous, physical, and demanding. Fire fighters must be in good physical shape to perform their job and duties.

Evaluation of Existing Skills and Knowledge

Before you can call yourself a competent member of an RIC, you will need to honestly evaluate yourself with the following questions:

- What is my level of training with basic skills (Fire Fighter I/II)? Is my foundation solid or does it need improvement?
- How well do I know my SCBA? Am I 100 percent confident that in an emergency situation I know the procedures that might save my life?
- Do I understand the concept and principles of air management?

- Do I understand fire behavior and what the smoke indicates about the fire? Do I realize how important this knowledge is to myself and the other members of the RIC?
- Do I understand the importance of calling a MAYDAY as soon as I get into trouble or do I wait for my low air alarm to sound when I am really in peril?

Each of these questions is pertinent to your level of expertise and abilities as a fire fighter and as an RIC member. We as fire fighters are skilled in what we do. When you add the RIC component to this, skills and abilities increase. RIC members need to be trained to the highest possible level FIGURE 1-2 . If you are the fire fighter in peril, your life depends upon the quality of the RIC. Would you want those saving you to be second best? Or would you want the most skilled, competent team to quickly rescue you? The choice is obvious.

What skills, knowledge, and abilities make a competent RIC? At a minimum, all RIC members must:

- Be trained and competent to the minimum level of Fire Fighter I/II (NFPA 1001, *Standard for Fire Fighter Professional Qualifications,* and NFPA 1407), with ongoing skills enhancement and development throughout their career
- Be physically and mentally capable of performing the job under stress and adverse conditions
- Understand their personal limitations and how they perform while using SCBA
- Understand fire and smoke behavior, building construction for the fire service, communications, and size-up; incident command system (ICS); and crew integrity and accountability
- Have been thoroughly trained in the skills and techniques of rapid intervention
- Understand a risk versus benefit analysis

In addition to these basics, the RIC Company Officer must:

- Be 100 percent committed to rapid intervention
- Be trained and competent to the minimum level of Fire Officer I (NFPA 1021, *Standard for Fire Officer*

FIGURE 1-2 RIC members need to be trained to the highest level possible.

Professional Qualifications), with ongoing skills enhancement and development throughout their career

- Possess strong leadership skills and exhibit a presence of command
- Be able to evaluate the team's capabilities before being committed to a rescue
- Possess strong communication skills
- Have an understanding of the company's role operating under an RIC activation and within the ICS
- Realize that your #1 job is to bring your company out safely

Rapid intervention and survival skills are beyond a fire fighter's basic skills foundation. Departments who train in the RIC discipline must ensure a solid education in the basic skills all fire fighters must possess. It will be upon this foundation that RIC and survival skills will be built and the capabilities of your team members will be judged.

Always remember, if the time comes and you are involved in an RIC activation to answer a MAYDAY, will you have the abilities and skills to make the rescue? A fire fighter in peril and his or her family hopes you do!

Wrap-Up

Chief Concepts

- Fire fighters must embrace the concept of RICs. Without it, we are putting ourselves at a greater risk.
- RICs must have a solid foundation of basic firefighting skills before learning enhanced skills.
- The importance of physical fitness cannot be overstated. Firefighting is already demanding, and rescuing a fire fighter in peril can be even more so, both physically and mentally.
- Fire fighters must understand the importance of calling a MAYDAY the moment they *think* they are in trouble.
- The fire service needs to embrace change. We are not invincible. We must realize that many innovations in the design of our PPE and SCBA have allowed us to be able to enter into much more hostile environments.
- Each of us is responsible for reducing the number of fire fighter fatalities. This can only be accomplished by being safer, training frequently, maintaining good overall health, and embracing the concept of rapid intervention.

Hot Terms

<u>Mutual aid</u> An agreement among emergency responders to lend assistance across jurisdictional boundaries. This may occur because an emergency response exceeds local resources, such as a disaster or a multiple-alarm fire, or it may be a formal standing agreement for cooperative emergency management on a continuing basis, ensuring that resources are dispatched from the nearest fire station regardless of which side of the jurisdictional boundary the incident is on.

<u>Rapid intervention</u> A team of fire fighters who are trained, prepared, and standing by at the scene of a fire to rescue a fellow fire fighter who is in peril.

<u>Rapid intervention crew (RIC)</u> A crew or company that is assigned to stand by at the incident scene, fully dressed, equipped for action, and ready to deploy immediately to rescue lost or trapped fire fighters when assigned to do so by the Incident Commander.

References

National Fire Protection Association (NFPA) 1001, *Standard for Fire Fighter Professional Qualifications*. 2013. http://www.nfpa.org/codes-and-standards/document-information-pages?mode=code&code=1001. Accessed October 11, 2013.

National Fire Protection Association (NFPA) 1021, *Standard for Fire Officer Professional Qualifications*. 2014. http://www.nfpa.org/codes-and-standards/document-information-pages?mode=code&code=1021. Accessed October 11, 2013.

National Fire Protection Association (NFPA) 1407, *Standard for Fire Service Rapid Intervention Crews*. 2015. http://www.nfpa.org/codes-and-standards/document-information-pages?mode=code&code=1407. Accessed February 11, 2014.

National Fire Protection Association (NFPA) 1500, *Standard on Fire Department Occupational Safety and Health Program*. 2013. http://www.nfpa.org/codes-and-standards/document-information-pages?mode=code&code=1500. Accessed October 11, 2013.

National Institute for Occupational Safety and Health (NIOSH) Fire Fighter Fatality Investigation and Prevention Program. 2006. *Death in the line of duty . . . Report #2005–05: Career Captain Dies After Running Out of Air at a Residential Structure Fire—Michigan*. http://www.cdc.gov/niosh/fire/reports/face 200505.html. Accessed December 23, 2013.

National Institute for Occupational Safety and Health (NIOSH) Fire Fighter Fatality Investigation and Prevention Program. 2008. *Death in the line of duty . . . Report #2008–34: Volunteer*

Fire Fighter Dies While Lost in Residential Structure Fire—Alabama. http://www.cdc.gov/niosh/fire/pdfs/face200834.pdf. Accessed December 23, 2013.

National Institute for Occupational Safety and Health (NIOSH) Fire Fighter Fatality Investigation and Prevention Program. 2009. *Death in the line of duty . . . Report #2007–28: A Career Captain and an Engineer Die While Conducting a Primary Search at a Residential Structure Fire—California.* http://www.cdc.gov/niosh/fire/pdfs/face200728.pdf. Accessed December 23, 2013.

National Institute for Occupational Safety and Health (NIOSH) Fire Fighter Fatality Investigation and Prevention Program. 2010. *Death in the line of duty . . . Report #2010-10: One Career Fire Fighter/Paramedic Dies and a Part-Time Fire Fighter/Paramedic Is Injured When Caught in a Residential Structure Flashover—Illinois.* http://www.cdc.gov/niosh/fire/reports/face201010.html. Accessed December 23, 2013.

Peterson, K.K., M.B. Witt, K.B. Morton, M.G. Olmsted, H.E. Amandus, S.L. Proudfoot, and J.T. Wassell. 2010. *Fire Fighter Fatality Investigation and Prevention Program: Findings From a National Evaluation.* RTI Press publication No. RR-0007-1003. Research Triangle Park, NC: RTI International. http://www.rti.org/rtipress. Accessed May 22, 2013.

U.S. Fire Administration (USFA) and Federal Emergency Management Agency (FEMA). 2013. U.S. Fire Administration Releases Annual Report on Firefighter Fatalities. Learning-Academic Materials and Programs (LAMP-POST) for U.S. Fire Administration. http://www.usfa.fema.gov/fireservice/firefighter_health_safety/firefighter-fatalities/reports/. Accessed October 10, 2013.

RAPID INTERVENTION CREW MEMBER
in action

You are the Training Officer of your department. The Department has never embraced the concept of RIC and decides to develop a program and start training the members. There is much opposition from older members of the department who do not want to buy into the concept. They feel it is a waste of time and effort, and cite the fact that they have never needed an RIC before. Their opposition causes younger members of the department to also not support the concept of an RIC. Many members refuse to attend training and refuse to believe in the importance of why the department needs rapid intervention training. As the training officer it is your responsibility to educate and train your department. It is also your responsibility to convince the members of the importance of the RIC concept.

1. How could you as the Training Officer get the fire fighters to buy into the concept of RIC?
    A. Tell them they have no choice, it is an order from the Chief.
    B. Discuss incidents that have occurred where RICs have made a difference and saved a fire fighter's life. Talk about how the outcome would have been tragic if not for a dedicated, well-trained RIC.
    C. Offer more pay to members wishing to participate.
    D. Threaten to punish members who do not accept the concept.

2. How can you convince fire fighters that an RIC staging outside a structure fire is the most important job on the fireground?

    A. Tell them it is the most important job on the fireground.
    B. Educate the fire fighters about the importance of being proactive and being in an active standby mode to perform important tasks that could help a fire fighter in peril escape.
    C. Discuss how important it is to have the additional size-up and recon of the RIC team and to keep the IC up-to-date on building and fire conditions.
    D. All of the above.

3. How could you prove the importance of an RIC to your department?
    A. Tell them that many other departments do the same.

B. You do not need to prove anything. You are the Training Officer, therefore it will just happen.

C. Make them think about their families and how they would feel.

D. Utilize research from NIOSH on fire fighter fatalities. Discuss how most LODD reports usually have common themes relative to establishing an RIC.

4. If someone says that there are not enough people to staff an RIC, how would you respond?

A. Develop a mutual aid agreement that provides for an RIC and train with that department.

B. Advise the members it is of no concern to them and will be handled by the Chief.

C. Hire more fire fighters.

D. Tell them that a dedicated crew will be assigned at every fire and that the number of available personnel for firefighting will be reduced.

5. How do you respond to the claim that a team could be assembled quickly if something were to go wrong?

A. Try it! Conduct a drill in which the members of the department are working hard and then have to stop suddenly, grab the appropriate equipment, and reenter. Do the rescuers have the appropriate air supply to conduct a search after working? Time the drill. The simulated downed fire fighter should have a minimum amount of air left in his cylinder. Then conduct the same drill with a dedicated crew properly geared up and equipped. Compare the results.

B. Tell the members of your department that they do not have what it takes to perform a rescue after working in a fire environment.

C. Research articles that have been written about the subject. Ask your members to read them and decide for themselves if they think they can assemble quickly enough.

D. Tell them they have no choice.

Planning for a Prepared RIC

© Photos.com

Knowledge Objectives

After studying this chapter you will be able to:

- State the four key attributes of a great rapid intervention crew (RIC) instructor. (p 12)
- List minimum basic requirements for RIC students. (NFPA 1407, 4.2.4, 4.3.1, 6.1, 6.1.1, 6.3 , p 12)
- List minimum basic requirements for RIC officers. (NFPA 1407, 6.2.1, 6.2.2, 6.3 , p 12–13)
- Understand the responsibilities and requirements for an RIC instructor. (NFPA 1407, 5.1, 5.1.1, 5.1.2, 5.1.3, 5.2, 5.2.1, 5.2.2, 5.2.3, 5.2.4, 7.3 , p 12–14)
- Understand the eight key RIC elements that fire fighters need to understand in order for an RIC training program to be successful. (p 12)
- Understand that it is important to teach the "whys" along with the "hows." (p 12)
- Understand the importance of safety when conducting RIC training. (p 13)
- Understand the importance of using a belay system when conducting RIC training and for what types of RIC skills it should be used. (NFPA 1407, 7.10.1, 7.10.2(1), 7.10.2(2) , p 13)
- Understand the importance of wearing your personal protective equipment (PPE) properly. (p 14)
- Understand the importance of training on all tools and equipment the RIC might use. (NFPA 1407, 7.14, 7.14.1, 17.14.2 , p 14–26)
- Explain what personal tools a fire fighter should carry, how to carry them, and why they are carried. (NFPA 1407, 7.14, 7.14.1, 7.14.2 , p 15–22)
- Differentiate between basic RIC company tools and situation-specific tools. (NFPA 1407, 7.14, 7.14.1, 7.14.2 , p 14, 22–26)
- Understand the three types of ropes used in rapid intervention and their applications. (NFPA 1407, 7.10(1) , p 25)
- Understand what a mechanical advantage is and its applications in rapid intervention. (p 27)

Skills Objectives

After studying this chapter, and with extensive practice and drilling, you will be able to perform the following skills:

- Tie a:
  - Water knot (NFPA 1407, 7.10(3) , p 15–16)
  - Handcuff knot (NFPA 1407, 7.10(3) , p 27–28)
  - Tie knots for RIC rescue. (NFPA 1407, 7.10(3) , p 26–27)
- Rig a 2-to-1 mechanical advantage system (NFPA 1407, 7.10(2) , p 27)

Additional NFPA Standards

- NFPA 1001, *Standard for Fire Fighter Professional Qualifications*
- NFPA 1021, *Standard for Fire Officer Professional Qualifications*
- NFPA 1981, *Standard on Open-Circuit Self-Contained Breathing Apparatus (SCBA) for Emergency Services*
- NFPA 1983, *Standard on Life Safety Rope and Equipment for Emergency Services*

You Are the Rapid Intervention Crew Member

As the Training Officer, you have been tasked with developing a comprehensive RIC and Survival training program for your department. The objective is to train all members in these skills. In order to do this, you must select other fire fighters to assist you in the training.

1. What qualities will you look for in potential instructors?
2. How will you make the training meaningful to your department's members?

Introduction

The key to any great crew of fire fighters is to have the right tools, equipment, and training to be prepared! With rapid intervention crews (RICs), this is not only key; it is imperative. The members of an RIC must be properly trained, have the right tools and equipment, and know how to use them. In order for a fire department to staff a qualified and capable RIC team, a comprehensive training program needs to be developed to ensure that all fire fighters have been properly trained and that the training continues with skills refreshers and updates.

National Fire Protection Association (NFPA) 1001, *Standard for Fire Fighter Professional Qualifications*, sets a standard for basic fire fighter training, *Fire Fighter I/II*. It is the *minimum* standard. All fire fighters who participate in RIC training should be trained to this minimum training level. It is imperative that all fire fighters who receive RIC training possess the knowledge and the skills to meet NFPA 1001, and that they have the physical capabilities to perform the basic skills along with the ability to actively participate in RIC training.

Fire Officers who will function as crew leaders need to meet the same minimum standards as all fire fighters and should meet the requirements of NFPA 1021, *Standard for Fire Officer Professional Qualifications*, as well. They must also be physically capable of performing all the required tasks and skills. Fire departments need to be proactive and include rapid intervention skills training in basic fire fighter training, and NFPA 1407, *Standard for Fire Service Rapid Intervention Crews*, provides the guidelines from which to organize and conduct RIC training.

Conducting Basic Rapid Intervention Training

First off, note that this is *basic* rapid intervention training; this text is intended to train all fire fighters in the basics of RIC, which will ensure a solid foundation from which to build. There are many more advanced techniques, skills, and ideas being taught and brought forward, and all fire fighters are encouraged to strive to better themselves and their skills by continually learning new methods. Fire fighters with basic training will learn that as rescues or recoveries become more technical in nature, the need for specialists comes into play. Some examples of more technical rescues might be building collapse, interior collapse where the fire fighter is buried or trapped under structural elements, or a rescue that involves a hazardous materials (HazMat) situation.

Training in this discipline needs to be meaningful, so that the student can relate the rescue skill being taught to a realistic scenario in which the skill would be used. For example, teaching the basic one-rescuer fire fighter drag is an important skill. What really makes it important, however, is the fact that you might be in a situation where you have to begin to move your partner to safety right away. When there is no time to wait for help, even though the RIC is activated and has entered the building, knowing this skill will have an immediate impact on your partner's survival.

Crew Notes

> Remember, teaching the "hows" does not build understanding unless you also teach the "whys."

Whenever RIC training is conducted, some basic guidelines need to be followed. Teaching the skills is only part of RIC training. For a rapid intervention and survival program to be successful, the eight key RIC elements that fire fighters need to understand also need to be addressed. These eight elements are:

1. Why we really need RIC and how your skills could save a fellow fire fighter
2. Risk assessment and management of the crews
3. How the crews will be activated and managed
4. Communications procedures
5. Incident accountability
6. Roles and responsibilities of all those involved in the RIC operation
7. Rescue versus recovery
8. The impact an RIC operation might have on your fire department

■ RIC Instructors

The four key attributes of a great RIC instructor are being passionate about rapid intervention, having physically participated in prior RIC training, being physically able to perform the skills and drills, and understanding how to teach and train fire fighters. For a successful RIC program, an RIC instructor should be qualified to deliver the RIC training and be responsible for ensuring that the training meets the NFPA 1407 standard. The instructor/student ratio should not exceed 5-to-1 for hands-on training, and instructors must ensure that all equipment that will be used for the training is ready for use and that the students are wearing their <u>personal protective equipment (PPE)</u> properly. An accountability system should be used, and students should always be monitored for safety, health, and welfare. During actual training sessions, rest, hydration, and rehabilitation should be taken into consideration, and this includes student access to medical evaluation and treatment, food and fluid replacement, and relief from weather extremes (heat/cold).

The goal is to properly and enthusiastically train our fire fighters, not subject them to health and welfare dangers that could be avoided. A good instructor needs to always remember that the students are there to learn the skills being taught. As Assistant Chief Robert Lamb of the Cheshire, MA, Fire Department summed up: "Don't teach me how smart you are, teach me what I'm here to learn!" As an RIC Instructor, your job is to make sure that the fire fighters are learning and are competent with the skills being taught.

■ Safety When Training

The skills and techniques of rapid intervention are very demanding and can be dangerous, so whenever any training is conducted, every safety precaution must be taken. These include ensuring the durability of the equipment being used, the capabilities and skills of the instructors setting up and operating belay lines, and awareness of the fire fighters' physical health needs, especially hydration. Rigging and running proper belay lines is a skill that should be done only by fire fighters trained and capable in this technical skill. **FIGURE 2-1** shows a proper belay, set and ready. A belay line should always be used if you want to conduct proper and safe training. Its use during training is *not* optional. All RIC skill drills where a fire fighter is playing the role of the downed fire fighter; in a drill that requires elevated emergency egress or vertical lift and lowering scenarios; or where a fire fighter is executing an emergency egress self-rescue, whether over a ladder or using a escape rope system that is an elevated egress, requires the use of a belay system. The belay system must include, at minimum, an escape belt properly secured to the fire fighter executing the emergency egress or playing the role of the downed fire fighter for the safety of all those training.

> **Safety**
>
> In this text we will be proactive and lead by example, showing belay lines being used during training for the safety of the fire fighters.

When practiced with full PPE including <u>self-contained breathing apparatus (SCBA)</u>, RIC skills can take a physical toll on those training. It is suggested that every RIC training scenario have an Incident Safety Officer (ISO) assigned and that all RIC instructors adopt a zero tolerance policy for unsafe practices.

FIGURE 2-1 A belay, properly set and ready.

The success or failure of any training program is in how it is delivered. Teaching RIC and survival skills requires that the fire fighters have a true understanding of why these skills are important. A good training program will encourage the fire fighters to not only buy into the concept of rapid intervention, but to make a commitment to keep learning new skills and techniques that can make a difference if a fire fighter in peril lives or dies.

■ The Very First Thing

In our basic training, we are taught about our PPE and how important it is, yet we still see fire fighters not wearing it properly. All fire fighters must learn to wear their turnout gear properly and as a complete ensemble. From the RIC perspective, this is important because you want to have all the protection possible and you need to have the gear function correctly. From the perspective of the downed fire fighter, this is important because not only do you need to have all the protection possible, but you also need to make sure things are where they are supposed to be if the RIC needs to make adjustments or remove the SCBA.

An excellent example of this are the SCBA straps. **FIGURE 2-2** shows a fire fighter who has adjusted his SCBA waist straps improperly by not putting the connector buckle in the middle of his waist; **FIGURE 2-3** shows the same equipment worn properly. If you, as a member of an RIC, found the fire fighter in Figure 2-2 and had a need to convert the waist strap to a drag rescue harness, it could be a time issue. In a smoky, hot environment, with little if any visibility, your first instinct would be to feel (not look) for where the buckle is supposed to be and, not finding it, you would then have to start feeling the strap from one side to the other trying to find the buckle. If the fire fighter was on his side, there is a good chance it would be under him, which means rolling him over just to locate the waist strap. Time is critical in an RIC operation, so wear your gear properly—it might help save your life!

Tools and Equipment for the RIC

When we discuss RIC tools and equipment, we need to break them down into three categories: (1) <u>personal basic tools</u>; (2) <u>RIC Company basic tools</u>; and (3) <u>situation-specific tools</u>. It is important that fire fighters understand the difference, needs, and applications of each set of tools. It is also important that the fire fighters are skilled and competent in the use of all the tools. This competence cannot be achieved by looking at pictures in a book or listening to a lecture, but rather by putting the tools in your hands and actually training with them. It might require building props, such as a wall breach prop, or using a vacant building to practice taking the building apart, breaching walls, venting the roof, taking the glass, and forcing doors; as opposed to using the building to conduct live fire exercises. Your skills using these tools need to be taught, developed, and practiced, so as to maintain them for future use.

■ Personal Basic Tools: What Is in Your Pockets?

Personal basic tools are the individual tools that every fire fighter should carry with them, and proactive departments consider these tools standard issue. All too often, however, some fire fighters will take this to the extreme by carrying everything they can think of or can fit into their pockets. Remember, every item adds weight and makes it more difficult to find what you need when you need it.

The tools in your pocket are divided into two categories: those needed in a burning building and those that might be needed at other incidents, such as a car accident or carbon monoxide call. A key difference in these scenarios is that in a burning building, you will have your gloves on and cannot take them off, so the tools need to be easily accessible and

FIGURE 2-2 Improperly worn SCBA gear.

FIGURE 2-3 Properly worn SCBA gear.

easy to operate. At a car crash you can momentarily, if needed, remove your gloves to find something in your pocket. It is a good practice, therefore, to always learn to deploy and use your tools with your gloves on **FIGURE 2-4**.

FIGURE 2-4 Learn to deploy and use your tools wearing your gloves.

Crew Notes

As you purchase and train with any tools, always learn to deploy and use them with your gloves on. This simple but often overlooked concept will make a difference in your ability to manipulate your personal tools and equipment when needed.

In a burning building, which pocket you carry your tools in will also make a difference. Consider that you are carrying your knife in your coat pocket and need it; are you going to undo your SCBA waist strap to get into your pocket? Instead, use your Bunker pants bellows pockets for tools and equipment that are interior firefighting-related and your coat pocket for tools that are not usually needed under fire conditions.

■ Personal Tools: The Basics Everyone Should Have

Every member of an RIC should have basic personal tools to make them more effective and prepared, including a radio, webbing, flashlight, knife, door chocks, wire cutters, and personal escape rope system or a personal escape rope with the proper anchor (carabiner).

Radio for Each Crew Member

Every member of the RIC should have a portable radio to ensure communication and safety. The RIC operation can be assigned a separate radio channel, and radios should be carried in a manner in which the fire fighter can easily receive messages and communicate as needed **FIGURE 2-5**. Carrying the radio in your radio pocket on your coat, without a lapel mike,

FIGURE 2-5 Radios should be carried in a manner that makes communication easy.

will require you to open the pocket, remove the radio, find the transmission switch, send your message, and reply, after which you will have to return it to the pocket. This is not fast, efficient, or effective.

Webbing

Every fire fighter should always carry 20 to 25 feet (6 to 8 m) of 1-inch (3-cm) tubular webbing, with 25 feet (8 m) being the ideal length. The applications of webbing on the fire ground are almost endless: you can drag, lift, carry, tie off, raise, and lower with webbing, just to name a few uses. Your webbing should be tied into a loop using a water knot. The water knot is used to join the two ends of the webbing to make one large loop, and it is this loop that will make your webbing versatile and function as a great tool **FIGURE 2-6**.

Tie your webbing into a loop using the water knot as demonstrated in **SKILL DRILL 2-1**:

1 Take the webbing and run it in between your fingers to make it flat. (**STEP ①**)

2 Take one end and make a loose overhand knot, leaving about 8 inches (20 cm) of "tail." (**STEP ②**)

3 Take the opposite end of the webbing and trace it back over and through the overhand knot on the other end. (**STEP ③**) Make sure to follow through completely, (**STEP ④**) and complete the trace by having about 8 inches (20 cm) of excess webbing pulled through. (**STEP ⑤**)

4 Grab both ends and pull tightly (**STEP ⑥**) to form the finished water knot. (**STEP ⑦**)

SKILL DRILL 2-1   Tying the Water Knot
NFPA 1407, 7.10, 7.10(1)

1 Take the webbing and run it in between your fingers to make it flat.

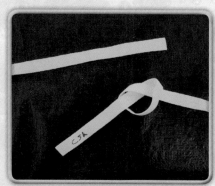

2 Take one end and make a loose overhand knot, leaving about 8 inches (20 cm) of "tail."

3 Take the opposite end of the webbing and trace it back over and through the overhand knot on the other end.

4 Make sure to follow through completely.

5 Complete the trace by having about 8 inches (20 cm) of excess webbing pulled through.

6 Grab both ends and pull tightly.

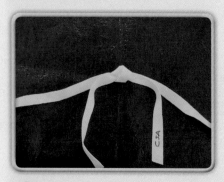

7 The finished water knot.

To carry your webbing and have it easily ready for deployment, it is recommended that it be carried in a roll with the water knot on the outside FIGURE 2-7 . The rolled webbing can be stored in an EMS glove to help keep it secure and ready for deployment. The advantage of preparing your webbing in this way is that it allows the webbing to be easily deployed under extreme conditions when you have no visibility. In a technical rescue, webbing is usually carried in a daisy chain, but to do so for an RIC rescue could become an exercise in futility. Remember that with a daisy chain, there is only one way to release the webbing for use. If you pull it the wrong way it will bunch up into one large knot and become useless for rapid deployment FIGURE 2-8 .

Crew Notes

To keep your webbing secured in a roll and ready for easy deployment, put an EMS glove over it.

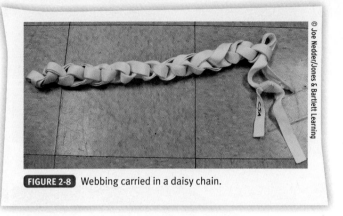
FIGURE 2-8 Webbing carried in a daisy chain.

FIGURE 2-6 A water knot is used to join the two ends of webbing, creating one large loop.

To deploy the webbing, simply grab it by the water knot and "snap" it out. As the webbing extends, grab it with your other hand. You now have total control of your webbing, and it is ready for use FIGURE 2-9 .

Flashlights

At minimum, fire fighters should always have a personal flashlight with them. The right angle light and the hand-held lantern are the most widely used types FIGURE 2-10 . It is suggested that, as a member of the RIC, you should have these two types of flashlights with you. Helmet-mounted lights are also available; however, these should be used with caution because the way the light and some of the brackets are mounted can lead to a wire entanglement hazard. If you choose to use a helmet-mounted light, seek one that is very low in profile and has a minimal exposure for entanglement. The right angle light is a good choice for a personal light, and is typically mounted to your coat. Think of your personal light as a backup or auxiliary light. The best light for an RIC member is the lantern-style light, as it will last longer, has a larger lens for more light, and is more flexible for the operation. Caution should be taken with lantern-type lights, in that they are usually supplied with a nylon strap and plastic buckles. Experience has shown that in extreme heat, the nylon straps will melt and can fail. A possible solution is to take the lantern and, by using a carabiner or snap hook, hang it from your SCBA waist strap or a truck belt FIGURE 2-11 .

Whatever light you choose to use, inspect it frequently and keep it charged to make sure it will work when you need it. Most of the rechargeable batteries used in these types of lights will lose a percentage of their charge every day. Check with the manufacturer for more information.

Knife

A good folding knife is a necessity for fire fighters. The folding knife's blade should lock in place when in use, so that the blade will not close on you and cause injury. The biggest issue with most folding knives is finding it in your pocket with your

FIGURE 2-7 Carry your webbing in a roll with the water knot on the outside for easy deployment.

A.

B.

C.

FIGURE 2-9 Deploy the webbing by grabbing the water knot and snapping it out. **A.** Webbing ready to deploy. **B.** Deploying the webbing. **C.** Webbing fully deployed.

FIGURE 2-10 The right angle and the lantern-type lights are most widely used as personal flashlights.

FIGURE 2-11 A lantern can be hung from a truck belt or SCBA waist strap.

gloves on. One way to overcome this is to purchase a folding knife that can be attached to a carabiner FIGURE 2-12. When you reach into your pocket with your gloves, finding the knife attached to the carabiner is much easier.

Crew Notes

Opening a folding knife in blacked out conditions with your gloves on is extremely difficult. If you purchase a knife that has a finger hole in the blade, you can create a grab point that will easily allow you to open the blade with your gloves by installing the Lamb Handle using a piece of threaded rod and two stainless acorn nuts.

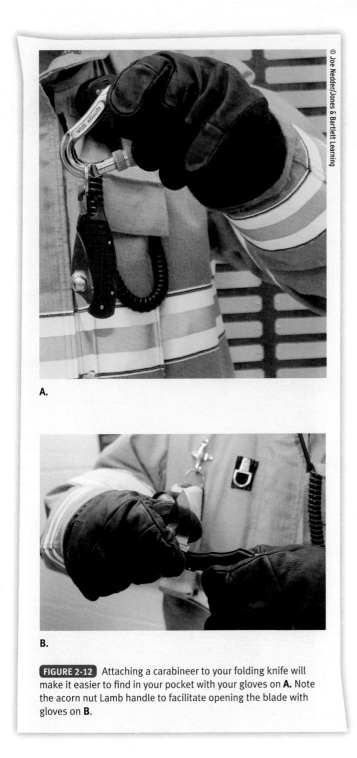

FIGURE 2-12 Attaching a carabineer to your folding knife will make it easier to find in your pocket with your gloves on **A.** Note the acorn nut Lamb handle to facilitate opening the blade with gloves on **B.**

Door Chocks

Door chocks are one of the simplest tools a fire fighter can carry, yet they are also one of the most overlooked. The most cost-effective and versatile chock is a wood wedge. The simple wood chock will do everything an expensive chock can do, plus it is easy to make and inexpensive. Chocks can be carried in your pocket or on your helmet, held in place by a piece of tire inner tube. The easiest way to make multiple wood chocks is by cutting 2 × 4s **FIGURE 2-13** .

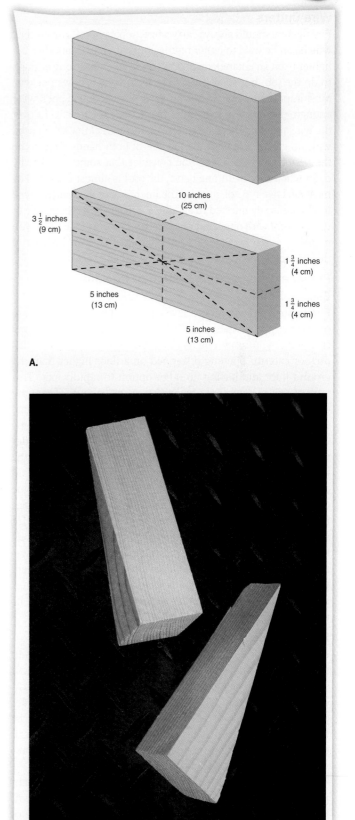

FIGURE 2-13 An example of **A.** the cut pattern to make wood chocks from a 2 × 4 and **B.** some ready-made chocks.

Wire Cutters

Fire fighters should always carry a functional pair of wire cutters, which can be used to either rescue yourself or assist another fire fighter from an entanglement. Cutters need to be strong enough to do the job, but small enough to fit in your bellows pockets. Most importantly, cutters need to have a jaw that will open wide enough, using only one hand to manipulate the tool, to fit over the wire and cut. There are cutters on the market that when used with no gloves and with the full use of both hands seem to be the best option, but when you consider that some jaws hardly open when used with one hand, the tool is almost useless. The most effective type of cutter is the lineman's or side cutter's type tool, commonly referred to as "dikes" **FIGURE 2-14**. Being able to use this tool with one hand is an important consideration as you might be severely entangled, in a cramped area, or because of the entrapment have the use of only one hand.

Personal Escape Rope System or Personal Escape Rope

Fire fighters should always have a personal rope to use in case of an emergency. The primary purpose of this system or rope is to assist you in an emergency rapid egress through a window or exterior opening if trapped on a floor higher than the ground floor and having no other option to rapidly exit. This type of exit is commonly referred to as a "bailout." There are two options available to accomplish this exit: (1) the personal escape rope system and (2) the use of a rated personal escape rope with a carabineer in what is called a rope or cylinder wrap. Early in the development of rapid intervention skills, the rope wrap was the most common method taught for a rapid egress emergency escape. Over the years, safer personal escape devices have been developed. The primary reason there has not been a massive move to purchase such escape systems for all fire fighters is the associated cost of this equipment—not all fire departments can absorb such costs. Regardless of what system your department has, if you are ever in a situation where you have no other option but to rapidly exit, the proper equipment and the skills gained through continued training in its use will determine your success or failure.

Personal Escape Rope Systems

In the past few years, there have been great improvements in the manufacturing of rope escape systems **FIGURE 2-15**. These systems are designed to assist the fire fighter in rapidly anchoring and exiting from an untenable situation. The systems are designed such that if the fire fighter loses consciousness or becomes incapacitated during the descent, the mechanical device portion of the system will secure to the rope, thus preventing the fire fighter from falling to death or injury.

These systems all require the use of an NFPA-compliant personal harness that secures the fire fighter to the system. Harnesses can be part of your PPE bunker pants at the time of purchase, a part of your SCBA, designed and rated as an escape belt, or a stand-alone rated escape belt that is worn over your coat or is part of your SCBA. Regardless of what method of harness used, it must be fully compliant and properly worn so as to ensure the safety of the user.

A.

B.

FIGURE 2-14 Two examples of wire cutters being operated with one hand: **A.** note how with one-handed operation, the jaws will not open adequately for emergency use and **B.** dikes, which if operated with one hand allow the jaws to open adequately for emergency use.

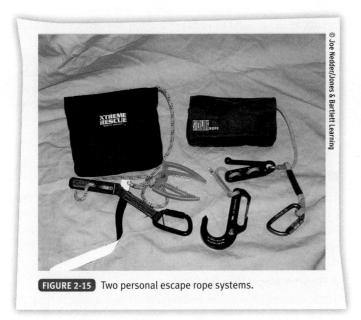

FIGURE 2-15 Two personal escape rope systems.

sizes and how much easier the larger carabiner will be to lock the rope in place. The rope should also be pliable and flexible enough to fit easily in your pant's bellows pockets, which will keep it ready for deployment. Storing the rope in your bunker pant's bellows pockets will keep it clean and protected. Further, because this is *your* personal rope, you are more inclined to protect it and maintain it so that it is always ready for deployment if needed.

Crew Notes

The life safety ropes that are used for the cylinder wrap method should ONLY use a carabiner for an anchor and never a hook! The hooks must be managed and held in place while rapidly exiting the window, and for the cylinder wrap, you must have both hands around the rope before you exit.

The system itself must be certified to NFPA 1983, *Standard on Life Safety Rope and Equipment for Emergency Services*, for escape. The system typically will have 50 feet (15 m) of ⅓-inch (7.5- to 8-mm) rope, and it should be static kernmantle, constructed of an aramid-type sheathing that is heat-resistant to 932°F (500°C), with a nylon core and round construction and resistant to abrasion and cuts. The rope should be pliable and flexible enough to easily fit and store in its intended pouch, bag, or bunker pant's pocket. The system will have a method of anchoring the rope securely to an unmovable object. The most common methods of anchoring are the hook and the carabiner. Whichever device utilized, it must be NFPA-compliant and designed to do what is needed and to be deployed rapidly to expedite your escape.

Personal Escape Rope for the Cylinder Wrap

As discussed, the cylinder wrap was the initial skill taught using your personal rope for escape. This method is extremely dangerous and if you lose consciousness while exiting, there is no avoiding falling to your death or severe injury. The method requires that you have a ⅓-inch (8-mm) personal rope that is NFPA 1983-rated—the same as the rope used for the escape systems—and that it should be 50 feet (15 m) long and have an NFPA-rated carabiner on one end. It is suggested that the carabiner be a ½-inch (13-mm) modified extra-large D type with a screw lock. This device will give you the most flexibility and is easy to deploy rapidly. The NFPA does call for the use of a "self-closing with a locking design carabiner" when in use for technical rope rescues; however, your ability to rapidly deploy your anchor will affect your exit time, and the use of a self-locking medium-size carabiner will slow you down. Imagine trying to take a medium-size carabiner, undoing the self-locking feature, opening the gate with your finger, and passing the rope through to anchor the escape line with your gloves on, under duress, and in a totally blacked out environment. Now try it with an extra-large D or modified D-type carabiner with a screw type lock left in the unlocked position. **FIGURE 2-16** shows the vast difference between the two

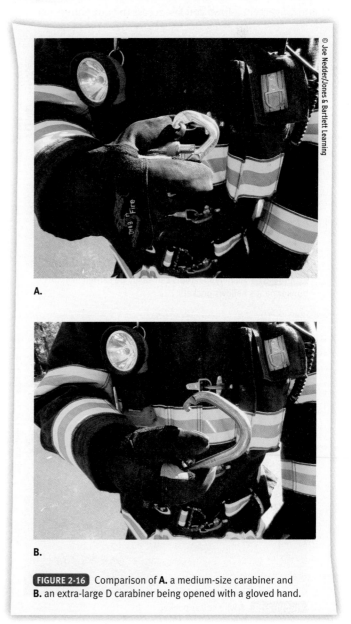

A.

B.

FIGURE 2-16 Comparison of **A.** a medium-size carabiner and **B.** an extra-large D carabiner being opened with a gloved hand.

The rope should be placed into the pocket using the "stuff" method, commonly used to load water rescue rope into the bags. A carrying option would be to place the rope in a storage bag that is designed to easily fit within the bellows pocket. Placing your personal rope in a bag that hangs on your SCBA is not suggested, as it becomes an entanglement hazard and is restricting and confining if a low-profile maneuver is required. Most importantly, your rope is not a shared piece of equipment, and as such, should be taken care of and maintained because your life might depend upon it.

All the personal tools outlined here can easily be carried in your bunker pant's bellows pockets. It is suggested that every department standardize the use of each pocket for commonality. One example would for the *left* pocket to be used for *life safety* (personal rope) and the *right* pocket be used for *rescue* (webbing, chocks, knife, chocks, and cutters) **FIGURE 2-17**.

A recent development has been the introduction of the leg pocket for carrying wire cutters. The pocket is placed on the pants near the ankle, on the outside of the leg **FIGURE 2-18**. This arrangement allows you fast access to your cutters without

having to feel around in your pocket and also reduces the load in the bellows pocket.

These are the basic personal tools that are needed for you to be most effective. If you choose to carry additional tools, such as adjustable wrenches, a screwdriver, or seat belt cutters, it is suggested that they be carried in your coat pockets, as they are less prone to be used during a rescue operation under adverse conditions.

■ RIC Company Basic Tools

The basic team tools an RIC member needs to have include the basic personal tools every fire fighter needs, plus a 200-foot (61-m) search rope, a thermal imaging camera (TIC) and spare battery, forcible entry tools, RIC rescue air unit for the down fire fighter, and a rescue rope bag.

A Search Rope in a RASP Bag

There are very few times you would *not* use a search rope in an RIC operation. Every Rapid Intervention Company should have their own Company search rope bag. The bag should be designed to be carried on your shoulder, with seat belt-type buckles to quickly release if needed. There are several styles of rope bags available, but not all were designed to function in an RIC application. In fact, we find that most search ropes are being stored and deployed in throw-style bags. To be functional in an RIC application, the bag must:

- Be designed to carry 200 feet (61 m) of ⅓- to ⅖-inch (7.5- to 9-mm) rope. Traditionally the fire service has used ½-inch (13-mm) utility rope for search. Using ½-inch (13-mm) rope is too heavy, too hard to manage, too hard to carry, and is not necessary. The ⅓- to ⅖-inch (7.5- to 9-mm) has become the nationally proven rope size for conducting RIC operations and rope searches in general.
- Have the ability to deploy for a left- or right-handed person.
- Have an identification tag attached to the bag and another matching identification tag attached to the rope at the anchor point.
- Have an attachment point inside the bag for the end of the rope.

Rope bags that meet these requirements are most commonly referred to as rope-assisted search procedure (RASP) bags, RIC rope bags, and Chicago-style bags **FIGURE 2-19**.

The rope should be compliant with the NFPA 1983 standard for escape rope, which is discussed previously. The rope should have one knot at 50 feet (15 m), two knots at 100 feet (30 m), and three knots at 150 feet (46 m), with the end of the line attached inside the bag marking the full deployment of 200 feet (61 m) **FIGURE 2-20**. When using the ⅓-inch (7.5-mm) rope, it is suggested that the distance indicator knots are "doubled up" to ensure that a fire fighter will feel the knot as it passes through his gloved hand.

Thermal Imaging Camera and Spare Battery

The use of a TIC during an RIC operation has greatly helped to improve the speed and efficiency in finding a downed fire fighter **FIGURE 2-21**. When the camera is deployed, always ensure that the battery is fully charged and that you

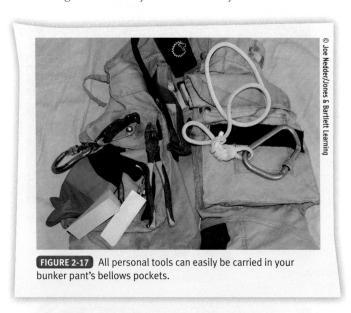

FIGURE 2-17 All personal tools can easily be carried in your bunker pant's bellows pockets.

FIGURE 2-18 A wire cutter pocket.

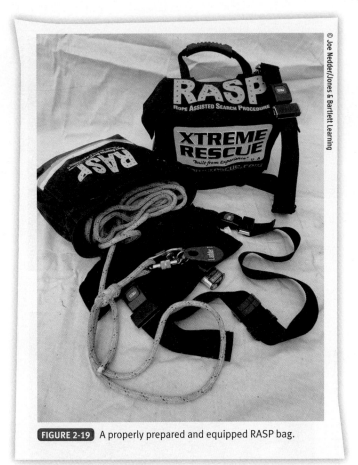

FIGURE 2-19 A properly prepared and equipped RASP bag.

FIGURE 2-20 Deployed rope showing knots marking 50-, 100-, and 150-foot (15-, 30-, and 46-m) distances and identification tags at the bag and at the anchor point.

FIGURE 2-21 A thermal imaging camera (TIC).

have a spare battery with you. Train and practice changing the battery in blacked out conditions with your gloves on, as that is the environment you will probably have to do it in.

Forcible Entry Tools

Of the four members on the RIC, two are usually designated as "tool men." The primary forcible entry tools every RIC must have are the Halligan bar and the flat head axe or another good striking/forcible entry-type tool such as a maul, a TNT tool, or the Lone Star Pig. When paired together, they are commonly referred to as "the irons" **FIGURE 2-22**. Additional forcible entry tools might include another set of irons, a rabbit tool, a maul, or a pick head axe, but of all of these, the irons are the most versatile. If you are operating with two tool men, having a second set of irons and using a maul-type tool instead of another flat head axe brings more versatility and capabilities to the team. The irons can be carried "married together" with a strap designed for quick deployment or by using other, nonconventional ways, such as using duct tape with a grab tab for quick release.

The RIC Rescue Air Unit for the Downed Fire Fighter

The RIC rescue air (also referred to as the RIT pack) for the downed fire fighter may be the most important of all the tools and equipment we bring with us **FIGURE 2-23**. The RIC air unit can also be utilized, *in an emergency situation*, by the rapid intervention company if a low air situation arises for a company member, allowing the company to exit safely. Remember, the downed fire fighter is in an environment where he cannot survive without a SCBA. To be prepared to deliver air, we need to have a method, and this method is the RIC rescue air unit, most commonly referred to as the RIC air bag. There are two ways to have RIC air: (1) the use of a manufactured RIC air unit from your SCBA provider, or (2) create an RIC air bag using an existing SCBA. Whatever your department's situation, a rapid intervention team must have the tools necessary to enter and deliver air.

A.

B.

FIGURE 2-22 Halligan bars married to **A.** a flat head axe and **B.** a Lone Star Pig.

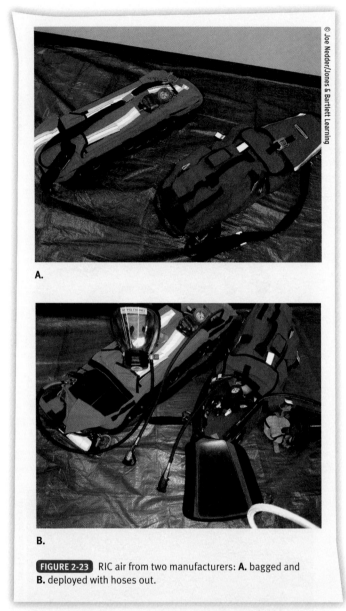

A.

B.

FIGURE 2-23 RIC air from two manufacturers: **A.** bagged and **B.** deployed with hoses out.

The Manufactured RIC Air Unit

The advantage of a factory-purchased RIC air unit is that it is specifically designed to deliver air to a downed fire fighter. It not only has a mask and mask-mounted regulator, but it also has a <u>universal air connection (UAC)</u>. The UAC is a NFPA-mandated device that allows one manufacturer's brand RIC air unit to provide air into any other manufacturer's brand SCBA, as long as the units are manufactured since the NFPA 1981, *Standard on Open-Circuit Self-Contained Breathing Apparatus (SCBA) for Emergency Services*, 2002 edition became effective. SCBAs with a UAC cannot be air donors to another SCBA, even if it is the same brand, with one exception: The MSA-brand SCBAs with a UAC (MSA refers to it as a URC, a universal rescue connection) are specifically designed to be a donor pack to another SCBA, if needed in an emergency as long as it has a UAC, regardless of brand. It is important to know your equipment and what it can and cannot do!

Today, most every SCBA manufacturer has developed an RIC air unit. These units are designed to incorporate the UAC connection found on every SCBA manufacturer since 2002 and to also provide a SCBA-specific regulator. These units are

Crew Notes

It is important to know your equipment! Some earlier factory-manufactured RIC air units *do not* have any audible alarm to indicate that the rescue air supply is getting low. Know what you have and train with how it works!

usually sold in a protective bag to safeguard the unit during deployment. They also have many other features that improve the performance and capabilities of the unit.

Creating an RIC Air Unit

If your department does not have a factory-manufactured RIC air unit and you are creating an RIC air unit using an existing SCBA, it is suggested that the SCBA be placed in a bag that is designed for this application, not in a gym or duffle bag. Bags for this application have the needed features such as carry handles, drag straps, places to attach to the downed fire fighter with a carabiner, and have two access points: one at the bottom of the SCBA to turn the air cylinder on, and the other at the top of the unit where the mask and the mask-mounted regulator are stored for deployment. These bags typically have pockets to store webbing and a few other items. By placing the RIC air pack in such a bag, the rapid intervention company can drag the unit along without worry that the SCBA will become exposed to contaminates or other items that might impede its operation. When creating an RIC air unit such as this, you must be aware that these units will most likely have a personal alert safety system (PASS) built in. Once the SCBA is turned on, the PASS device will go into pre-alert activation when left motionless for more than 30 seconds, followed by full alert if not tended to immediately. As an RIC member, you will have to try and manage this distraction as activated PASS devices might indicate another fire fighter down or in distress! One option is to utilize specific units that have had the straps removed so that they cannot be worn as a SCBA. The battery that powers the PASS can then be removed, as the unit is impossible for a fire fighter to wear. Any SCBA that a fire fighter can wear should never have any safety feature compromised or deactivated, including removing the batteries, as this type of action can seriously jeopardize a fire fighter's life if such device is needed.

Crew Notes

Always keep the rope bag separate from the RIC air bag. Trying to put it all in one creates something that is unnecessarily heavy. The rope or RASP bag needs to be carried and deployed by the RIC Officer, while the RIC rescue air unit is deployed by the crew's "airman." Placing all the resources in one bag will disrupt team positions and responsibilities.

Safety

Every position on the team has specific jobs and equipment responsibilities. This is what will make the team rapid and efficient.

Rescue Rope Bag

The three types of ropes used in rapid intervention include the company members' personal escape rope, the RIC team's search rope (RASP Bag), and the <u>life safety rope</u> for hauling **FIGURE 2-24** . The rescue rope bag holds a life safety rope that is used for specific RIC situations such as lifting a fire fighter up through a hole in the floor (Nance drill) and hauling, lifting, or lowering a downed fire fighter. Each of these RIC techniques uses the rope for a life safety rescue situation, and the rope that will best do the job must meet the NFPA 1983 standards for life safety rope: 1/2-inch (13-mm) or 7/16-inch (11-mm) static kernmantle rope. This thicker rope versus using 1/3- or 2/5-inch (8- or 9-mm) rope provides a more secure grip for the rescue fire fighters and will help prevent the rescuers from wrapping their hands around the rope to get a better hold. Wrapping the rope around your hand is dangerous and should never be done **FIGURE 2-25** .

The life safety hauling line should be 75 to 100 feet (23 to 30 m) long and carried in a rope bag for easy deployment. The bag should also contain extra-large carabiners with screw locks

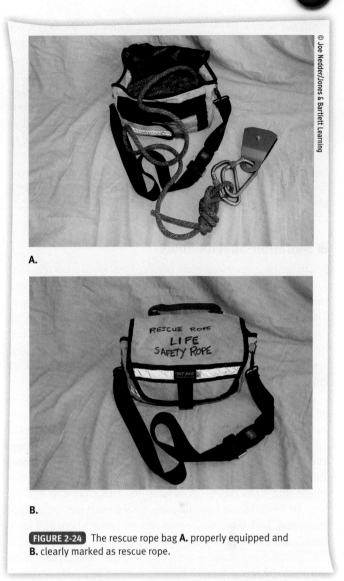

A.

B.

FIGURE 2-24 The rescue rope bag **A.** properly equipped and **B.** clearly marked as rescue rope.

FIGURE 2-25 Never wrap rope around your hand.

© Joe Nedder/Jones & Bartlett Learning

and a pulley with a carabiner attached for 2-to-1 mechanical advantage situations. The lead out end of the rope should have a figure eight on a bight with a 1/2-inch (13-mm) extra-large carabiner with a screw lock attached. There are various types of bags on the market. Consider using a RASP or Chicago-style bag to carry the hauling rope. The biggest advantages of this type of bag are that it has a shoulder strap with seat belt-style buckle that makes it easy to carry and release and deploy, instead of trying to lift it over your helmeted head with your SCBA on. The bags usually have a second pouch for the additional carabiners and rope pulley, and the bags are designed to be used in RIC operations. If these or other types of bags are used, they need to be clearly marked as "Life Safety Rescue Rope" so as not to be taken in error as the RASP rope bag or a utility rope. One way might be to clearly label them and then to use a colored tag that designates it as rescue rope.

■ Situation-Specific Tools

So far, we have reviewed basic personal and team tools required for the RIC. These are the tools that the crew will gather and bring with them when the RIC reports to the Command post. The 360-RECON, discussed in the chapter *RIC On Scene: Preactivation and Actions While Staged*, will determine the need or requirements for any situation-specific tools. The need for specialty tools can also be driven by an "event" of significant calamity, such as a wall or roof collapse.

There is a difference between having the tools ready to go and having them available. Personal and RIC company basic tools should be with you, ready to go, when the team is assigned and staged. The tools for a situation-specific rescue should be available to you, meaning that you know where the tools are stored and can get them quickly if needed. You might find that, based upon your 360-RECON, it would be wise to have certain specific tools staged with the team right away, and if so, it would be proactive to gather them and have them staged as well. We are not advocating, however, that the RIC gather every possible tool for any situation and place it on a 30- by 30-foot (9- by 9-m) tarp labeled RIC—this type of action is a waste of human resources and effort and would be discouraging to the company assigned to the RIC.

Crew Notes

Get what you need and *know* where the tools you might need are located.

Tools for specific situations can include almost everything that the fire department has, including:

- Rabbit tool: You are in a hotel, senior housing, or apartment building and must force several doors to conduct your search. A rabbit tool will expedite forcing the doors.
- Mauls and sledgehammers: You must enter a big box store, with masonry block construction, to locate a missing fire fighter in the B/C corner of the building. With limited door access and with a potential of a rope search 800 to 1,000 feet (244 to 305 m) into the store

to reach that area, the use of sledgehammers to breech the wall near the B/C corner would expedite entry and possible rescue.

Other possible tools that the RIC should have access to, but might not need to stage except for a specific situation, include:

- K-12 saw
- Chain saw
- Vent saw
- Sawzall (battery powered)
- Ladders
- Hauling ropes
- Bottle jacks
- Cutting torches
- Hydraulic tools
- Portable lighting
- Pry bars
- Pike pole
- Cribbing

This is just a partial list of some tools that might be needed, and there are many more that can be added to the list for specific rescue situations.

Safety

A key part of an RIC is to always plan ahead when considering possible tools and equipment that might be needed for a specific situation. Be proactive!

RIC Rope Knots and 2-to-1 Mechanical Advantage Systems

A competent RIC must know and understand fire service rope and basic knots. They must also know how to rig and use some basic mechanical advantage systems that might be required to rapidly make an RIC rescue. When operating under extreme conditions, it is important to be able to perform these without hesitation.

■ RIC Knots

The basic skills of ropes and knots are taught as part of the Fire Fighter I and II curriculum. Many of us, however, fail to remember how to tie basic knots because of the lack of use and practice. All fire fighters need to know the specific types of rope and their use, and have the skills required to tie basic knots. Continued practice and training is critical to maintaining the *basic skills* of knot tying. In rapid intervention, knowing and being able to tie certain knots is critical to the operation. Every member of every RIC must be able to tie the knots most often used in RIC, which include the figure eight, the figure eight on a bight, the figure eight follow-through, the open clove hitch, the closed clove hitch, and the handcuff knot **TABLE 2-1**. All of these knots, with the exception of the handcuff knot,

TABLE 2-1	Suggested Knots and RIC Applications
Knot	**Sample Application Examples**
Figure Eight	Stopper knot tied at the end of a rope if used with a descent or belay device so as to stop the rope from completely passing through the device.
Figure Eight on a Bight	Used to attach a carabiner to a rope during training for a belay line or to a life safety rope for dragging or extraction.
Figure Eight Follow-Through	Used to create an anchor point for a life safety rope, belay line in training, or hauling rope.
Open Clove Hitch*	Used to quickly secure a rope around an object, especially during a RASP search. It is not considered a life safety knot and cannot be used for anchor points.
Closed Clove Hitch* Handcuff Knot*	Used for hauling tools and equipment. Used to secure an unconscious fire fighter to a rope while performing an extraction (Nance drill) from a lower level rescue, such as a fire fighter who fell through the floor.

*These are not considered life safety knots.

are taught as basic fire fighter skills and can be reviewed in your *Fundamentals of Fire Fighter Skills* textbook. Your ability or inability to carry out and execute any basic skills will reflect poorly in an emergency situation. Fire departments, fire fighters, and training officers must be diligent to ensure that our basic skills are always sharp and maintained!

Handcuff Knot

The handcuff knot is a very important RIC knot that is used to either drag or lift a downed fire fighter. It is easy to tie and deploy, and its most common use has been to lift a fire fighter up through a hole in the floor when no other means of escape is available. The knot is placed on the forearms of the downed fire fighter, over his turnout coat and as close to the elbows as possible. When the knot is tightened and the lift begins, the sleeve material will bunch up and help secure the knot.

The handcuff knot, which is used when dragging or lifting a downed fire fighter, is demonstrated in **SKILL DRILL 2-2** :

1. Take the left side of the rope and cross it over the right to make a loop about 6 to 8 inches (15 to 20 cm) across. (**STEP** ❶)
2. Repeat step 1, making a second loop to the left of the first loop. An easy way to remember this to think "left over right" and then repeat "left over right." (**STEP** ❷)
3. Take the loop on the right side and place about one-half of it over the loop on the left. The result will look like a pretzel. (**STEP** ❸)
4. Take the left edge of the right loop and put it under the left edge of the right loop, and at the same time take the right edge of the left rope and put it over the right edge of the right loop. (**STEP** ❹)

5. Pull the two ropes through, tightening the center of the knot. (**STEP** ❺)
6. Once the knot is placed over the forearms of the downed fire fighter, the knot is pulled as tight as possible.

■ The 2-to-1 Mechanical Advantage Systems

The use of a mechanical advantage system has some application in rapid intervention; however, it must be emphasized that rapid intervention is just that—RAPID! Sometimes there is no time to create complicated or elaborate systems to assist in the removal of the fire fighter. Evaluate the situation and determine what skills are needed to get the downed fire fighter out quickly. There are situations when using a 2-to-1 mechanical advantage system will assist in the rescue, such as dragging a large fire fighter over a long open space.

The 2-to-1 mechanical advantage will help to reduce the load weight by one-half, making it easier for the fire fighters doing the hauling to move the downed fire fighter. The basic components of the system are the same items found in the rescue rope bag: at least two ½-inch (13-mm) extra-large carabiners (non-auto locking), a rope pulley, and 100 feet (30 m) of ½-inch (13-mm) or 7/16-inch (11-mm) static kernmantle rope. **FIGURE 2-26** shows the basic components and how to rig this system for hauling along a floor. Once the rig is in place and attached to the down fire fighter, one or two fire fighters can haul (or pull) on the rope and drag the fire fighter to their location.

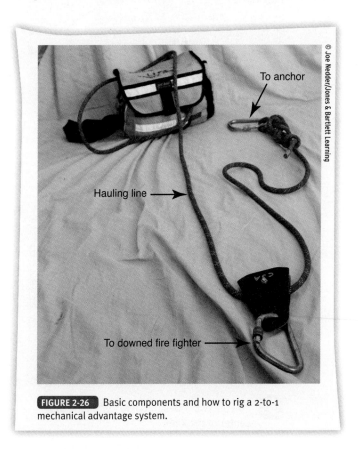

To anchor

Hauling line

To downed fire fighter

© Joe Nedder/Jones & Bartlett Learning

FIGURE 2-26 Basic components and how to rig a 2-to-1 mechanical advantage system.

SKILL DRILL 2-2

Tying the Handcuff Knot
NFPA 1407, 7.10

1 Take the left side of the rope and cross it over the right to make a loop about 6 to 8 inches (15 to 20 cm) across.

2 Repeat step 1, making a second loop to the left of the first loop. An easy way to remember this to think "left over right" and then repeat "left over right."

3 Take the loop on the right side and place about one-half of it over the loop on the left. The result will look like a pretzel.

4 Take the left edge of the right loop and put it under the left edge of the right loop, and at the same time take the right edge of the left rope and put it over the right edge of the right loop.

5 Pull the two ropes through, tightening the center of the knot.

Chief Concepts

- Members of an RIC must be properly trained, have the right tools and equipment, and know how to use them.
- A successful RIC training program will address the eight key elements.
- When training in rapid intervention skills, every safety precaution must be taken, including belay lines where applicable, for the safety of all involved.
- Fire fighters must learn to wear their turnout gear and SCBA properly to ensure they are protected and to make it easier for an RIC rescuing a downed fire fighter to locate straps, buckles, and grab points of the victim in low visibility.
- Fire fighters should only carry personal tools that are needed in a building fire in their bunker pant's pockets. Carrying too many personal items can add unneeded weight as well as making access to tools difficult.
- The RIC members must be able to use their personal and company tools with gloves on.
- An RIC must understand that tool selection should be based on the ability to use them in zero visibility situations and with gloved hands.
- The minimum tools required for an RIC team include:
  - Radio for each member
  - Hand light for each member
  - 200-foot (61-m) search rope (RASP bag)
  - Thermal imaging camera with spare battery
  - Forcible entry tools
  - RIC rescue air unit
- It is imperative that once assigned to the position of RIC, the crew be proactive and conduct a 360-RECON to determine the need for specialized rescue equipment based on building type, construction, and access.
- Any specialty tools deemed necessary should be staged for easy access of the team.
- Any training with RIC tools or techniques must be performed in a safe manner, making safety the number one priority.

Hot Terms

360-RECON A 360-degree view of a structure for the purposes of gathering information such as fire conditions, building construction, access and egress, and hazards found around the building. Commonly conducted by an RIC after arriving on the scene.

Door chocks Small wedge-like devices used to hold a door open, usually made from wood, plastic, or metal.

Flashlight A hand-held battery-powered light source. Usually the light source is a small incandescent light bulb or a light-emitting diode (LED).

Forcible entry tools Tools used by fire fighters to gain access to areas that are blocked or locked.

Handcuff knot A knot used to secure the arms/wrists of a victim that is typically utilized when hauling an unconscious fire fighter from a lower level.

Hazardous materials (HazMat) Any materials or substances that pose an unreasonable risk of damage or injury to persons, property, or the environment if not properly controlled during handling, storage, manufacture, processing, packaging, use and disposal, or transportation.

Knife A cutting tool with an exposed cutting edge or blade, hand-held or otherwise, with or without a handle.

Knot A fastening made by tying together lengths of rope or webbing in a prescribed way, used for a variety of purposes.

Life safety rope Static kernmantle rope in lengths of 75 to 100 feet (23 to 30 m) and typically ½ or 7/16 inch (13 or 11 mm) diameter, kept in a rope bag for easy deployment, and used for hauling or lowering a victim.

Mechanical advantage system A system usually consisting of ropes and pulleys used to assist in the hauling of a heavy object. Systems can be pre-rigged for fire service use as in the use in RIC applications. Mechanical advantage is the ratio of the output force produced by a machine to the applied input.

Personal basic tools Tools that every fire fighter should carry on his/her person. The tools should be limited to the following: webbing, flashlight, knife, door chocks, wire cutters, and a personal rope with carabiner or a personal escape rope system.

Personal escape rope system An NFPA-rated escape system kept on the fire fighter that has 50 feet (15 m) of rope, an anchor device (hook or carabiner), and a mechanical device to control the escape and decent, designed for easy deployment to assist a fire fighter in escaping a situation in which his or her life is in imminent danger. The entire system must be NFPA-compliant.

Personal protective equipment (PPE) Gear worn by fire fighters that includes the helmet, gloves, hood, coat, pants, SCBA, and boots. PPE provides a thermal barrier that protects fire fighters against intense heat.

Radio A battery-operated hand-held transceiver used to communicate information.

Rescue rope bag A dedicated bag for life safety rope, which is 75 to 100 feet (23 to 30 m) of ½-inch (13-mm) or 7/16-inch (11-mm) static kernmantle rope; is used for hauling and lowering; and is equipped with extra-large carabiners, one which is attached at one end of the rope with a figure eight on a bight knot. The bag may also contain a pulley and extra carabiners to rig a

2-to-1 mechanical assist. This bag is used exclusively for the extraction and rescue of fire fighters.

RIC company basic tools Tools used by the RIC for the purposes of finding, assisting, and removing a downed fire fighter. The minimum tools needed include a portable radio, 200 feet (61 m) of search rope, a thermal imaging camera, forcible entry tools, and an air supply for the downed fire fighter.

RIC rescue air unit A fully functioning SCBA in a protective carrying device that is able to deliver air to a downed fire fighter. The device can simply be a SCBA or a specific manufacturer's device designed for RIC operations.

Rope-assisted search procedure (RASP) bag A 200-foot (61-m) search rope designed to enable the RIC to move quickly and maintain an awareness of where they are (distance in) on the line.

Search rope A guide rope used by fire fighters while conducting searches in a structure that allows them to maintain contact with a fixed point for easy exit or to allow easy access for others.

Self-contained breathing apparatus (SCBA) A respirator with an independent air supply used by fire fighters to enter toxic or otherwise dangerous atmospheres.

Situation-specific tools Tools or equipment needed to address specific situations, such as a tool needed for a specific task determined by the type, size, and construction of a building.

Thermal imaging camera (TIC) Electronic devices that detect differences in temperature based on infrared energy and then generate images based on that data. These devices are commonly used in obscured environments to locate victims.

Universal air connection (UAC) A device installed on all SCBA and a requirement per NFPA, it allows a direct air fill to a SCBA cylinder from another SCBA/RIC emergency air system.

Water knot A knot commonly used to tie webbing. This knot is used to create a webbing drag strap that is part of the fire fighter's personal basic tools. It is constructed by tying an overhand knot in one end of the webbing and then following through the knot with the other end of the webbing.

Webbing High-strength nylon or polyester material woven in the same fashion as rope, but made to be flat. Can be constructed in a single layer (flat) or in a tube shape (tubular). When tied with a water knot, can create a drag strap.

Wire cutter Hand-held device used to cut many types and sizes of wire.

References

International Association of Fire Chiefs, National Fire Protection Association. *Fundamentals of Fire Fighter Skills, Third edition.* Burlington, MA: Jones & Bartlett Publishing; 2014.

National Fire Protection Association (NFPA) 1001, *Standard for Fire Fighter Professional Qualifications.* 2013. http://www.nfpa.org/codes-and-standards/document-information-pages?mode=code&code=1001. Accessed October 11, 2013.

National Fire Protection Association (NFPA) 1021, *Standard for Fire Officer Professional Qualifications.* 2014. http://www.nfpa.org/codes-and-standards/document-information-pages?mode=code&code=1021. Accessed October 11, 2013.

National Fire Protection Association (NFPA) 1407, *Standard for Fire Service Rapid Intervention Crews.* 2015. http://www.nfpa.org/codes-and-standards/document-information-pages?mode=code&code=1407. Accessed February 11, 2014.

National Fire Protection Association (NFPA) 1981, *Standard on Open-Circuit Self-Contained Breathing Apparatus (SCBA) for Emergency Services.* 2013. http://www.nfpa.org/codes-and-standards/document-information-pages?mode=code&code=1981. Accessed October 14, 2013.

National Fire Protection Association (NFPA) 1983, *Standard on Life Safety Rope and Equipment for Emergency Services.* 2012. http://www.nfpa.org/codes-and-standards/document-information-pages?mode=code&code=1983. Accessed October 14, 2013.

You are assigned to the RIC at a building fire. You are the company officer and have gathered the basic tools for an RIC deployment. You perform a 360-RECON of the structure and note that the exterior of the structure is made of solid brick. The building is 150 × 150 feet (46 × 46 m) and four stories tall. The only access points to the structure besides the front door on the A side are windows with metal bars across them. You note there is no ground level access/egress on sides B, C, and D. You further note the windows are very narrow and are above the first floor level. To complicate matters further, there is no access for a ladder truck to set up.

1. Based on the 360-RECON, what concerns regarding access and egress to the structure do you have?
   A. Access is not an issue. The main door on side A will be sufficient.
   B. The windows could easily be open with prying tools if needed. No additional concern is warranted.
   C. This building has significant access/exit issues and considerations for additional access and egress should be considered and prioritized.
   D. Old brick buildings do not pose a significant fire hazard due to the noncombustibility of brick; therefore, the fire will most likely not be major and access will not be an issue.

2. What additional equipment beyond the basic RIC equipment might you need for access to the structure?
   A. Saws with metal blades.
   B. Ground ladders.
   C. Sledge hammers.
   D. All of the above.

3. What proactive measures could the RIC do that could potentially allow for easier escape from the structure?
   A. No proactive measures are needed.
   B. Ladder the windows and remove the metal bars while maintaining a state of readiness.

   C. Stand by with an additional hoseline in case the fire intensifies.
   D. Widen the front door.

4. Why would utilizing rope-assisted searches be so important in a structure like this?
   A. With limited access and egress, it is important to tie off to a safe area to ensure easy escape.
   B. The rope will just get in the way in a brick building and should not be used.
   C. With the potential of additional equipment needed, the rope will have to be left outside at the point of entry.
   D. Another crew could use the rope and follow your company in a search if needed.

5. Based on your 360-RECON, what additional resources, if any, would you feel is important in the event the RIC was activated?
   A. Access to this building could be extremely difficult. Additional RIC should be requested.
   B. Additional engine companies should be called to assist with extinguishment.
   C. Further resources are cost-prohibitive and should only be used when absolutely necessary.
   D. No additional resources are deemed necessary. You have a well-trained crew that can handle any situation.

RIC On Scene: Preactivation Considerations and Actions While Staged

Knowledge Objectives

After studying this chapter you will be able to:

- Know the rapid intervention crew (RIC) team assignments and the tools needed for each assignment. (NFPA 1407, 8.1.1 , p 33–34)
- Understand the purposes of the RIC rescue air and the preactivation responsibilities of the airman. (p 33–34)
- Know what actions the RIC should take upon arrival. (NFPA 1407, 8.1.1 , p 34)
- Understand why the brief initial report meeting with the Incident Commander (IC) is important. (p 34)
- Know how to conduct a 360-degree RECON as a company, and what we need to look for and understand. (p 34–38)
- Know what a proactive RIC should be doing preactivation. (NFPA 1407, 8.1.1 , p 34–41)
- Know the incident and personnel indicators that might foretell the possible activation of the RIC. (p 38)
- Understand how to select tools specific to the situation. (NFPA 1407, 8.1.1 , p 39)
- Understand the importance and variables of RIC rescue air. (NFPA 1407, 7.9 , p 39–40)
- Recognize the importance of radio protocols and procedures during an RIC operation. (p 41)
- Understand the importance of RIC specific communications procedures. (NFPA 1407, 7.1(3) , p 41)
- Know and understand the importance of RIC team management and the responsibilities of the RIC Operations Group Supervisior and the RIC Company Officer. (p 41–43)

Skills Objectives

There are no skill objectives for Rapid Intervention Crew candidates. NFPA 1407 contains no Rapid Intervention Crew Job Performance Requirements for this chapter.

Additional NFPA Standards

- NFPA 1981, *Standard on Open-Circuit Self-Contained Breathing Apparatus (SCBA) for Emergency Services*

You Are the Rapid Intervention Crew Member

Your company arrives at a structure fire with heavy fire showing. Upon arrival, your company is designated the RIC. There are numerous companies working, and the IC has just called for a second alarm.

1. What actions can you as the RIC take that will assist in preparing your company in the event of a MAYDAY and RIC activation?
2. What assignments should the RIC Officer make and what tools and equipment are immediately needed?
3. What indicators should we be watching and listening for that might indicate a potential for an RIC activation?

Introduction

When a company is assigned to be the RIC, they must immediately become <u>proactive</u> for this important combat position, meaning they must be prepared to respond and react to hostile conditions while conducting a search and rescue operation for downed or trapped fire fighters. The word proactive is frequently used in the fire service, but what does it really mean? Within the rapid intervention discipline, a proactive officer or company is one that tends to prepare for an occurrence rather than react to an event after the fact. It is a position of anticipation! The RIC needs to be ready to go at a moment's notice, and staying proactive will help them to be mentally prepared. This, coupled with the skills in which they have trained, will make them an effective team.

The RIC assignment can happen in different ways. Your assignment can be preplanned as part of the run card assignment, it can be assigned while en route, or it can be assigned while on scene.

Crew Member Assignments

Once you have the RIC assignment, the rapid intervention Company Officer should assign the company members their responsibilities (i.e., RIC rescue airman, tool man, etc.). RIC assignments are based upon the number of fire fighters in the company, but the ideal team size is four (Officer and three fire fighters). For a crew of four, the assignments, tools, and equipment required include basic personal tools (radio, webbing, flashlight, knife, door chocks, wire cutters, and personal escape rope system as described in the chapter *Planning for a Prepared Rapid Intervention Crew*). In addition, each team member has specific tool and equipment assignments: the RIC Officer (fire fighter #1), with a thermal imaging camera (TIC), a spare battery, and a search rope carried in a <u>rope-assisted search procedure (RASP) bag</u>; fire fighter #2 (tool man, forcible entry), with the irons, a 30-inch (76-cm) Halligan bar, and an 8-lb (4-kg) flat head axe or other good striking tool; fire fighter #3 (tool man, forcible entry), with a <u>hydro-force-style tool (rabbit tool)</u> with an 8-lb (4-kg) maul-type tool to open multiple doors in buildings, such as apartments and offices, or if the rabbit tool is not required, a second set of irons; and fire fighter #4 (airman), with the <u>RIC rescue air unit</u> properly bagged and ready to deploy **FIGURE 3-1**.

FIGURE 3-1 A properly equipped RIC team.

© Joe Nedder/Jones & Bartlett Learning

Crew Notes

The airman position has numerous other critical responsibilities besides just bringing the RIC rescue air unit. Once the team is staged, it is his responsibility to inspect the RIC rescue air unit, including:

- Checking the cylinder and making sure it is 100 percent full.
- Checking all hoses and connections and making sure everything is secured and operational.
- Inspecting the <u>mask-mounted regulator (MMR)</u>, if so equipped, and making sure that the donning switch is in the closed position and the purge or bypass valve is closed.
- Making sure the hose for the <u>universal air connection (UAC)</u> is in proper position for quick deployment and the rubber cover is on the connector to protect it. The connector should also be inspected to make sure it is clean and that any air in the hose from prior charging of the system is bled off.
- Making sure the extra mask is in place and ready to go.

The airman should also be very familiar with the bagged unit, knowing what flap to open to charge the system and what flap to open to retrieve the mask, MMR, or UAC. An airman's lack of inspection or familiarity with this most critical unit can have a negative and catastrophic effect on the RIC and the downed fire fighter if the unit needs to be quickly put into use.

The RIC rescue air unit's primary purpose and function is to deliver air to a downed or trapped fire fighter who is low or out of air when found by the team. The RIC rescue air unit can also be utilized *in an emergency situation* by the RIC if a low air situation or SCBA emergency arises for a company member, allowing the company to exit safely. It should never be used to extend the RIC's air supply so that they can go longer or deeper into the building. The RIC is not only responsible for finding the distressed fire fighter, but also for getting themselves out alive and unharmed. This requires that the company's air status be monitored and the company exits before becoming dangerously low on air, potentially requiring another RIC to extract them.

On Arrival

Once assigned or upon arrival on the scene, the RIC needs to always stay together and move as a company. If you are preassigned, upon arrival have your crew gather their equipment and then, as a company, report to the Command Post and Incident Commander (IC). If an RIC Operations Group Supervisor has already been assigned and is *in place*, the RIC will report to this person. If you are assigned after your arrival or later in the incident, gather the needed equipment and tools *first*, and then report for a face-to-face meeting.

■ Getting the Basic Information, the Brief Meeting With Command

Regardless if you are designated RIC en route or while on scene, the RIC Officer must report to the IC and get a brief initial report of what is happening **FIGURE 3-2**. To be effective as an RIC if activated, you need to understand the fireground operation. This can be accomplished by having a brief face-to-face meeting with the IC. This meeting need only be a few short moments, yet it is necessary for the RIC Officer to have

an understanding of the operation. What we are asking the IC to do is "paint a picture" of what is going on:

- We have a… (provide an incident size-up).
- Engine 1 is doing… (provide information of how many companies are operating, their assignments, and where they are operating).
- As you can see… (provide a progress report).

If the position of RIC Operations Group Supervisor is staffed with the arrival of the RIC company, the RIC Operations Supervisior along with the RIC Company Officer should meet with the IC for this important briefing. Once the information is obtained from the IC, acknowledge it, and then ask where the RIC should be staged. The RIC Officer should then share the information he has just obtained with his crew, so that all members of the RIC are properly briefed and prepared and have a solid understanding of the operation.

With the information gathered from the brief meeting, the RIC is now better prepared and has a solid understanding of the operation. This is critical information to have if a MAYDAY is declared and the RIC is activated. For example, if a MAYDAY is called from a fire fighter assigned to Ladder 1 trapped under a ceiling collapse, it would be advantageous for the RIC and the RIC Operations Group Supervisor to know that Ladder 1 was assigned to assist Engine 3 with extinguishment and overhaul on floor 2, C/D corner. This initial information would assist in knowing the target location for beginning the rescue (floor 2, C/D corner), and that there is a potential of additional fire fighters in peril (two companies were assigned to that area).

What a Proactive RIC Is Doing While Staged

A capable rapid intervention crew, once staged, realizes that they are a very active position. To stay proactive, the 360-degree RECON or 360-RECON (RIC size-up) should be performed. After the 360-RECON is completed, the RIC should monitor the operation and continue to conduct an ongoing size-up, gather tools specific to the incident and what the building might require, ensure that the team is prepared to deliver emergency air to a downed fire fighter, monitor radio traffic, and create access points for fire fighter emergency egress. During this process the RIC Officer will continue to provide team management.

360-RECON and RIC Size-Up

It is important to always take into consideration the size of the building beforehand. Many large commercial buildings, malls, big box stores, and residential buildings, such as condos and apartments, are not easily walked around. Let common sense prevail! Before an RIC leaves its staging area for the RECON, the IC or the RIC Operations Supervisor needs to be informed. The team needs to remember they are still the RIC and should have their basic equipment with them while conducting the RECON **FIGURE 3-3**. Always be prepared to activate!

FIGURE 3-2 The RIC Officer being briefed by the IC on the fireground operation.

© Joe Nedder/Jones & Bartlett Learning

FIGURE 3-3 An RIC performing the 360-RECON with equipment.

■ What the RIC Should Observe and Pay Attention To

The purpose of the 360-RECON is to help the RIC to stay proactive and be better prepared by having a good understanding of the building that stands before them, the overall operation, the smoke, and the fire behavior. During the 360-RECON, the RIC should be observing:

- Fire situation
- Smoke
- Building construction
- Fireground operations
- Possible hazards
- Building variables
- Access/egress

Fire Situation

Where is the fire? Is it being contained or is it spreading? If the fire is spreading, where is it going and how? Is it by interior convection or exterior overlapping? *The importance of fire behavior cannot be overstated.* Understanding fire behavior will assist you in observing what is happening. The nature and science of fire behavior is a key component in being a competent fire fighter **FIGURE 3-4**. Not understanding what is happening might lead you into a situation that is more dangerous than you realize if and when the RIC is activated. An excellent and helpful way to "observe" the fire in the interior is by utilizing the TIC. The TIC will assist the RIC from outside the building to determine where the fire is and if the fire has spread.

Smoke

Observe the smoke coming from the building. Look at the volume, velocity (pressure), density, and color **FIGURE 3-5**. Is it indicating fire spread or containment? Is it thick, dark smoke exiting under extreme pressure, indicating a possible flashover situation? Does the volume and density indicate a lot of fuel is burning? Are the engine companies making progress with the extinguishments? Are you seeing an increase in pressure and volume as steam is generated? Is the smoke color beginning to turn light gray and white, indicating the water is doing its job and beginning to turn to steam? The aspects of smoke indicate many things, including fuel load, early or late stage burning, a deep-seated fire within the building's structure, possibilities

FIGURE 3-4 Fire in a building.

FIGURE 3-5 Heavy smoke from a fire.

of flashover, and successful knockdown and extinguishment. Understanding smoke and the message it is sending enables the RIC to better evaluate the situation during size-up and be better prepared, if activated, for the situation they are entering.

Building Construction

Look at the building and determine the type of construction. Fire fighters and fire officers need to know and understand the five classifications of building construction **FIGURE 3-6**. How a building is constructed will affect the growth and path of travel

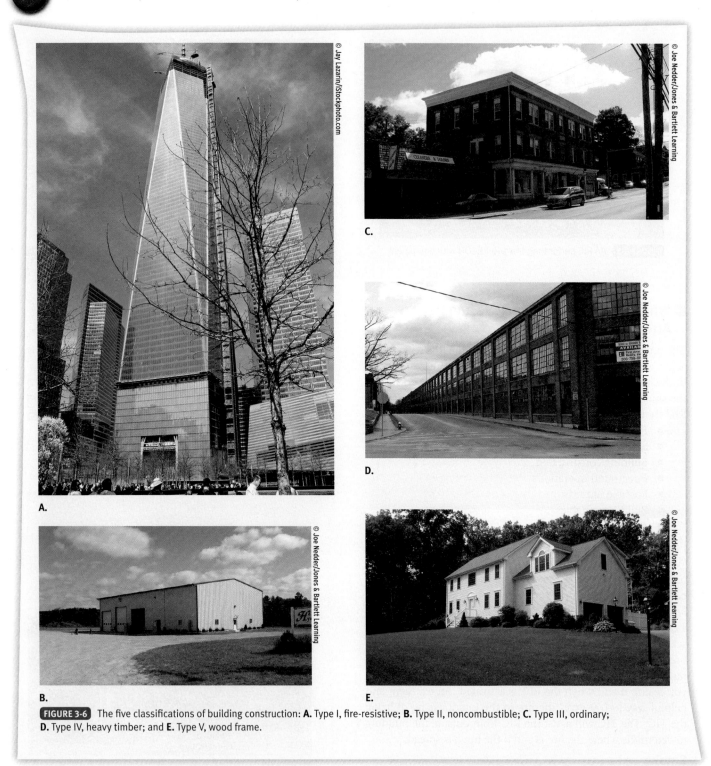

FIGURE 3-6 The five classifications of building construction: **A.** Type I, fire-resistive; **B.** Type II, noncombustible; **C.** Type III, ordinary; **D.** Type IV, heavy timber; and **E.** Type V, wood frame.

of the fire and how it might collapse or experience structural failure. Collapse is a significant event. For an RIC, understanding building construction is something you must have expertise in.

For example, if you are standing in front of a Type III ordinary construction building, perhaps a three-story taxpayer in an urban or suburban downtown area, ask yourself these questions: Are there common walls or a common cockloft? What other access do you have besides the front door? Are the floors above commercial or residential? Do you observe any structural modifications? What about entry points? Is the building in an urban downtown, with alley access to the rear, where you will

encounter a heavily fortified rear door? All of these questions are pertinent to the situation at hand. They can affect the fire spread and the lives of the fire fighters operating on scene.

Often, by observing a building you can reasonably determine a probable floor plan. For example, if you are looking at a Type V, wood frame ranch-style house, you can be reasonably sure that the front door leads to a living room and behind that is the dining area and kitchen. The bedrooms are usually down the hall with two in front and one in the rear. Between the kitchen and the rear bedroom will be the bathroom. It is this type of knowledge that will make the team more capable, safer, and possibly more efficient in the rapid intervention operation. The same is true for other types of buildings. A Type III ordinary building, for example, that is three stories tall, will most likely have stores, shops, or restaurants on the first or ground floor and offices or apartments on the (second) floor above. Your knowledge of building construction, as it pertains to the fire service, can and will make a difference to the safety and survival of your crew and your ability to reach a downed fire fighter.

Fireground Operations

Fireground operations involve what is currently happening. As a member of the RIC, the following types of questions need to be asked: Are the companies that are working making progress? Do you know how many companies are in the building, what their assignments are, and where they are approximately located? How many hoselines are in the building? If you are in a more rural area, what is the water supply situation? Is there a chance there might be a loss of water to the companies operating within the building? Has the vent crew completed its tactic, or have they been slowed down by access, weather, roof conditions, or lack of personnel? Is the vent working as anticipated? Has a search crew been committed to the operation? How much geography do they have to cover, what are the fire conditions where they are going, and do they have adequate personnel? If the search crew is above the fire, has a ladder or ladders been thrown to the floor they are operating on for emergency egress? How long have the crews been in the building? Are we seeing crews starting to rotate out due to low air? Is the fire knocked down and are we in overhaul mode? You need to have a good grasp of the situation of who is doing what and where. This type of knowledge is critical to making the crew more efficient, focused, and targeted if activated.

Possible Hazards

As you RECON around the building, operate in a discovery mode. Look for things that might not be visible to the IC in the front of the building. Consider building modifications and how they might affect the structure, the fire load, and any possible rescues. Ask the following questions: What is the condition of the building right now? Where is the fire and what structural components are burning or have been exposed to the extreme convection heat and smoke currents? Is it in danger of partial collapse? Are lightweight or truss construction components involved in fire? Is the building in derelict condition? Is it under major renovation? Answers to these types of questions will give you a better assessment of the condition of the building.

Also look for things that could increase the danger, such as propane tanks adjacent to the building—are they in danger of being impinged upon by the fire? Look for other hazards too, such as security fencing that could slow down access, security dogs, retaining walls that have a significant drop off, a portion of the building under renovation, barred windows, or anything that could create a problem for the crews operating or could further complicate an RIC activation. If presented with obstacles and hazards such as barred window or locked gates, do they need to be removed or opened? Proactive actions such as this will help to "soften" the building for a possible emergency entry

or egress. Have you found the entry point for the utilities? Is there natural or propane gas? Is the electrical service overhead or underground? Has the power been turned off at the street? Remember, anything that could slow down or complicate an RIC's ability to gain access or egress should be noted.

Building Variables

Watch for and note any variables to the building. Is the main building one construction type and another part of the building a different construction type? What about the number of stories? Is the front (A side) different than the rear (C side)? Is there a change in grade from the front to the rear—is there a walk-out cellar or basement on side C? This is a very common occurrence and must be noted. What if it was never observed or noticed, and a MAYDAY is called for a "fire fighter down on floor 2" and a ladder is requested to the second floor rear (C side) to extricate the downed fire fighter? To reach the second floor, most fire fighters would use a 24-foot (7-m) ladder. But when they go around back, they discover a change in grade and a walk-out basement, in which case the 24-foot (7-m) ladder will not reach to do its job and another, taller ladder would have to be quickly obtained. If it had been known in advance that the rear of the building was a walk-out basement, the call for a ladder to that sector would have been answered with a 35-foot (11-m) ladder. The delay in getting the correct ladder might affect the survivability of the downed fire fighter!

On February 14, 1995, three Pittsburgh fire fighters died in a situation similar to this. The building appeared to be a two-story structure from the A side, but had four stories in the rear. When the fire fighters got in trouble, no one was aware of the difference in grades, and as companies tried to rescue them, there was confusion about the number of levels in the building. The FEMA report (USFA-TR-078/February 1995) states: "This incident also reinforces a concern that has been identified in several fire fighter fatality incidents that have occurred where there is exterior access to different levels form different sides of a structure. These structures are often difficult to 'size-up' from the exterior, and there is often confusion about the levels where interior companies are operating and where the fire is located. In these situations, it is particularly important to determine how many levels are above and below each point of entry and to ensure that the fire is not burning below unsuspecting companies."

Access/Egress

When a MAYDAY is called and the RIC is activated, a decision has to be made on what is the best, quickest, and safest entry point for the team. As you conduct your 360-RECON, look for additional access and egress points, including doors, windows, fire escapes, and loading docks. How accessible are they? Have any of these entry/egress points been softened up? Do they need to be? What tools would be needed to gain entry at that point? Is there an overall lack of access/egress with this building? For example, the C side of an industrial building might include only a few man doors set up for emergency egress from within, with no exterior hardware showing. This presents a forcible entry situation if access is needed in that general vicinity **FIGURE 3-7**.

As you walk around the building, think proactively about what actions might be necessary to gain entry at any given

FIGURE 3-7 An egress door with no exterior hardware.

point on the building and formulate a plan for gaining access. Do we need to force locks, doors, and gates now, keeping the doors closed? This type of action, to break locks, open gates, and create access, all without creating ventilation paths is called softening the building. These actions should always fall within the IC's Incident Action Plan or be a preapproved standard operating guideline. If not, always seek permission from the IC before taking such actions. Are the walls masonry blocks? If so, what will it take to breech and enter this type of wall? Is there good access on all four sides? What ground ladders need to be raised for possible rescue or evacuation of fire fighters (and civilians)? Are there overhead wires or building variations that might interfere with the raising of a rescue ladder?

Other things to consider are making sure that doors are secured open. If garage doors have been raised, secure them in place so that they cannot close on crews operating within the structure. Also, on what floor and what sector of the building have ground ladders been raised? Is there at least one ladder raised to every floor fire fighters are operating on? Review the grades of the property and plan what size ladders will be needed to reach the different floors on all sides of the structure in the event the RIC is activated and quick access is needed.

After 360-RECON: The Next Steps

Once the 360-RECON is completed, the RIC members should report back to their assigned supervisor, either the IC or the RIC Operations Group Supervisor, and share their observations. If you have found things that the IC or the RIC Operations Group Supervisor should be aware of (different grades, different number of floors front and rear, lack of access, propane tanks), tell them face-to-face. If you have found things that can affect the safety of those working the operation, tell the Incident Safety Officer face-to-face. This type of sharing of information and observations will only strengthen the incident's overall operational safety. Once the information has been shared, the RIC should report to their designated staging area.

■ Monitor the Operation and Continue an On-Going Size-Up

The Company Officer and the entire RIC should now continue to monitor the operation. Watch what is happening, monitor crew rotations, and pay attention to the building and its deteriorating condition. Are companies exiting, changing cylinders, and going right back to work, or are they rehabbing? Fatigued companies might be exposed to an increased danger level. Every bit of information you can gather might make a difference on the success of an activated team. What is happening with the operation? Are more lines needed? Are companies making progress? Is the fire in control of the incident?

There are some indicators that can be used to anticipate the possibility of an RIC activation, and for the purpose of this book they are broken into two categories: incident indicators and personnel indicators.

- **Incident indicators:** a prolonged burn time, extensive structural damage, poor or no ventilation, which can have an adverse effect on a coordinated fire attack (venting is done in coordination and to support the entry of the engine company), sagging floors, bulging walls, and interior collapse. Other incident indicators are lightweight construction being compromised, structural components being compromised, excessive water in the building, multiple floors involved and not under control, ineffective tactics, implementing the wrong strategy, and the initial staff being overwhelmed by the incident.
- **Personnel indicators:** the lack of training (and it shows!), not enough help and fatigued fire fighters, the weather's affect on the well-being of the fire fighters, SCBA air consumption, companies not maintaining crew integrity, and a lack of or a poorly organized accountability system.

Crew Notes

As the RIC officer, you should continue to monitor the operation frequently and conduct your own situational analysis. Are the operating crews knocking down the fire or is it still growing? Anticipate possible entry points if your crew is activated and consider the fire conditions you would be leading your crew into. *Pay attention and stay proactive!*

■ Gather Tools Specific to the Incident and Building

When it comes to gathering tools for the RIC, consider what is the best strategy for tool staging. Basic tools should already be in place. The 360-RECON might identify specific issues or possible problems that require additional or specialty tools, which should also be gathered and staged.

There is a school of thought that promotes the concept that the RIC should spread a tarp out at every fire and place every possible tool they might need on it. Some manufacturers actually sell tarps with the letters "RIT" printed along the edge of the tarp. The single advantage of gathering every possible tool needed is that everything is in one location. The disadvantages, however, include:

- It will take the crew a significant amount of time to locate, gather, and place every tool possible, and while doing so, they are not staged.
- It will take the crew at least as long to replace all the tools, which will delay returning the company to service.
- In inclement weather (rain, snow), the tools and equipment are exposed to the elements.

These types of tasks are an excellent example of why most fire fighters resent the RIC assignment—it is seen as busy work.

An alternate tool-gathering strategy, however, is based upon the initial size-up of the situation on arrival and the 360-RECON, and involves gathering only those tools that best fit the situation and building type of the incident before you. In utilizing this type of tool strategy, a Stokes basket can be used to gather the additional tools (if needed) and stage them with the team. If the RIC is activated, these tools can be easily deployed. This type of strategy is based upon learning to be observant, to anticipate, and to always be proactive.

What Additional Tools Do You Need?

With the knowledge gained from the 360-RECON, any specialty tools that might be needed should be gathered, or at least ensure team members know where they are located. Specialty tools are tools that are specific to the incident in front of you. For example, if a masonry block wall was discovered during the 360-RECON that might need to be breached, sledgehammers will be needed and should be gathered and staged. If the front entrance is locked with a roll down gate, access could be obtained using a K-12 style saw with a metal cutting blade. Quick access could also be obtained using the hydraulic cutters (especially the electric models), normally used for vehicle extrication to cut through the gate. It is important to know if the apparatus these tools are carried on is on scene. If a ladder is needed quickly, know what pieces of apparatus still have ladders available and their sizes. Being proactive means finding out this information before it is needed. Quick access to the tools needed could make the difference between the success and failure of a rapid intervention operation.

■ Ensure the Team Is Prepared to Deliver Emergency Air

There are very few MAYDAY situations where the downed fire fighter's air supply is not limited. The RIC is responsible for delivering an emergency air supply to the downed fire fighter.

This can be done in four ways: (1) provide a similar SCBA for a regulator swap; (2) delivery of air via an RIC rescue air unit using the UAC or a regulator swap; (3) execute a mask swap; or (4) utilize the manufacturers' built-in buddy breathing system. The delivery of emergency air, however, is not necessarily as simple as it might seem. In order to know what method of air delivery to utilize, the answers to some very important questions are needed:

1. Does the downed fire fighter's SCBA have a UAC fitting?
2. Does the RIC rescue air unit have a UAC high-pressure hoseline?
3. Can the RIC deliver a similar SCBA where a simple regulator swap can provide the needed air?
4. Is it a mutual aid fire fighter? If so, is it the same SCBA as yours? If not, does the SCBA have a UAC?
5. Does the situation allow for a SCBA face piece (mask) swap? Indicators that would allow a mask swap are: the environment is not super-heated smoke or gas, either of which could cause immediate danger to the respiratory system or death, and most importantly, the distressed fire fighter's breathing is strong enough that he can "open" the regulator with a breath.
6. If both SCBAs are equipped with an identical buddy breathing system, is there time to extend and connect? Will tethering the downed fire fighter to another fire fighter's SCBA or RIC air pack inhibit or slow down the rapid exit?

As basic as all this might sound, it is imperative to know and understand. Many departments are fortunate to have all the same model SCBA. Some department's SCBAs, however, are not 100 percent similar. They might be the same brand, but a different model or generation. The RIC must be prepared to deal with this.

The Universal Air Connection

In 2002, National Fire Protection Association (NFPA) 1981, *Standard on Open-Circuit Self-Contained Breathing Apparatus (SCBA) for Emergency Services*, required that all new SCBAs purchased have a UAC; SCBAs purchased before that date, however, are NOT required to have the UAC. The UAC is designed so that every SCBA is interoperable with an RIC rescue air supply unit, allowing air to be transferred and equalized from the RIC rescue pack to the cylinder of the downed fire fighter, regardless of brand FIGURE 3-8 . This means that, if you have a *factory manufactured RIC rescue air unit*, you can, via the UAC, assist a downed fire fighter with air supply, regardless of brand or model. It is important to understand that the UAC is a high-pressure line, and the air must come from a factory manufactured RIC rescue air unit. Currently, one SCBA manufacturer, MSA, has the ability for its SCBA to also be an RIC UAC donor. Check with the manufacturer of your department's SCBAs to understand the most current technology available and what capabilities the SCBA units in your department possess.

The UAC is not designed to fill another cylinder; rather, it equalizes the pressure (not volume) between the donor unit and the receiving unit. It, in essence, works similar to a cascade system and in the majority of cases will, at best, provide the downed fire fighter with approximately one-half of a cylinder. If your department does not have an RIC rescue air unit, you still

© Joe Nedder/Jones & Bartlett Learning

A.

B.

FIGURE 3-8 Two SCBAs **A.** from different manufacturers showing the UAC connections and **B.** a closeup of the UAC connections.

need to provide the fire fighter with air. In this event, you will need to bring the downed fire fighter a SCBA compatible to the one he is wearing, which will allow you to do a regulator swap.

What If the SCBA Is a Different Brand?

In areas where different brand SCBAs are used by mutual aid departments, it is important for RIC members to know what brand/model is used and some of the basics about those units. By being proactive and knowing what your mutual aid departments use for SCBAs, if a member of a mutual aid department calls a MAYDAY, you will have the proper units staged with the team and be able to provide that fire fighter with emergency air. Proactive also means knowing how these units operate. By cross-training with departments you mutual aid with, you will be competent with their equipment and they with yours.

Some believe that if they do not have a UAC or the same SCBA as your department, the RIC should bring them what you have and swap out the SCBA. Stop and think about this! How long would it take to change out the SCBA and mask of the downed fire fighter, especially under extreme conditions? The answer is "too long." If the downed fire fighter is having a medical emergency, every second counts. Spending 3 to 5 minutes trying to change SCBAs is not a viable option. It is important for a competent RIC to know what SCBAs are in use

on the fireground, how they operate, and to have the correct unit(s) and equipment available with the team if needed.

Buddy Breathing

The term buddy breathing has been around for almost as long as the SCBA has been in use. Buddy breathing is two (or more) fire fighters sharing the same air supply. Many years ago, before the introduction of positive-pressure SCBAs, fire fighters would remove their mask and share it with another fire fighter who had run out of air to escape. Given today's fire behavior, extreme conditions, and the high heat and toxic smoke faced in a structure fire, under no circumstances should a rescuer buddy breathe by sharing his mask or MMR with the downed fire fighter. This position has been supported by the National Institute for Occupational Safety and Health (NIOSH) for many years. The removal of the regulator from your mask, or worse yet the removal of your mask, will expose you to an immediately dangerous to life or health (IDLH) environment of extreme toxic and heat danger!

NFPA 1981 standard, 2013 edition, recognizes that SCBA manufacturers today have developed systems that allow two fire fighters to share their air while not at any point exposing themselves to an IDLH environment. In this new standard, this type of system is referred to as an Emergency Breathing Safety System (EBSS), and the NIOSH refers to such systems as Emergency Escape Support Breathing Systems (EESBS).

Today, most SCBA manufacturers offer an EBSS. This is a system that connects one SCBA to another via a special hose and couplings. Regardless of what each manufacturer calls it, you are sharing your air so it is still "buddy breathing." As a member of an RIC, you need to think very carefully about using your *personal air* before deploying and using such systems. Your RIC company should be equipped with an RIC rescue air unit, which is the primary way you will deliver air to the downed fire fighter. Note: if your RIC rescue air unit is equipped with this type of system, it can be used as an alternative to a regulator swap.

Consider this: if you have exhausted half of your air supply locating or getting to the downed fire fighter, and you now buddy breathe, your personal air supply is now one-fourth of the cylinder, as the downed fire fighter is now using the other one-fourth of the cylinder. Depending on your air consumption rate and the size of your SCBA cylinder, you might personally have 3 to 6 minutes of air left—is this enough to get you out safely? Probably not, and now, in this example, there might be two fire fighters without air. Also, utilizing a buddy breathing system physically tethers you to the downed fire fighter with approximately 3 feet (1 m) of hose, increasing the difficulty of evacuation and escape, especially if it is necessary to crawl or climb or if the distressed fire fighter tethered to you experiences severe fatigue or becomes unconscious. There are places and circumstances where an EESBS is an excellent option to have and deploy, but using your air in a rapid or extended intervention rescue is not the place. The job of the RIC is to locate and evacuate the downed fire fighter rapidly, and jeopardizing your air supply will not assist in the mission.

Brand-Specific SCBA Features

Many of the SCBAs being manufactured today have specific features regarding supplying air to a downed fire fighter. Consult with the manufacturer to learn about these features and incorporate what you learn into your RIC training.

Monitor Radio Traffic

During the 360-RECON and continuing once you are back at the staging area, all members of the RIC should constantly monitor radio traffic. Listening to company progress reports, updates, and additional assignments will help you to know what is happening, who is operating where, and, most of all, how the overall operation is progressing. When listening to what is being said and how it is being said, does the tone of the voice calling suggest urgency or are people having trouble sending and receiving messages? Notice if the radio traffic is out of control or if it is disciplined in manner. It is not unusual for radio traffic to have a significant decrease in quality when things go wrong—numerous transmissions trying to be sent at the same time, being unable to get a message through, and unimportant traffic getting priority can be a warning sign that the communications protocols and practices on that fireground are not working. Repeated calls for charging a line or getting assistance, or urgent calls for missing stairs, holes in floors, or deteriorating conditions can also provide advance notice of the likelihood of RIC activation.

■ Fireground Radio Communications and Preplanning the Radio Protocol if RIC Is Activated

Exceptional radio communications during a rapid intervention activation is a necessity. Failure to achieve this level of radio communications could be a warning sign of failure. The entire issue of fireground radio communications is being discussed at the highest levels, and manufacturers are working to find technical solutions to ensure that the radios will work when needed. However, it is important to focus on what team radio protocols to use now and not wait for technology advances for answers. The discussions revolve around two basic questions when a MAYDAY is called: (1) does the downed fire fighter change channels to an RIC frequency, or (2) does the downed fire fighter stay on the fireground channel with the RIC operation and all other fireground suppression companies move to another channel?

Realistically, a downed fire fighter who calls the MAYDAY and is potentially trapped, lost, disoriented, and in heavy smoke should *never* be asked to change radio frequency channels. Upon receiving and acknowledging the MAYDAY, the IC should declare radio silence before moving anyone to another channel. When an IC declares radio silence, and it is immediately abided by all members, it will improve the chances of confirming the MAYDAY and obtaining additional rapid intervention information.

> ### Safety
>
> A downed five fighter who calls the MAYDAY and is potentially trapped, lost, disoriented, and in heavy smoke should *never* be asked to change radio frequency channels.

In the chapter *Activation of the RIC and Organizational Considerations*, the differences between a rapid intervention and an extended intervention will be discussed. For any extended intervention operation, the fireground suppression operations should be moved to another channel. Moving to another channel is not simple or easy, and it is highly urged that it be coordinated with the personnel accountability report (PAR) that must be conducted. The extended intervention operation will then remain on the original channel, which will ensure that the downed fire fighter has on the correct radio frequency and that the rescue operation has a dedicated frequency on which to operate and possibly communicate with the downed fire fighter.

Moving the fireground operations to another channel provides three benefits: (1) it provides a clear channel to the rapid or extended intervention operation; (2) it limits those hearing the RIC operations transmissions, and therefore fire fighters will be less prone to freelance or mutiny on their company officers thinking that they "hear something;" and (3) it moves significant emotional transmissions that the downed fire fighter might be sending. It is not uncommon in past incidents for fire fighters who are trapped and dying to transmit very disparaging statements and requests to be heard by everyone. Hearing these types of transmission is difficult and emotional for all, so moving the majority of the operation to another channel limits those listening.

Creating Egress/Access Points

While staged and under the direction of command, if manpower is shorthanded, the RIC can be used to create emergency egress and/or access points to support fire fighter safety and survival. They can also be used to position ladders as directed by Incident Command for fire fighter emergency egress or to "ladder the building" (positioning ladders to all floors where companies are working) to ensure that emergency egress is provided to every floor fire fighters are working on that ground ladders can reach. Utilizing an RIC to achieve tactical considerations other than rapid intervention actions, however, can fatigue the company, cause them to lose focus on the incident, and put them in a position where, if activated, there might be a significant delay in activation.

Team Management: The RIC Operations Group Supervisor and RIC Company Officer

The management of an RIC is a very important consideration on every fireground. Traditionally, and especially in smaller departments, the initial RIC is managed exclusively by the Company Officer reporting directly to the IC. Upon activation of the RIC, the IC has to staff an RIC Operations Group Supervisor position. This takes time and is really playing catch-up during a MAYDAY operation. A solution for this, and to increase the effectiveness of any RIC and possible operation, is to staff the position with the arrival of the RIC as a part of the initial action plan.

■ RIC Operations Group Supervisor

Getting a MAYDAY call for a fire fighter in trouble is one of the worst nightmares an IC could ever have. It can take the entire operation and turn it upside down! One of the most important things that can be done on a fireground with any RIC activation and a fire not yet under control is to control the fire. The question becomes how can an IC maintain a good fireground operation and run a rapid intervention activation at the same time? The answer is, he cannot. The jobs of fire containment and rescuing the downed fire fighter are two jobs divided into the Incident Command and the RIC Operations Group Supervisor.

The IC continues to be responsible for the overall fireground operation actively happening. He knows the strategies and tactics being used and is in charge overall. The RIC Operations Group Supervisor position is best assigned when the RIC is assigned at an incident. Having an RIC Operations Group Supervisor with the staging of the RIC provides optimal accountability, control, and communications with the RIC both preactivation and in the case of the team being activated.

The term Group Supervisor is used to keep within the terminology of the National Incident Management System (NIMS) and the Incident Management System (IMS). In reality however, this position is best filled with a Chief Officer *who is RIC trained* **FIGURE 3-9** . Not having RIC training will put this Chief immediately into a weak position, as he is not familiar with the skills and concepts of RIC; has no idea what efforts are needed by a team to accomplish their goals; does not speak the "language" of the RIC; and by not being trained or experienced in RIC Operations, he will not be able to properly stay proactive and anticipate RIC strategies, tactics, and needs. If a Chief Officer is not available or has little or no RIC training, the position should be filled by the most qualified officer available.

The immediate staffing of this position during preactivation will create a stronger command structure as indicated:

- From the initial RIC 360-RECON, the RIC Operations Chief is interacting with the RIC company members.
- The RIC reports to someone who has the responsibility and focus for the RIC operation.
- As the incident progresses, and while in preactivation, the RIC Operations Chief can anticipate possible activation scenarios and discuss them with the RIC. This planning and preparation can help to organize several MAYDAY strategies, and from this, the company can alter or add to the RIC equipment cache, communicate with any other command positions freely, and keep the RIC focused and informed.

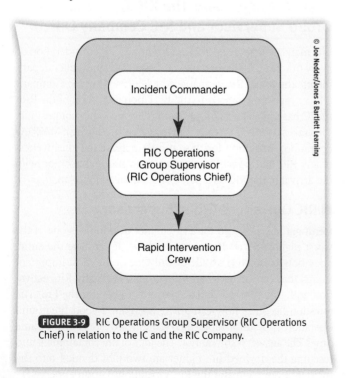

© Joe Nedder/Jones & Bartlett Learning

FIGURE 3-9 RIC Operations Group Supervisor (RIC Operations Chief) in relation to the IC and the RIC Company.

- In the event of a MAYDAY or RIC activation and working with and through the IC, the RIC Operations Chief immediately takes over as the "IC" for the rescue operation. Although he is not called a Commander or RIC Command, he is responsible for the entire RIC operation. Most likely he will be referred to as RIC Ops or RIC Operations. Being up to par with the incident by having been in this position and planning preactivation, the RIC Operations Chief will reduce the chances of becoming overwhelmed. He will start prepared and proactive. It is this type of forward thinking that can and will make a difference in the end results of an RIC operation when fire fighters' lives are at stake.

■ RIC Company Officer

During the staging of an RIC team, it is the Company Officer's responsibility to manage his company. This includes, but is not limited to, making position assignments and making sure each member has taken the needed actions to ensure his position is properly prepared and tooled up; maintaining crew integrity and accountability; keeping all members focused on the situation before them, anticipating what might go wrong, and how they will react as a team; and assessing the capability of the team.

In addition, it is important for the RIC Officer to assess what his team is capable of, to understand any limitations on a member's part, and if activated, to plan how the team will react. Remember, the Company Officer's primary job is to safely lead the team in and out intact and unharmed. While the RIC is in preactivation, the Company Officer should frequently interface with the RIC Operations Chief to discuss changing conditions; anticipate the team's activation; request team needs; and, with the entire company, monitor the radio traffic and observe, by listening and looking, how the incident is progressing.

Prolonged Preactivation Operations

Ideally it is best to always keep an RIC intact without any change out. For a "bread and butter" operation, single alarm, this is very achievable. The advantage of keeping one crew intact is that they are ready, they know the situation, and they are prepared to react. However, reality tells us that as a fireground operation grows, the timeline of the event also grows. Expecting one crew to standby as an RIC for an extended period and through multiple alarms will cause mental fatigue; anxiety, because the crew wants to actively "go to work;" and, in the case of extreme weather (heat or cold), the need for some rehabilitation.

Before an IC can rotate an RIC out with a replacement company, important actions need to happen. First, the replacement crew must be physically at the RIC staging area. Next, the Officer from the active crew needs to share the knowledge he has gathered and gained with the Officer of the replacement crew. Crew members must review where the specialty tools are located and why they were selected, and any SCBA interoperability issues should also be discussed. Information regarding egress points and ladder placement needs to be shared. This enables the replacement company to be "good to go" and just as prepared as the crew it is replacing. When the transfer happens, the new RIC should report to the IC or

RIC Operations Chief that they are staged and ready. It is also advisable that, if possible, the new crew conducts a quick 360-RECON in order to visually establish the fire situation.

When Does the RIC Position Stand Down?

The RIC should be kept intact and in place as long as fire-fighting activities are going on within or around the structure.

This includes during overhaul and the removal of lines and equipment from the building. It has not been an uncommon event to experience partial interior collapse (ceilings and floors) during the overhaul process. Even though the fire has been extinguished, visibility has been improved, and the smoke condition and fire gases have mitigated, having an RIC ready to go in case of a fire fighter in distress is not only advisable, but recommended.

Wrap Up

Chief Concepts

- When a company is assigned to be the RIC, they must immediately become proactive for this important combat position.
- The RIC needs to be ready to go at a moment's notice. Staying proactive will help them to be mentally prepared.
- Within the rapid intervention discipline, being proactive refers to an officer or company that tends to prepare for an occurrence *rather* than react to an event after the fact. It is a position of anticipation!
- Ideally, an RIC will have four members whenever possible.
- Remember, as the RIC, you are responsible not only for finding the distressed fire fighter, but also for getting yourselves out alive and unharmed.
- Once assigned or upon arrival on the scene, the RIC needs to always stay together and move as a company.
- The RIC Officer must report to the IC or RIC Operations Chief and get a brief report of what is happening as soon as possible.
- It is always best to assign an RIC-trained Chief as the RIC Operations Group Supervisor (RIC Operations Chief).
- The 360-RECON and RIC size-up should be a priority as soon as the face-to-face meeting with the IC has been completed.
- The purpose of the 360-RECON is to help the RIC to stay proactive and be better prepared by having a good visual understanding of the building involved, the overall operation, and the smoke and fire behavior.
- Understanding fire behavior will assist you in observing what is happening at the scene.
- The RIC must not only observe fire conditions, but also monitor smoke conditions and behavior, monitor the fireground operations, monitor radio traffic, identify any possible hazards, and if possible, identify what type of construction the building is.
- Always continue to monitor what is going on and resize-up the situation as conditions change. The RIC should

base their tool/equipment selection on the size-up of the structure.
- The RIC should always be prepared to deliver emergency air to a downed fire fighter. This is probably the single most important function to increase survivability for a downed member.
- The RIC must ensure the compatibility of emergency air to the SCBA types utilized on the fireground.
- In the event of RIC activation that will be an extended operation, all radio traffic for suppression operations should move to another channel, while the RIC stays on the same channel as the distressed fire fighter.

Hot Terms

<u>Buddy breathing</u> Sharing one's air with another via a system on some SCBAs.

<u>Hydraulic cutters</u> A powered rescue tool consisting of at least one moveable blade used to cut, shear, or sever material.

<u>Hydro-force style tool (rabbit tool)</u> Hydraulically operated hand tool used to force doors.

<u>K-12 style saw</u> A rotary cut off saw used for cutting through wood, metal, or concrete. The saw blade can be changed to a specific blade to cut the desired material.

<u>Mask-mounted regulator (MMR)</u> The SCBA low-pressure regulator that attaches to the breathing mask. For most manufacturers, this regulator is removable and only mounted when the fire fighter goes on SCBA air.

<u>Personnel accountability report (PAR)</u> An accountability report or roll call taken in place.

<u>Proactive</u> Within the rapid intervention discipline, being proactive refers to an officer or company that tends to prepare for an occurrence *rather* than react to an event after the fact. It is a position of anticipation!

Wrap-Up, continued

Rope-assisted search procedure (RASP) bag Typically 150 to 200 feet (46 to 61 m) of less than ½-inch (13-mm) rope used to assist in searching large areas.

RIC Operations Group Supervisor This position is best filled with a Chief Officer who is RIC trained.

RIC rescue air unit Self-contained air supply used to provide air to a downed fire fighter. Can be an SCBA or a specialty unit bought specifically for RIC air use.

Universal air connection (UAC) A device installed on all SCBA and a requirement per NFPA, which allows a direct air fill to a SCBA cylinder from another SCBA/RIC emergency air system.

References

National Fire Protection Association (NFPA) 1407, *Standard for Fire Service Rapid Intervention Crews*. 2015. http://www.nfpa.org/codes-and-standards/document-information-pages?mode=code&code=1407. Accessed February 11, 2014.

National Fire Protection Association (NFPA) 1981, *Standard on Open-Circuit Self-Contained Breathing Apparatus (SCBA) for Emergency Services*. 2013. http://www.nfpa.org/codes-and-standards/document-information-pages?mode=code&code=1981. Accessed October 14, 2013.

U.S. Fire Administration/Technical Report Series. *Three Firefighters Die in Pittsburgh House Fire*. Pittsburgh, PA; USFA-TR-078/February 1995. http://www.usfa.fema.gov/downloads/pdf/publications/tr-078.pdf. Accessed October 10, 2013.

RAPID INTERVENTION CREW MEMBER
in action

Your engine company has been assigned to the position of RIC while responding to a confirmed working structure fire. As the Company Officer you have the responsibility of gathering as much information as possible while responding. You also must assess your crew capabilities and assign the crew their positions. Being a proactive RIC cannot be overstated.

1. What is the first thing the Company Officer should do once he has been assigned RIC?
   A. Assign the company their positions.
   B. Report to the staging officer.
   C. Gather the necessary tools for deployment.
   D. Conduct the 360-RECON while the rest of the company prepares for deployment.

2. What should the brief face-to-face conversation with the IC include?
   A. What the situation involves.
   B. What companies are assigned and where.
   C. The current status of the situation.
   D. All of the above.

3. What are the benefits of conducting a 360-RECON?
   A. It makes the RIC feel important.
   B. RIC can be thought of as a boring job and the 360-RECON will help to keep the crew from being bored.
   C. Vital information can be gathered about the type and size of the building, as well as the conditions of the fire.
   D. It allows the IC to concentrate on suppression activities.

4. What responsibilities does the Company Officer have while staging his crew for RIC?
   A. Making position assignments and making sure each member has taken the needed actions to ensure his position is properly prepared and tooled up.
   B. Maintaining crew integrity and accountability.
   C. Keeping all members focused on the situation before them, anticipating what might go wrong, and how they will react as a team.
   D. All of the above.

5. Why is it important to assign an RIC Operations Group Supervisor as quickly as possible on any fireground?
   A. It increases the effectiveness of any RIC and will provide a qualified immediate Group Supervisor for an RIC activation.
   B. It gives another Chief Officer a position to fill.
   C. The IC has someone who can assist him during fireground operations.
   D. If there is an RIC activation, the IC has someone to assist him running the RIC operation.

Activation of the RIC and Organizational Considerations

Knowledge Objectives

After studying this chapter you will be able to:

- Define a MAYDAY versus emergency traffic. (NFPA 1407, 7.4(1), 7.5(1), 7.5(2) , p 46–47)
- Understand how to activate and manage an RIC operation. (NFPA 1407, 7.1(2), 7.1(5), p 47–48)
- Understand the importance of an incident accountability system on all firegrounds. (NFPA 1407, 7.1(4), 7.1(6), 7.1(7), 7.1(8), p 48)
- Know the difference between a rapid intervention operation and an extended intervention operation. (NFPA 1407, 7.1(9), p 47–48)
- Describe the difference between a rapid intervention and an extended intervention. (p 47–48)
- Define the term personnel accountability report (PAR) and explain what it does. (NFPA 1407, 7.1(4), 7.1(7), 7.1(8), 7.5(2), p 49)
- List six actions an Incident Commander (IC) will take during a rapid intervention crew (RIC) activation for a MAYDAY. (NFPA 1407, 7.2, p 48–50)
- List seven ways an IC can remain proactive. (p 48–50)
- Understand the importance and procedure of securing a dedicated communications channel for RIC operations (NFPA 1407, 7.11(3), p. 49)
- Understand the importance of protecting the downed fire fighter during an RIC activation, (NFPA 1407, 7.4(6), p 50)
- Describe the role of the RIC Operations Group Supervisor. (NFPA 1407, 7.2, p 50–51)
- List the six key actions of an RIC Operations Group Supervisor. (NFPA 1407, 7.2, p 50–51)
- Describe why the position of an RIC Operations Group Supervisor is best filled by a Chief Officer and what the qualifications are for this position. (NFPA 1407, 7.2, p 50)
- Describe the number one job of the RIC Company Officer. (NFPA 1407, 7.2, p 51)
- List seven performance standards for the RIC Company Officer. (NFPA 1407, 7.2, p 51–52)
- Describe risk management. (NFPA 1407, 7.1(1), p 52–53)
- List four examples of proper feedback. (NFPA 1407, 7.1(3), p 52)
- Describe the term CAN and how it will help the IC to anticipate. (p 53)

Skills Objectives

There are no skill objectives for Rapid Intervention Crew candidates. NFPA 1407 contains no Rapid Intervention Crew Job Performance Requirements for this chapter.

Additional NFPA Standards

- NFPA 1521, *Standard for Fire Department Safety Officer*
- NFPA 1561, *Standard on Emergency Services Incident Management System*

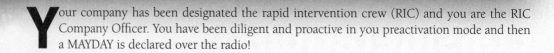
Your company has been designated the rapid intervention crew (RIC) and you are the RIC Company Officer. You have been diligent and proactive in you preactivation mode and then a MAYDAY is declared over the radio!

1. With the activation, what is your number one job as the RIC Company Officer?
2. What must you do as the RIC Company Officer to execute a rapid and safe search?
3. What type of feedback is important to provide to the RIC Operations Commander?
4. As you search for the missing fire fighter, what can you do in regard to risk management?

Introduction

When a MAYDAY is declared on a fireground, it is a downed or distressed fire fighter pleading for help! We need to be capable and ready to go, or the end result might be a <u>line-of-duty death (LODD)</u>.

In this chapter, we will look at the MAYDAY process, the protocols for activating the RIC, and the tremendous responsibilities placed upon not only the crew, but also the Incident Commander (IC) and his staff. The integration of the RIC operation into the <u>Incident Command System (ICS)</u> will enable the command staff to manage the process in an organized manner. A MAYDAY also places burden and strain on all fire fighters operating on the fireground. Everyone operating on the fireground has specific responsibilities to support a MAYDAY. It is important to understand and realize that not all MAYDAYs end with a life being saved. Risk assessment and management plays a key role in this type of operation and will help determine if a transition from a rescue mode to a recovery mode is needed.

Fire fighters are frequently exposed to tremendous emotional hardships. A MAYDAY that ends with severe injury or LODD will tax the emotional capabilities of the department more than can be imagined. We must be able to seek <u>critical incident stress debriefing (CISD)</u> assistance quickly for our personnel.

Activation of an RIC is not a simple matter. This chapter will assist you in planning your decisions and actions before they are needed.

■ History of the Term MAYDAY

Frederick Stanley Mockford developed the term MAYDAY in 1923. Mockford, a British senior radio officer at Croydon Airport in London, was asked to find a word that could be used to indicate distress and could be easily understood by all pilots and ground personnel in the event of an emergency. Much of the air traffic Mockford worked with was between Croydon Airport and an airfield called Le Bourget in Paris. He determined that in French the term "come and help me" was "Venez m'aider." Taking the word "m'aider" he proposed the word "mayday." It was adopted and quickly became a term used internationally.

MAYDAY calls were specifically to announce a situation where an aircraft, vessel, vehicle, or person was in grave and imminent danger and that immediate help and assistance was required. To this day, when the term is used, the accepted protocols are:

1. When a MAYDAY is called, no other radio traffic is permitted on that frequency except to assist in the emergency.
2. A MAYDAY call may only be made when life or craft is in imminent danger of death or destruction.
3. The call is given three times in a row: "MAYDAY, MAYDAY, MAYDAY." This prevents it from being mistaken for a similar sounding phrase under noisy, loud conditions and to help to distinguish the actual MAYDAY call from a message or radio traffic about a MAYDAY.
4. When a MAYDAY is called, additional information such as the name of the vessel, position, nature of the emergency, assistance required, and number of people on board is also provided.

As you read and study this chapter, you will find that this internationally accepted term and the established protocol fits perfectly into the fire service, if used properly and when needed.

Declaring a Fire Fighter Emergency

What constitutes a MAYDAY? How is it different from "emergency or urgent traffic?" What term should we use and does it really matter? Many things are standardized in the fire service: apparatus terms, what we call different equipment (i.e., LDH, "the irons," etc.). Standardization and common-use terminology helps us all to be consistent, communicate (via training seminars, periodicals, text books) with other fire fighters from different areas and understand what is being said, gives us the ability to work together at a mutual aid fire, and provides uniformity to the industry.

■ MAYDAY!

The term MAYDAY should be reserved for fire fighters in extreme distress calling for immediate help! In essence, given that we work in a toxic, smoke-filled environment with a limited air supply, the term MAYDAY can actually be equated to a fire fighter calling for immediate help or he might die. When a fire fighter calls a MAYDAY, all other radio traffic should immediately cease. Nothing has higher radio priority than a MAYDAY call. Examples of extreme fire fighter distress are:

- Separated from the company and lost
- Trapped or entangled in wires or debris

- Running out of air or having run out of air due to a self-contained breathing apparatus (SCBA) failure
- Fire fighter is having a medical emergency (unconscious, heart attack) and a company officer or partner calls the MAYDAY
- Trapped or caught in a structural collapse

Emergency Traffic

The term emergency or urgent traffic should be used on the fireground when you need to get a message out indicating an imminent hazard or a situation that is of immediate danger. Examples of emergency traffic include:

- Fire fighters exiting the building very low on air
- Reporting a blown or uncharged hoseline
- Structural hazards that put fire fighters in immediate danger, such as a hole in the floor, missing stairs, or interior collapse
- Rapid fire spread, flashover, rapidly changing conditions, and other fire behavior indicators that change the situation in a drastic way
- To report missing civilian occupants to working companies
- To report finding missing occupants

The call "emergency traffic" is of high priority. All other radio traffic should stop until the emergency message gets through and is acknowledged. All fire fighters operating on the fireground should stop and listen to the message. It might be reporting a hazard or situation that could put you in greater harm's way if you were unaware.

MAYDAY, MAYDAY, MAYDAY! Now What?

When a MAYDAY is called, many actions need to be taken to react quickly and efficiently. The roles we will look at include the IC, the RIC Operations Group Supervisor (or RIC Operations Chief), the RIC Company Officer, all Company Officers operating at the incident, and all fire fighters operating at the incident.

Safety

If an RIC is not on scene and ready, they are not an RIC! Forming a crew when a MAYDAY is called is in no one's best interest, least of all the fire fighter in trouble!

Crew Notes

To be consistent with the terminology of the ICS, we are using the term RIC Operations Group, which will typically be run by a supervisor or RIC Operations Group Supervisor (or RIC Operations Chief). Although it is recommended that this position should be filled by a Chief Officer, for the purposes of this book, the terms RIC Operations Chief, RIC Operations, and RIC OPS will be used.

Activation of the RIC

The term rapid intervention has been used to describe most any type of fire fighter rescue, regardless of the complexity or length of time needed. In reality, there is a difference between a rapid and an extended intervention. We must understand these differences and accept that not all intervention rescues will be done rapidly and, in some cases, successfully. **TABLE 4-1** delineates the differences between a rapid and extended intervention operation.

Rapid Versus Extended Intervention Operations

Generally, when the RIC must deliver emergency air to a downed fire fighter, the rescue operation has really gone from a "rapid" to an "extended" intervention operation. Anytime the RIC activation for a MAYDAY becomes an extended intervention operation, the risk levels for all involved dramatically increase and the need for competent resources greatly increases as well.

TABLE 4-1	Rapid Versus Extended Intervention Operations

Rapid intervention operation:

- Usually limited to one SCBA cylinder
- Found quickly, obvious rescue (i.e., a grab and go!)
- RIC assists the fire fighter victim with self-rescue (i.e., light entanglement, disorientation in a small structure)
- RIC positions a ladder for a fire fighter in need of rescue (i.e., the fire fighter or company has been trapped by fire and needs to quickly exit a window)

Extended intervention operation:

- Organized search for a missing, trapped, or distressed fire fighter with an unclear location
- Use of a search (RASP) rope
- The fire fighter in trouble is in need of SCBA air immediately
- Multiple RIC teams have been activated or will be needed
- Entanglement
- Extrication (floor or ceiling collapse)
- Advanced life support (ALS) intervention (the fire fighter has experienced a traumatic injury or has a significant medical emergency and is possibly unconscious or unable to assist in his own rescue, including an initial MAYDAY radio communication)
- Initial RIC has consumed their SCBA supply and another team is needed to continue the search or assist in the rescue/removal
- Multiple fire fighters are in need of rescue (i.e., ceiling collapse trapping a company or floor collapse)

Not only is there a demand for personnel to have great skill sets with tools, but also there could be greater demands for technical rescue knowledge. For example, if there is an interior structural collapse trapping a fire fighter, the RICs operating must have an understanding of aspects including secondary collapse, structural path loads, shoring, and much more. If you and your department are not proficient in technical rescue aspects, you must identify in your preplans and run cards available resources that can be brought in quickly. Unfortunately, delays in getting this type of assistance do not bode well for the survival of the fire fighter(s) who are in the MAYDAY situation.

For an extended intervention, the organizational aspects will stay the same: the RIC Operations Group Supervisor (or RIC Operations Chief) reports to the IC, the crews are still identified as RIC teams, but it is understood that the operation will most likely be an extended operation. The length of time that qualifies as "extended" is not known, but you must understand the differences and provide the needed personnel and equipment. Based upon the information received, the IC must determine if it is a rapid or extended intervention, and if the IC is unsure, plan for the extended intervention and gear down if it is not needed.

Rapid Versus Extended Intervention Radio Protocols and Company Accountability Considerations

A rapid operation, as described in Table 4-1, or for example, if a fire fighter has a medical emergency and receives assistance from his own company right away to get out, typically will not require a personnel accountability report (PAR) or that the entire operation be moved to another radio channel. Extended operation rescue situations, however, as detailed in Table 4-1, will require that the IC conduct a PAR. Its purpose will be to determine if the MAYDAY is isolated or if additional fire fighters or companies are in trouble.

Role and Responsibilities of the IC

When a MAYDAY is called, the IC must cease all other activities and answer the call for help. A simple radio communication is "command answering the MAYDAY." The distressed fire fighter will hopefully be capable of providing you their location, problem, and identification. Once this information has been obtained and acknowledged, tell the fire fighter that help is on the way, and prompt them to activate their personal alert safety system (PASS) device. It is important to remember that the distressed fire fighter needs reassurance that help is coming. How the fire fighter calls for help, the acronyms used to send the pertinent information, and the radio procedures will be discussed in the chapter *Rapid Intervention and Self-Rescue*.

The IC must take specific actions, including:

- Notifying the RIC Operations Group Chief to activate the RIC.
- Notifying the fire alarm or dispatch that a MAYDAY is in progress.
- Notifying all companies operating of an RIC activation.

If it is a *rapid* intervention, all companies working should be notified that there is an activation of the RIC. Many times

in a rapid intervention, the rescue might be a quick "grab and go," and the fire fighter is safely outside before there is even time for an announcement to all the companies working or an attempt to move radio channels. The key here is to identify the situation and take the appropriate actions that will not overcomplicate the situation.

If it is an *extended* intervention, along with the items listed previously, the IC should:

- Tell the companies to standby for a PAR.
- Conduct a PAR.
- Move the fireground channel to a different radio frequency, leaving the downed fire fighter and the RIC operation on the channel in current use.

In addition, the IC should:

- Ensure that ALS is on scene.
- Get extra help.
- Commit resources to protect the fire fighter in distress and the RIC.

Notify the RIC Operations Chief to Activate the RIC

After the IC has answered the MAYDAY call, the RIC needs to be activated by notifying RIC Operations. A competent RIC will have heard the MAYDAY call and started to gear up. Provide the RIC with the information obtained. Who is missing, and from which company? What was their assignment? Where (location, if it is known) is the fire fighter calling the MAYDAY? Is the fire fighter lost, stuck, trapped? What is the problem and situation? What is the status of his air supply? Much of this information can be obtained by maintaining a good accountability system, from which it can be determined which companies are operating, what their assignments are, where they should be located, and how many fire fighters are in the company. Such information will help the RIC to give them focus and a possible target area for the search.

Notify Fire Alarm or the Dispatcher

The IC also needs control of the airwaves. If the department is reasonably disciplined in good radio procedures and protocols, radio silence can be declared. Alternatively, notify the fire alarm, dispatcher, or whoever is responsible for base transmissions that there is a MAYDAY RIC activation in progress and have them announce that there is an RIC activation at your fireground location. An excellent way for the dispatcher to clear the airwaves to make the transmission is to use an emergency traffic tone, or if your department does not have this, they can activate the standard fire department tones used for call notification to the fire personnel and announce that there is a MAYDAY. Regardless of what method is used, the IC needs to gain control of all radio traffic immediately.

Notify All Companies Operating of the RIC Activation

If RIC Activation Is a Rapid Intervention

Notify all companies working on the fireground that there is an RIC activation, which will serve two purposes. First, it will, with good discipline, help to maintain radio silence or at least limit the radio transmissions to only those urgently necessary, such as safety concerns, deteriorating fire conditions, or feedback from the RIC.

If RIC Activation Is an Extended Intervention

Notify all companies working on the fireground that there is an RIC activation and to stand by for a PAR. A PAR is an accountability report or roll call taken in place. Personnel accountability reports (PARs) can be used not just during an RIC activation, but at other times as well. For example, when a company exits the building, the officer should make sure he "has PAR," which means that all the fire fighters on that company are accounted for.

When the IC notifies the companies to standby for a PAR, each officer should quickly determine that his crew is intact. At this point there are two methods to conduct the PAR. The first, primarily used in smaller organizations with limited staffing, is for the IC to conduct a PAR on the radio or to delegate the PAR to the Incident Safety Officer (ISO). The second option, usually done with organizations that have ample staffing on the fireground, is for the PAR to be done face-to-face. Company officers report to Division or Sector Chiefs, who report to Operations Chiefs, who then report to the IC. If this is possible, it will greatly reduce radio traffic and free the IC and/or ISO from the task, allowing them to focus on other critical actions needed.

For the purposes of this example, we will focus on a radio roll call for PAR. Begin the PAR or roll call, and each company operating should be contacted and should reply "all members accounted for" or they "have PAR" if all members are accounted for. If they do not have PAR, they need to issue a MAYDAY for another missing fire fighter. The need for a PAR during the RIC activation will help determine that no other fire fighters are in trouble or distress. *Never assume* that no one else is in trouble.

Conducting a PAR

A MAYDAY is received from a fire fighter on Ladder 1, reporting that he is trapped under a ceiling collapse. From our accountability system, it is known that L-1 (officer and two fire fighters) was sent to assist Engine 6 (officer and two fire fighters) operating on the second floor of the B/C corner of the building.

- **Command:** to all Companies operating, standby for a PAR. (Allow a few moments for the officers to check their crews.)
- **Command:** to Engine 1
- **Engine 1:** we have PAR
- **Command:** to Engine 12
- **Engine 12:** all members accounted for
- **Command:** to Ladder 1 (*and then you hear this message*)
- **Ladder 1:** Command, I can't find my crew!

From the accountability system, you know that L-1 has an officer and two fire fighters. One MAYDAY from L-1 has been received, but you now have two fire fighters from L-1 in trouble! Make sure you continue the PAR.

- **Command:** to Engine 6 (*no answer*)
- **Command:** to Engine 6 (*still no answer*)

From the accountability system, you know L-1 was assisting E-6 with containment and overhaul. The initial MAYDAY from L-1 reported a fire fighter trapped under a ceiling collapse, L-1 was working with the three members of E-6, and we have

no answer from E-6. The question now becomes: is the crew of E-6 in trouble or is it a radio issue? For safety's sake, you need to assume that the crew of E-6 is also in trouble. From the original one MAYDAY received, after the PAR, it has been determined that there are possibly five fire fighters in distress. The size and scope of the RIC operation has just changed dramatically!

Conducting a PAR in place is a critical element of an RIC operation when the problem is consistent (i.e., collapse, rapid fire spread, structural failure) with putting other fire fighters at greater risk and will assure that you know how many fire fighters are in trouble.

Moving to a Different Radio Channel

It is important to have a preplan for radio procedures during an RIC activation. Moving a fireground operation to a different channel can be difficult and confusing. Did everyone move? Did everyone hear the transmission to move to channel "X?" In conjunction with this, should the PAR be conducted on the new channel or the one you started with? These are all legitimate questions, and the answers can and will impact your operation.

There are two options with regard to which channel the PAR should be called:

Option 1. As soon as you can, notify the dispatcher or fire alarm that you are moving the fireground operation to another channel (you must designate which channel). Have the dispatcher make an announcement to all companies working at your location to immediately move to channel "X." The RIC operation will stay on the channel the fireground had previously been operating on. This will ensure a clear channel and allow the RIC, the RIC Operations Chief, and the downed fire fighter to communicate without being impeded by the operational radio traffic.

Option 2. Keep all companies on the initial fireground channel, and as you conduct the PAR have the company switch channels upon giving their company's PAR. This method will control the channel switchover and make sure that it is done in an orderly and safe manner.

The second option has the advantage of providing the IC with more control over the situation. For example, with the first option, the IC has fire alarm or dispatch order everyone to immediately switch to channel "X." His intent is to conduct the PAR on the new channel, thus keeping the first channel open for the RIC rescue operation. However, if one or two company officers fail to move to the new channel in their desire to "assist" with the RIC operation or if they move to the wrong channel, those companies not on the correct channel will not hear PAR being called on the radio and will not answer the IC. The IC now needs to determine if those companies are also in trouble and additional help is needed, or if those companies are on a different channel. Either way, control of the situation has begun to break down, and precious time is being wasted trying to figure out what is going on.

Ensure ALS Is On Scene

When a downed fire fighter is found and rescued, his or her survivability possibilities might be in question. It is important that the best possible emergency medical services (EMS) capabilities are available on scene, which means having paramedics or the highest EMS capabilities your region offers standing by and multiple ambulances available on scene. Given that this is an RIC operation and that fire fighters are now operating in an even higher danger mode, the availability of quality EMS, at the ALS level if available, is paramount.

Get Extra Help!

While all this is going on, the IC needs to think ahead. One thing that usually is synonymous with an RIC activation is the need for additional help. Be proactive, anticipate the need for a lot more help, and request it. This might be in the form of striking an additional alarm, calling the dispatcher and requesting specific companies, and/or requesting mutual aid. Whatever procedure your department uses, the need for additional help is imminent, and needs to be requested immediately.

Commit Resources to Protect the Fire Fighter in Distress and the RIC

The IC must evaluate the situation and dangers that the downed fire fighter is exposed to and what dangers the RIC is entering into. Required resources to protect the RIC need to be identified and committed. At the same time, it is critical that the IC maintain the firefighting efforts to control and extinguish the fire. Firefighting positions must still be maintained along with deploying or redeploying resources to protect the RIC operation. Fire control and extinguishment will affect the RIC operation and rescue by helping to remove that element of danger.

Role and Responsibilities of the RIC Operations Group Supervisor or Chief

The RIC Operations Group Supervisor is in charge of and responsible for the entire RIC deployment. This includes the crew(s) currently operating and the staging of additional RICs. The RIC Operations Group Supervisor is also responsible for maintaining contact with the crew(s) and getting updates and feedback from the company officer(s). The Operations Group Supervisor should be very proactive, which can be achieved through six key actions:

1. Managing the operation
2. Developing a plan and a backup plan
3. Staging and prepping additional teams
4. Anticipating special tools needs
5. Providing feedback to the IC
6. Communicating with the distressed fire fighter. Because of the responsibilities of position, it is strongly suggested that it be filled by a Chief Officer with RIC training and knowledge.

■ Managing the Operation

An RIC operation is a very intense situation—everyone wants to be involved in the rescue of the fire fighter! The RIC Operations Chief must take charge and show command presence. Constantly evaluate what progress, if any, is being made; stage, prepare, and update more teams; and anticipate as best you can. For example, if the downed fire fighter is found, how quickly can we activate another RIC to assist in the rescue? Will they be looking for egress from a window? Are the additional RIC teams in staging prepared to aggressively throw a ladder to the designated window and carry the fire fighter down? The crews will be looking to the RIC Operations Chief for direction and guidance. This position is a key component to any RIC activation. As such, it cannot be overstated that the RIC Operations Chief must have prior training in rapid intervention, including the skills required, in order for this critical operational position to function properly.

■ Developing a Plan and a Backup Plan

Think quickly, anticipate, and be prepared to change as the situation or conditions warrant. *Stay proactive!*

- **Observe.** Observe the building and watch for any changing conditions for the better or worse. Watch for suppression progress: are they getting it or not? Is the situation getting out of control?
- **Monitor.** Monitor the radio transmissions from your crew. Based on the feedback you are getting from the crew, is the RIC operation progressing, or have conditions made the overall situation worse or untenable? Monitor the "time in" of the RICs. Remember the RIC also has limited air and that they are operating under extreme stress and possible anxiety. How will this affect their air supply and active work-time? Select time benchmarks and call the RIC, prompting them to check their air supply. A situation in which an RIC runs short or out of air is not needed!
- **Consider.** Consider the survivability factors of the downed fire fighter. When the MAYDAY was called, what was his air status? How long has it been and what do you think his air status is now? Has the fire fighter been without any breathable air for 20 or 30 minutes?

Thinking about these considerations will assist in doing a risk versus benefit analysis. Sometimes the situation is going to be hopeless—past events have proven that. The question at hand *at that given moment* might be the survivability status of the fire fighter you are looking for versus the deteriorating conditions that the RICs might be encountering. If the answer is that the downed fire fighter could not have survived the given situation, and the crews are getting into greater danger with the possibility of things going wrong for them, it is time to back the crews out.

■ Staging and Prepping Additional Teams

Most RIC activations will utilize numerous crews. Past instances indicate that, although there is no hard and fast rule, being prepared to utilize four or more crews is not out of the question. The RIC Operations Chief works with the IC and requests more help that can be pre-staged as additional RICs. The sooner you can assemble and stage additional crews, the better you will be able to react to requests for additional help or to replace a crew that needs to back out. While staged, these crews need to be prepped. They need to know what has transpired, the situation as it stands currently, and what, if any, progress is being made. The crews that are staged should be focused and monitoring the RIC operations radio channel.

Think about radio designations for your additional crews. If the company that has already been activated is the RIC, what do you call the other crews that are staged? Is it RIC-2,

RIC-3, or is it something else? Always be proactive! Training fire fighters in advance will make the real thing go more professionally and with fewer distractions.

■ Anticipating Special Tools Needs

Consider and anticipate what other special tools you might need. This is resource planning. Assign one of the staged teams to locate and gather these tools and stage them. If there is something you anticipate needing and it is not on scene, notify the IC and request it. Refer back to the chapter *Planning for a Prepared Rapid Intervention Crew* for more information on tools that should be available to an RIC operation.

■ Providing Feedback to the IC

The RIC Operations Chief should make every effort to provide the IC with adequate feedback and updates as to the rescue operations status. Remember that the IC is trying to stay focused on his job, but yet has extreme concern for the downed fire fighter. Maintain contact with the IC and provide him with pertinent feedback, such as RIC progress, if the fire fighter has been found, worsening conditions making the RIC operation questionable, or if the RIC is exiting with the downed fire fighter.

■ Communicating With the Distressed Fire Fighter

It is important that the RIC Operations Chief tries to maintain contact with the downed fire fighter and provide assurance that help is coming. Listen and perhaps console. Ask them to take survival actions, such as banging a tool to help the RIC locate and rescue them.

Role and Responsibilities of the RIC Company Officer

The RIC Company Officer plays a significant role at any emergency scene. Their primary role is to achieve their tactical assignment by leading and directing and getting the crew home safely. The RIC Company Officer's first priority is to bring the RIC out alive, and their second priority is to locate the downed fire fighter and begin the rescue. The RIC Officer position requires a person who understands and is capable of leadership, who has the proper RIC skills and training to fulfill the mission, and who truly understands risk assessment so as to bring his company back safely.

■ MAYDAY and the RIC Company Officer

Upon hearing the transmission of a MAYDAY call, the RIC Officer needs to immediately become proactive by giving the order for the RIC members to gear up and get ready, *even before the team is activated*. You know that a MAYDAY has been called, and as the RIC, the chances are likely that you will be activated. So, why wait to start gearing up? A good RIC Officer will stay within the workings of the IMS and not "activate and enter" on your own, but wait for the order. However, by being proactive, you are ready to go.

By "gearing up," the RIC should be in full personal protective equipment (PPE), masked up, tooled up, but not on air. It has been demonstrated that it often takes an RIC 3 to 4 minutes to activate! This is *totally unacceptable* and must be addressed. Your goal as the RIC Officer is to have your team ready to go in 30 seconds, which can be achieved through training. For the most part, SCBA-donning training has not included rapid emergency donning as the RIC company.

This type of donning is very different, in that we are not familiar with having to do it rapidly and with great precision. Gearing up under a MAYDAY and RIC activation, all while wearing the SCBA harness, entails removing your helmet; putting on the face piece, hood, helmet (with chin strap secured), and gloves; making sure that your cylinder valve is completely open; picking up your tools; and then as a company moving to the entry point as directed. The crew is now ready to make entry. The only thing left is for the fire fighter to go on air, which should not be done until entry into an immediately dangerous to life or health (IDLH) atmosphere. Even though this might sound basic and simple, try it as a group. You might be surprised by what the results are. It is suggested that fire departments begin to include this type of donning training.

Anticipating and being proactive can make a difference to the survivability of the distressed fire fighter. The Company Officer must have the skills and knowledge to lead the RIC. When activated, it is expected that the Company Officer has made sure that his crew is properly equipped.

■ Activation of the RIC

When the RIC is activated, they will be given information that pertains to the MAYDAY and their assignment, including the name and company of the fire fighter in distress, what the emergency is, and the probable location. The officer should quickly share what information he has with the team and give the task assignments: for example, "We have a MAYDAY from a fire fighter on Ladder 1, last known location was second floor, B/C corner. Reported trapped under a ceiling collapse, low on air. We are making entry on the A side and up the stairs. At the top, we will be conducting a left-hand search pattern." Once everyone is briefed and knows what the crew's objective is, the crew can now make entry.

■ RIC Company Officer Performance Standards

There are seven performance standards that a capable RIC Company Officer needs to remember.

1. **Show leadership!** You are the officer of an activated RIC. The crew is now looking to you for leadership and command presence. The RIC officer must understand this and exhibit this. Keep your crew focused on the mission and keep them moving. Do not overestimate your crew's skills. Good intentions do not save fire fighters in trouble—skills and courage do.

2. **Remember your #1 job.** Lead your crew in, manage them, and be aggressive in your search for the distressed fire fighter, but always remember that your number one job is to bring your crew out safely.

3. **Consider special tool needs.** Utilize your size-up and the situational analysis you have been conducting to select any special tools that might be needed and make sure they are with the crew.

4. **Accountability.** Keep your crew together, and work and operate as a team. The Company Officer must maintain crew integrity and accountability. Remember that you are operating in a hazardous, dangerous environment, and the intensity of finding a fire fighter in trouble has compounded the stress factor.

5. **Keep sizing up.** Do not focus only on finding the downed fire fighter. The Company Officer needs to also continue to size up the situation. Stop, look, and listen frequently; be aware of what is happening around you; and avoid tunnel vision. Watch the actions of the smoke. Is it intensifying? What about the pressure and density? Is it indicating possible preflashover conditions? What about the heat conditions? Are you being driven onto your stomach to get low and even that is not low enough? Remember, extreme heat is one of the warning signs of a possible flashover—make sure you know the other signs! Always be aware of the fire conditions around you and be prepared to take one step back so you can proceed forward.

6. **Remember, you and the RIC have limited air!** The officer needs to be 100 percent aware of the crew's air consumption. Periodically call for an air status from the crew members. The RIC cannot overextend with its air supply; the officer must ensure that the crew or individual crew members do not run out of air. When a SCBA low air alarm sounds, this is not the signal to back out. It is a warning sign of low air—*very soon to be no air!* A good RIC officer will make sure his crew members are not put in this position. He is always aware of the crew's air supply and starts exiting before they are in a low air or, worse, a no air situation! Always remember the officer's main job is to get the crew out alive, and that they should never put themselves or the crew into a position that another RIC is needed to rescue them.

Crew Notes

The RIC rescue air unit can, in an emergency, also be used to assist an RIC company member low on air. This action is not intended to allow the member to go deeper or work longer, but rather to assist the company in getting back out without running out of air.

7. **Provide feedback.** An important responsibility of any Company Officer is to provide adequate feedback. In the case of the RIC, feedback needs to be given periodically to the RIC Operations Group Chief and includes:
   - Reporting fire condition changes
   - Progress of the team (i.e., "company has made the second floor")
   - The RIC has found the downed fire fighter
   - Company air status

- Backing the company out and why
- Situational analysis and requesting more help
- Anything that is pertinent to the rescue that will help the RIC Operation Chief stay proactive and anticipate needs and next steps

Risk Management: Is the Benefit Worth the Risk?

The concept of <u>risk management</u> plays an important role with any RIC activation. What is risk? The dictionary defines it as "to expose to a chance of loss or damage." In the fire service, a more descriptive definition would be "the likelihood of exposure to injury or loss (death)." Risk management is a tool used in both preplanning and on scene to evaluate and reduce fire fighters' exposure to injury, loss, or death.

Risk is part of the job. Whenever fire fighters respond to any emergency scene, they are put at risk. The difference is the level of risk they are exposed to. A competent IC will always use a risk management tool at every scene to evaluate the dangers or potential dangers, and in doing a risk assessment, the IC can control the amount of risk the fire fighters are exposed to. Managing risk will also allow the IC to attempt to keep the risk at acceptable levels. In physics you learned "that for every action there is a reaction." The same principle applies to fire fighting. The IC must recognize and understand the consequences of certain predictable acts. When a risk assessment is conducted there are three rules that can help to guide you:

1. Risk a lot to save a lot (saving a life)
2. Risk a little to save a little (victim is dead)
3. Risk a little to save property (calculated minimal risk is good safe practice)

Safety

Understanding risk management and using it can make a difference if an RIC is activated. Stay proactive!

Always consider the severity of the situation (fire, building conditions, length of time the building has been burning, collapse potential) versus the probability of events if crews are placed in unnecessary danger. In order to stay proactive, an IC needs to:

- Remain calm, think clearly, and act decisively
- Conduct your size-up and reevaluate frequently
- Remember that for every action there is a reaction; plan accordingly!
- Know your limitations and seek help if needed
- Know the limitations of your resources, such as water supply and personnel
- Always be aware of the "time into the incident"
- Work with and within the Incident Command System (ICS)
- Get more help on scene *before* you need it
- Always use a personnel accountability system that gives you the information you need, when you need it
- Remain focused on the incident

- Stay within your span of control
- Utilize other command positions, such as the ISO and the RIC Operations Group Supervisor
- Always *anticipate*

Risk management is a very important skill that all fire service officers need to understand and use, the application of which can help prevent an RIC activation. It is important to seek out continuing education and training to further your knowledge and capabilities of risk management and risk assessment on the fireground.

Crew Notes

An excellent acronym to help you anticipate is CAN:
C = CONDITIONS, are they improving or deteriorating?
A = ACTIONS, are your actions supporting the strategies and tactics needed?
N = NEEDS, are you meeting the needs of the incident?

Safety

Risk a lot to save a lot, risk a little to save a little, and most importantly, risk nothing to save nothing!

Integration of Rapid Intervention Into the ICS

The ICS is a system that defines the roles and responsibilities of those operating at emergency incidents. National Fire Protection Association (NFPA) 1561, *Standard on Emergency Services Incident Management System*, provides the needed information on the system implementation, functions, and command structure. All emergency incidents need to be managed under an ICS utilizing the ICS.

For many midsize to smaller organizations, the "report to" position for the RIC within the ICS has typically been the IC. In the chapter *RIC On Scene: Preactivation Considerations and Actions While Staged*, the benefit of assigning an RIC Operations Group Chief with the initial assignment of the RIC company is discussed. Regardless of how you choose to set up your organizational chart, it is clearly to everyone's benefit that an RIC Ops Chief be used for all RIC activations for rescue of a fire fighter in distress. An operations organizational chart might look something like the samples shown in **FIGURE 4-1** or **FIGURE 4-2**. **FIGURE 4-3** shows suggested operations organizational charts

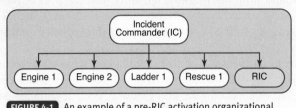

FIGURE 4-1 An example of a pre-RIC activation organizational chart for a smaller operation demonstrating the RIC reporting directly to the IC.

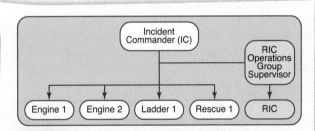

FIGURE 4-2 An example of a pre-RIC activation organizational chart demonstrating the RIC reporting to an RIC Operations Group Supervisor.

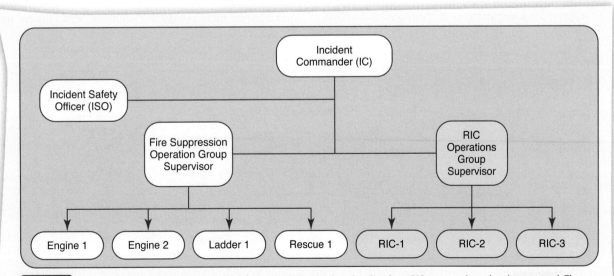

FIGURE 4-3 An example of an organizational chart during an RIC activation showing three RICs, one activated and two staged. The actual situation you are faced with will dictate the number of RICs needed.

during an RIC activation. Check with your jurisdiction prior to creating an organizational chart when planning for an RIC operation.

The ISO and Rapid Intervention

As an emergency incident grows in severity, so does the responsibility on the IC. One way to help to relieve some of the demands upon the IC is to use an ISO. The ISO is a key position, and for the purposes of this text, we will focus on this position's role on an active fireground. On any fireground, you need to always think safety first. The first of the three fireground priorities is always *life safety*, and the lives of the fire fighters should be the priority. On the fireground, the ISO is as capable as the IC in understanding the operations ongoing; however, the key element of this position is that it is a *position of anticipation*. The ISO should know what each company operating is doing, and what could potentially happen as a result of what they are doing now. A good ISO understands that the building talks to him and will understand what it is "telling" him. Is the fire free-burning? Are the fire fighters getting water on it? What about the smoke, what is it telling you? Are structural elements engaged in fire and is there a possibility of structural failure or collapse?

The ISO is responsible for operational safety and is in a position of great authority. He will conduct his own independent size-up, with the *anticipation* of possible problems. Fireground accountability as it relates to personnel is critical at all emergency scenes. The ISO should ensure that the department's accountability system is being used and used correctly at all incidents. He should also make sure that a rehab sector is set up if the incident warrants or requires it. Neither the rehab sector nor the accountability officer needs to report to the ISO. A key factor of a competent ISO is that he is constantly interfacing with the IC providing relevant information as needed.

Much of what has been described here is contained in NFPA 1521, *Standard for Fire Department Safety Officer*, and it is strongly suggested that departments properly staff and utilize

this position for the life safety of all those operating on scene. For more in-depth knowledge of the true role of an ISO, refer to *Fire Department Incident Safety Officer* by David Dodson.

■ Using the ISO Without an RIC Operations Group Supervisor

Many departments do not have the staffing to fill the position of RIC Operations Group Supervisor with the initial response and staffing of the team. When an incident is in its early stages, the IC has a lot of decisions to make, actions to implement, and companies to assign. Relieving the IC by having the RIC report to the ISO position, upon arrival, can greatly enhance safety on the fireground. For example, upon arrival, the assigned RIC reports to the ISO, usually located at or near the Command Post to better interact with the IC. The RIC gets its briefing from the ISO, who then instructs them to conduct a 360-degree RECON. The RIC continues to report to the ISO while staged, sharing observations and information.

If there is a MAYDAY and the RIC is activated, the ISO will become the RIC Operations Group Supervisor, and the position of the ISO must be replaced immediately. The downside of this type of organizational move is that the ISO's replacement will not be in a position to know what has been happening and what the companies are doing. The ISO is a major position on any fireground, as is the RIC. If staffing is an issue, look to start (preactivation) the RIC reporting to the ISO. The IC can then appoint an RIC Operations Group Supervisor as soon as a qualified person is available. If there is an RIC activation prior to an RIC Ops Supervisor being put in place, the ISO can move to the RIC Operations Group Supervisor position. The IC must replace the ISO as quickly as possible. During the transition, the IC is responsible for operational safety. **FIGURE 4-4** and **FIGURE 4-5** shows what this would look like organizationally.

Figures 4-4 and 4-5 are intended as a guide to show how RIC will fit into the National Incident Management System (NIMS). As any incident grows in size or complexity, the Incident Action Plan, the Command Staff, and the organizational chart will grow and become more complex. Command Staff Officers need to

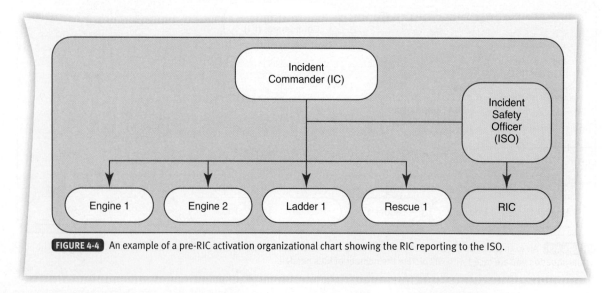

FIGURE 4-4 An example of a pre-RIC activation organizational chart showing the RIC reporting to the ISO.

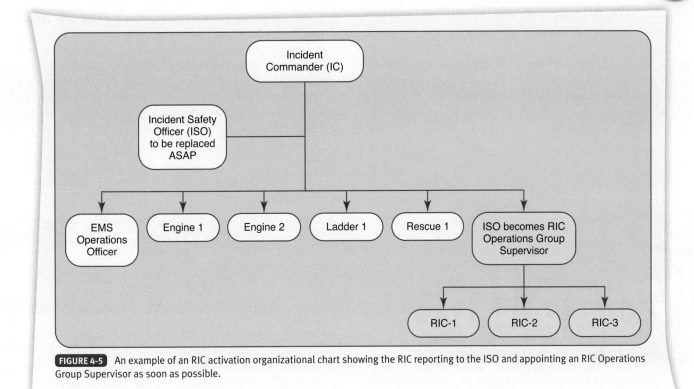

FIGURE 4-5 An example of an RIC activation organizational chart showing the RIC reporting to the ISO and appointing an RIC Operations Group Supervisor as soon as possible.

be proficient and familiar with the organizational considerations and Command Staff positions that need to be fulfilled during a rapid or extended intervention operation.

Post-Rapid Intervention Incident Actions

The process of postincident actions and assistance for the fire fighters is a very complex and important matter. Every department should establish protocols and procedures for such actions, and additional information can be obtained from numerous articles, books, and from the Learning Center at the National Fire Academy in Emmetsburg, MD.

When an RIC activation occurs, the levels of adrenaline and anxiety on the fireground go up dramatically. The results of an RIC activation will be one of the following:

1. Distressed fire fighter self-extricates, and the RIC can stand down.
2. The RIC locates the downed fire fighter and a rapid exit is made with the fire fighter injured or possibly unharmed. He is transported to the hospital, treated, and released when recovered.
3. The RIC locates the downed fire fighter, unconscious or in a full code, and exits as quickly as possible. The fire fighter is handed over to EMS and extreme measures are taken to revive the fire fighter, who is transported to the hospital where he is declared dead.
4. Every effort is made to locate the fire fighter, but conditions deteriorate to the point that it is unsafe for the RICs and other companies operating within the

structure. The time line shows that the risk greatly exceeds the benefit, as the downed fire fighter has been without air for a period of time that is nonsurvivable. The decision is made to evacuate the building, and the RIC operation now will become a prolonged recovery operation when it can safely commence.

Regardless of the end result, your first consideration will always be the mental welfare of your fire fighters. It might have been a self-extrication and successful escape, but is the fire fighter all right? Has his brush with danger affected him mentally and/or physically? What is your procedure to provide assistance? *The answer to this question is important.*

If the end result of the RIC deployment is death or severe injury, there are many considerations and actions needed. Most jurisdictions and states have protocols in place, but there are additional items that need to be addressed, including:

- The mental well-being of the fire fighters. CISD counseling will be needed immediately.
- Notification of the family of the fire fighter. Do you have a process and procedure in place? Who will make the notification?
- You need to control the information going out. Use a Public Information Officer (PIO) to interface with the Press.
- Ensure effective communications within your organization.
- In the event of a prolonged search and recovery:
  - Develop a relief schedule for crews searching.

- Maintain control of the site.
- Continue to evaluate the structural integrity and safety of the building.
  - What is your department's protocol for interfacing with the family of the injured or deceased? Do you understand the protocols for death benefits?
  - Does your department have protocols in place for a LODD?

The post-deployment events will be a difficult time for all. This is a time for strong leadership and command presence, with compassion and sensitivity. Rapid intervention is not just hands-on skills! For an RIC deployment to work, strong, capable leadership and a command and control system that everyone works within is required. It is important to possess a strong understanding of all the roles and positions that need to be fulfilled along with the skills to fulfill them!

Wrap Up

Chief Concepts

- Frederick Stanley Mockford developed the term MAYDAY in 1923.
  - In French, the term "come and help me" was "Venez m'aider."
  - The word m'aider was termed "MAYDAY."
- When a MAYDAY is called, no other radio traffic is permitted on that frequency except to assist in the emergency.
- The call is given three times in a row: "MAYDAY, MAYDAY, MAYDAY."
- A MAYDAY should be reserved for fire fighters in extreme distress calling for immediate help.
- The term *emergency* or *urgent traffic* should be used on the fireground when you need to get a message out, indicating an imminent hazard or a situation involving immediate danger.
- When a MAYDAY is called, the IC must cease all other activities and answer the call.
- Once the IC acknowledges a MAYDAY, it is important to tell the fire fighter that help is on the way, and prompt them to activate their PASS device.
- The IC needs to control the airwaves when a MAYDAY is declared.
- Move the fireground operation to a different radio channel, if possible, for an extended intervention operation.
- Get extra help! An RIC operation can become a very manpower-intensive operation.
- The RIC Operations Group Supervisor is in charge of and responsible for the entire RIC operation.

- The RIC Operations Group Supervisor must take charge and show command presence.
- A competent IC will always use a risk management tool at every scene to evaluate the dangers or potential dangers the fire fighters are exposed to.
- Risk a lot to save a lot (saving a life).
- Risk a little to save a little (victim is dead).
- Risk a little to save property (calculated minimal risk is good safe practice).
- Risk nothing to save nothing!

Hot Terms

Critical incident stress debriefing (CISD) Counseling designed to minimize the effects of psychological/emotional trauma on those at a fire or rescue incidents who were directly involved with victims suffering from particularly gruesome or horrific injuries.

Emergency traffic Used on the fireground when you need to get a message out; indicating an imminent hazard or a situation that is of immediate danger.

Incident Command System (ICS) The combination of facilities, equipment, personnel, procedures, and communications operating within a common organizational structure that has responsibility for the management of assigned resources to effectively accomplish stated objectives pertaining to an incident or training exercise.

Line-of-duty death (LODD) Death of a fire fighter while on duty or operating in his or her official capacity.

Risk management A tool used in both preplanning and on scene to evaluate and reduce fire fighters' exposure to injury, loss, or death.

References

Dodson, D.W. *Fire Department Incident Safety Officer, 2nd ed.* Cengage Learning: Independence, KY; 2007.

National Fire Protection Association (NFPA) 1407, *Standard for Fire Service Rapid Intervention Crews.* 2015. http://www.nfpa.org/codes-and-standards/document-information-pages?mode=code&code=1407. Accessed February 11, 2014.

National Fire Protection Association (NFPA) 1521, *Standard for Fire Department Safety Officer.* 2008. http://www.nfpa.org/codes-and-standards/document-information-pages?mode=code&code=1521. Accessed October 14, 2013.

National Fire Protection Association (NFPA) 1561, *Standard on Emergency Services Incident Management System.* 2014. http://www.nfpa.org/codes-and-standards/document-information-pages?mode=code&code=1561. Accessed October 14, 2013.

RAPID INTERVENTION CREW MEMBER
in action

You are the IC at a working fire. You have the situation well in hand and all companies are working, making what appears to be good progress on the fire, when over the radio you hear, "MAYDAY, MAYDAY, MAYDAY; fire fighter Jones, division 2 trapped, partial ceiling collapse, send help!"

1. As the IC, what is the very first thing you should do once a MAYDAY is declared?
   A. Notify fire alarm communications.
   B. Activate the RIC.
   C. Answer the call for help and prompt the distressed fire fighter to activate his PASS device.
   D. Find out if the fire fighter is really in need of help or is just panicking.

2. Why is it important to advise the distressed fire fighter that help is on the way?
   A. So he or she can relax.
   B. The downed member will need assurance that help is on the way.
   C. It is not necessarily important.
   D. All of the above.

3. Why is it important to gain control of the radio traffic?
   A. The IC must maintain radio control so as to monitor the downed fire fighter and maintain good communications with him.
   B. To minimize other crews from talking over the member who is in trouble.
   C. Fireground operations need to still take place while the rescue of a fire fighter is in progress. These functions should not interfere with the downed member.
   D. All of the above.

4. Why is getting a PAR so important?
   A. The IC needs to account for all of the fire fighters on the fireground to determine if any other members are missing or in trouble.
   B. For liability purposes.
   C. To determine how many ambulances may be needed.
   D. It is a requirement of the ICS system.

5. Why is additional help needed when a MAYDAY is declared?
   A. The RIC could potentially need additional help in removing the fire fighter.
   B. There could be more than one fire fighter in peril.
   C. The RIC could potentially become victims themselves.
   D. All of the above.

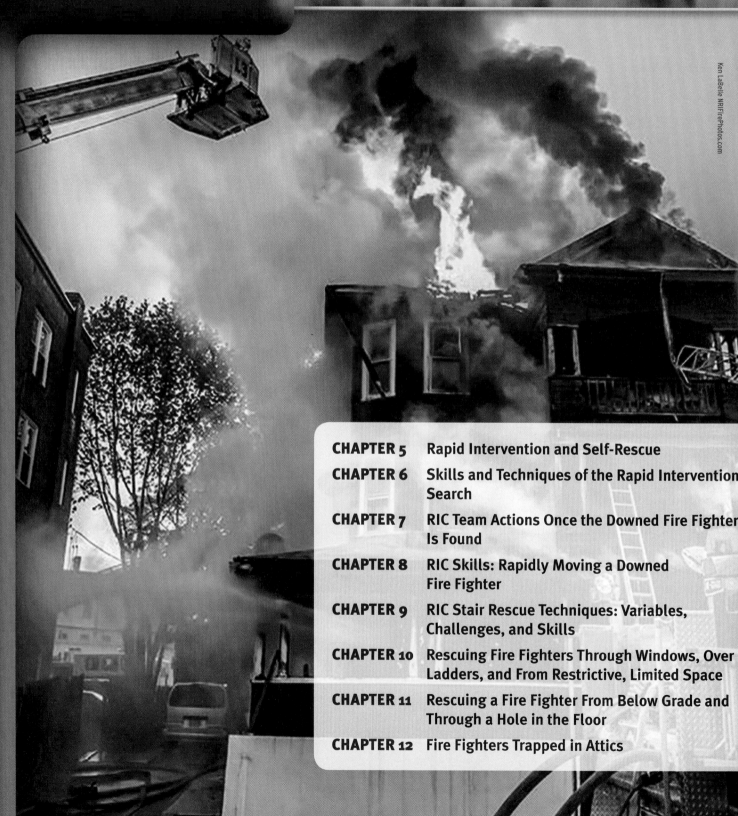

SECTION II

RIC Skills

Ken LaBelle NRIFirePhotos.com

Rapid Intervention and Self-Rescue

© Photos.com

© Joe Nedder/Jones & Bartlett Learning

Knowledge Objectives

After studying this chapter you will be able to:

- List the five keys to help avoid getting yourself in trouble. (p 60)
- Know why, how, and when you call a MAYDAY and how it can make a difference in your survival. (NFPA 1407, 7.4(1), 7.4(8), 7.5(1), 7.5(2), 7.13.1(1) , p 60–63)
- Define the location, unit, name, assignment, resources (LUNAR) and location, identification, problem (LIP) acronyms. (NFPA 1407, 7.5(2) , p 62–63)
- List the four things wall breaching will give you egress for. (NFPA 1407, 7.7(1), 7.7(2), 7.7(3) , p 64–72)
- Removal of a fire fighter from debris. (NFPA 1407, 7.7(5) , p 73–74)
- Define low voltage wire, list the four primary types of low voltage wire, and understand what they have in common. (p 74–75)
- List the three first steps of any entanglement escape. (NFPA 1407, 7.7(3), 7.7(4), 7.7(5), 7.13.1(1) , p 74–75)
- List four conditions that warrant a rapid emergency egress escape. (NFPA 1407, 7.13.1(1), 7.13.1(3) , p 75–77)
- Understand the importance of ladder positioning. (p 77–78)
- Know how to execute an emergency rapid escape over a ladder. (NFPA 1407, 7.4(8), 7.13.2 , p 77–82)
- Know the proper rope type and mechanical escape system needed for an emergency rapid escape and how to carry and deploy. (NFPA 1407, 7.13.2 , p 82–89)

Skills Objectives

After studying this chapter, and with extensive practice and drilling, you will be able to perform the following skills:

- Declare a MAYDAY. (NFPA 1407, 7.4(1), 7.5(1), 7.5(2) , p 60–63)
- Breach various walls. (NFPA 1407, 7.7(1) , p 64–67)
- Reposition your self-contained breathing apparatus (SCBA) to low profile yourself. (NFPA 1407, 7.7(2) , p 67–72)
- Locate and use a hoseline to find your way out if lost or separated. (NFPA 1407, 7.13.1(1) , p 72–73)
- Attempt to escape from an entanglement. (NFPA 1407, 7.7(3), 7.7(4), 7.13.1(2) , p 74–76)
- Perform a rapid emergency egress, by rope and/or ladder, to escape untenable conditions. (NFPA 1407, 7.10.2(2), 7.13.1(3), 7.13.2 , p 77–89)

Additional NFPA Standards

- NFPA 1981, *Standard on Open-Circuit Self-Contained Breathing Apparatus (SCBA) for Emergency Services*
- NFPA 1983, *Standard on Life Safety Rope and Equipment for Emergency Services*

Your company is moving in on a fire under extreme conditions. Visibility is zero. As the hoseline is being advanced you lose contact, become disoriented, and cannot find your crew.

1. Should you call a MAYDAY?
2. What actions can you take to self-rescue?
3. At what point are you really in danger?

Introduction

Learning how to avoid getting oneself into trouble, along with the skills to help self-extricate and declare a MAYDAY are very important. Here we will discuss why fire fighters get in trouble and how to avoid it, and we will present and teach basic survival techniques and skills that can make a difference in your ability to survive. It is important that as you study and train for these skills, extreme safety precautions are always in place and used. Remember: Train safely at all times!

Safety

Remember—train safely at all times!

Fire Fighter and Self-Rescue Survival Skills

Fire fighter <u>survival skills</u> can be used to not only save you as an individual, but also to rapidly allow a company to egress to safety. Here, we will focus on the following basic survival skills:

- Declaring a MAYDAY!
- Wall breaching and passing through the hole.
- Using a hoseline to find your way out.
- Entanglement survival.
- Rapid emergency egress techniques through windows.

The first step in any survival training is to always be prepared in the event of an emergency. This is a proactive approach to survival. Survival is not just the skills, but also the mental preparation and the ability to recognize situations where survival skills need to be activated immediately. The five keys to avoid getting into trouble are:

1. Avoid disorientation. Know where you are in the building—what floor, your approximate location (i.e., A/B corner)—at all times and your company's assignment.
2. Maintain <u>crew integrity</u>; stay together, work together, and exit together.
3. Do not overextend your air supply and do not overestimate your skills and abilities.
4. Know and maintain your basic skills.
5. Maintain your own interior size-up, watching the fire behavior and conditions, along with the smoke behavior.

Declaring a Fire Fighter Emergency

In this text, we will discuss the process and skill of calling the MAYDAY. Fire fighters should also understand the term MAYDAY, when fire fighters should declare a MAYDAY, and the activation of the rapid intervention crew (RIC) to initiate the <u>rescue</u> of the endangered fire fighter, all of which were discussed in the chapter *Activation of the RIC and Organizational Considerations*.

All fire fighters need to recognize and understand that when they think they are lost, trapped, or stuck, they are! Fire fighters operate within an <u>immediately dangerous to life and health (IDLH)</u> environment with a very limited air supply; therefore, any avoidable delay in calling for help can make a difference in whether you live or die. The facts are indeed facts, and they must be faced realistically.

■ How Calling a MAYDAY Can Make a Difference to Survival

For this example, assume the fire fighter is wearing a 30-minute cylinder. As such, it is accepted knowledge that an average fire fighter working hard will exhaust his air supply in 12 to 16 minutes. In this scenario, a fire fighter working with his company has separated from them. He immediately tries to make contact with the crew, but is unable to. He then tries to find them or a landmark that might lead him out to safety. This incident has happened 8 minutes after the crew made entry.

It is taught that, at this point, the fire fighter should declare a MAYDAY; however, most will not. Instead, they will continue to try and find their way out or to reunite with their company. When the self-contained breathing apparatus (SCBAs) low air alarm sounds, the fire fighter has 25 percent of his cylinder's capacity left, meaning 4 minutes of air (25 percent of 16 minutes). The fire fighter now declares a MAYDAY, finally realizing that he in fact is lost and in trouble. In 2013, the National Fire Protection Association (NFPA) revised the NFPA 1981, *Standard on Open-Circuit Self-Contained Breathing*

Apparatus (SCBA) for Emergency Services. Among the changes was a new requirement that the SCBA low air alarm now sound at 33 percent of the cylinder's capacity. However, for this example, we will still use the alarm sounding at 25 percent air capacity remaining, as the vast majority of SCBAs in service as of this writing are pre-2103 standard.

Crew Notes

The low air alarm in a 30-minute cylinder *does not* indicate 5 minutes of air remaining! This false information does not apply to today's SCBAs. The low air alarm indicates you have 25 percent (or 33 percent, if you are using a SCBA manufactured to the 2013 Standard) of your cylinder left at whatever rate you have been breathing. For most fire fighters, this is 3 to 4 minutes of air remaining.

With just 4 minutes of air remaining, it will take on average 2 to 3 minutes to declare the MAYDAY, get acknowledged from the Incident Commander (IC), have the RIC activated, and make entry. The fire fighter now has 1 minute of air remaining in his cylinder. How long will it take this first team to locate the lost fire fighter? Whatever number you pick is a guess, because each circumstance will be different. For this example, assume 8 minutes, since that is how long it took the fire fighter to get lost originally. Therefore, when the RIC found him, he had been without any air for over 7 minutes. In all probability, if smoke and heat conditions are severe, he is not breathing, his heart has stopped, and he is dead.

The initial RIC, after calling for a second team to start in to assist, quickly begins to extract him as rapidly as possible. When the second RIC intercepts the first team, the downed fire fighter is handed off to the second team to complete the rescue. Again, for example, say it takes 8 minutes to exit with the downed fire fighter, at which point the paramedics go to work, making their noblest efforts. However, after 16 minutes without breathable air, going into respiratory and cardiac arrest, the fire fighter in all probability is dead.

If, however, the MAYDAY had been called when first lost, the fire fighter would have saved 4 minutes of air, and when found, would have been without air for approximately only 3 minutes. In this case, there is a significant chance, depending on conditions that he might be unconscious but still ever-so-slightly breathing. The RIC can now quickly reestablish his air supply and rapidly extract him. There is an excellent chance that this fire fighter might have survived if he had called a MAYDAY immediately.

As fire fighters, we must not get caught up in the technical debates of exactly when a fire fighter would succumb to the smoke. There are too many variables, such as what poison gases might be in the smoke. Remember, there is always carbon monoxide, and this will cause altered mental status, loss of consciousness, respiratory arrest, and then cardiac arrest. Time lines might differ, but the message remains the same: calling a MAYDAY when you first get in trouble will extend your breathable air during the initial RIC operation and will increase your chance of survival.

The NFPA 1981 Standard, 2013 Edition, and How It Will Affect Us

The release of the 2013 edition of NFPA 1981 dramatically changes the way fire fighters have been taught to understand the audio warning of low air in their SCBAs. As fire departments begin to purchase new SCBA units, fire fighters will experience having two different units on the fireground—one sounding low air at 25 percent and the other at 33 percent. Unless fire departments replace 100 percent of their SCBAs that are in service at one time, this "confusion" will continue for many years. As fire fighters and as part of a competent RIC, we must always be aware of what equipment we are using and what is in use on the fireground. Imagine being part of a company with fire fighters wearing SCBAs from the different standards. If a low air alarm sounds, do you know what model you are wearing? Is it 25 percent or 33 percent? The best way to deal with this mixing of SCBAs is through continued training. In departments with SCBAs from both standards, we need to ensure that fire fighters clearly identify what model they will be wearing as part of their daily inspection or donning sequence. This proactive action might make a difference in your survival.

Most departments in the United States work with mutual aid companies from other jurisdictions and communities. As part of an RIC, it is important to know what SCBA they have. Fire fighters' lives depend on you. Know your equipment!

How to Avoid Getting Into a MAYDAY Situation

When a discussion comes up about fire fighters in distress, peril, or in need of calling a MAYDAY, the first thing someone usually will say is, "We need to teach them how not to get in trouble!" This is true, but they also need to be taught to be proactive in their approach to avoiding a fire fighter emergency. This can be accomplished in many ways, including:

1. Basic skills—fire fighters need to know all the basic skills and keep them fresh by use and continued training throughout their career.
2. A solid understanding of fire and smoke behavior, along with frequent continued training and refresher classes on this subject. The science of fire and smoke behavior cannot be underestimated—it's important to know and understand it before entering a burning building!
3. Understand and accept that a fire fighter can get in trouble, and if so, needs to call a MAYDAY without delay.
4. A solid understanding of building construction (how it is built, how it burns, and how it falls down).
5. The ability to conduct your own ongoing size-up of the incident you are involved in.
6. The availability and *use* of a fire fighter accountability system, and with this, crew integrity. Every fire fighter is responsible for the safety and welfare of the company he is assigned to and, as such, maintains the integrity of the crew at all times.
7. Strong leadership at all levels and a solid understanding of the principles of risk management.
8. The use of the Incident Command System (ICS) at all incidents.

9. The training of all fire fighters in rapid intervention and survival skills, and the staffing of a capable RIC at all incidents.

10. Excellent fireground radio communication skills and the reduction of useless radio transmissions.

Crew Notes

An excellent resource to learn more about risk management is the National Institute for Occupational Safety and Health (NIOSH), *ALERT Preventing Deaths and Injuries of Fire fighters Using Risk Management Principles at Structure Fires* (Publication Number 2010-153).

■ Identifying the Emergency and the Need to Call a MAYDAY

Examples of extreme fire fighter distress include being separated from the company and lost; being trapped or entangled in wires or debris; running out of air or have run out of air due to a SCBA failure; or a fire fighter having a medical emergency (unconscious, heart attack) and a company officer or partner calls the MAYDAY. All these are examples of true emergencies, and the need for calling a MAYDAY is apparent. You cannot be in a state of denial at this point. Hesitancy or reluctance to call the MAYDAY

could very well make the difference in the fire fighter's ability to survive or not. In a hostile environment with a limited air supply, calling for help will assist in your survival!

To call a MAYDAY, the first and most important initial step is to try to remain as calm as possible. The MAYDAY procedure is demonstrated in **SKILL DRILL 5-1**:

1 Using your radio, transmit and call "MAYDAY, MAYDAY, MAYDAY!" (**STEP 1**) Learn to use the radio so that your transmission can and will be understood. Do not shout, do not place the radio right on the voice amplifier if so equipped, and wait until after you have keyed the mike to begin talking.

2 Once your MAYDAY has been acknowledged and answered, provide the IC important pertinent information, including who you are, your company, where you are, and what the problem is, by using either the LUNAR or LIP acronym.

3 After the MAYDAY transmission and the LUNAR or LIP report, you should activate your personal alert safety system (PASS) device. (**STEP 2**)

4 If you are able to move, and depending on the circumstances, attempt self-rescue by seeking a window or safe haven. (**STEP 3**)

5 If you are unable to move, lie on your side in the fetal position with your PASS device emitter facing outward.

SKILL DRILL 5-1   Calling a MAYDAY!
NFPA 1407, 7.4(1), 7.5(1), 7.5(2)

1 Using your radio, transmit and call "MAYDAY, MAYDAY, MAYDAY!" Once your MAYDAY has been acknowledged and answered, provide the IC important pertinent information, including who you are, your company, where you are, and what the problem is by using either the LUNAR or LIP acronym.

2 After the MAYDAY transmission and the LUNAR or LIP report, you should activate your personal alert safety system (PASS) device.

(Continued)

SKILL DRILL 5-1   Calling a MAYDAY! (Continued)

3 If you are able to move, and depending on the circumstances, attempt self-rescue by seeking a window or safe haven. If you are unable to move, lie on your side in the fetal position with your PASS device emitter facing outward.

© Joe Nedder/Jones & Bartlett Learning

Crew Notes

The first and most common acronym is LUNAR, which stands for Location, Unit number, Name, Assignment, Resources needed. The alternative acronym is LIP, which stands for Location, Identification, Problem. This acronym is simple and accomplishes the same mission as LUNAR, and was developed by District Chief Michael O. McNamee (Retired) of the Worcester, MA, Fire Department. After the tragedy of the December 3, 1999 Cold Storage Warehouse fire, in which six Worcester fire fighters died in the line of duty, Chief McNamee worked ardently to create skills and procedures to help prevent fire fighter injury and death. The LIP acronym is an excellent example of developing something for the fire service that is concise, brief, and accomplishes the mission. Which one should be used is determined by the level of distress that the fire fighter is in and if he or she is able to send a full message. If the level of distress is extreme and we just know the fire fighter's location and company, and we have a good accountability system being used at the incident, we will most likely have enough information to help the RIC company locate and rescue the fire fighter. An example of either a LUNAR or LIP report might be: "This is fire fighter Johnson, Engine 3, I'm trapped and entangled 2nd floor B/C corner, low on air, send help!"

During the process of waiting for help, you need to make every effort to remain as calm as possible and control your breathing. While lying on your side, turn your flashlight on and

Safety

It is critical that you know your SCBA and its operation intimately. Different brand air packs have different locations for the PASS emitter: some face forward, some face to the rear, and some emit from both the front and the rear **FIGURE 5-1**. Know what you wear. Your life depends upon it!

© Joe Nedder/Jones & Bartlett Learning

FIGURE 5-1   SCBA with a rear facing PASS emitter.

have the beam facing upwards towards the ceiling **FIGURE 5-2**. Continue to radio with your air status, changing conditions, or if you think you hear fire fighters nearby. Use your tool by banging it on the ground or against something to help draw attention to your location **FIGURE 5-3**. A large part of survival is being proactive in your own rescue.

FIGURE 5-2 While lying on your side waiting for rescue, turn your flashlight on with the beam facing upward toward the ceiling.

FIGURE 5-3 Help draw attention to your location by banging your tool on the ground or against something.

Wall Breaching

If you are a fire fighter operating in the burning structure and are blocked by fire or a partial ceiling collapse, breaching the wall will create another path of entry or escape for you. Wall breaching is an important skill to learn, as it provides access to both victims and trapped fire fighters, and also provides egress for escaping with a victim, a downed fire fighter, via a door or window in another room when your primary way out is blocked by fire or collapse, and when no other avenue of escape is available due to deteriorating conditions.

To breach a wall can be very labor-intensive when performed in full personal protective equipment (PPE). Fire fighters need to practice and develop this skill while wearing full PPE and on air, so that they will be better prepared if it ever became necessary for them to execute the skill under real conditions.

To effectively breach a wall, you must have developed the skills and you must always have a tool with you. Most interior walls in residential construction are wood 2 × 4-framed walls. You may also find metal stud framing, especially in commercial construction. In some instances, you might find interior masonry walls. The framing is covered with various types of wall sheathing products, including wood lathe and plaster (typically found in older construction) and drywall. Each of these wall-sheathing products presents a different problem to accomplish the breach.

Other considerations and issues when breaching a wall include encountering electrical wires and outlets, air conditioning, heating ducts and pipes, HVAC ductwork, plumbing, waste pipes, and vent pipes **FIGURE 5-4**. In addition, you must also consider what might be on the other side of the wall you are going to breach. Are you breaking through a wall only to encounter the back end of a refrigerator, stove, or toilet? Have you broken through into kitchen or bathroom cabinets? Any of this is possible. To help avoid this type of issue when you enter the building, try and imagine a floor layout during your personal size-up process. Where is the kitchen and bathroom? When you make entry into the structure, keep your orientation, and remember what you saw during the size-up.

■ Breaching an Interior Framed Wall for Emergency Egress or Access

The skills of breaching *any* wall begin with the same pre-breach sequence.

- Try and determine where you are within the structure.
- Notify the IC that you intend to wall breach and why.
- Size up the wall you want to breach, and look for electrical, telephone, or cable outlets that would indicate

FIGURE 5-4 When breaching a wall, wires, plumbing pipes, and various ductwork may be encountered.

wires in that general vicinity. If you find an outlet, move a few feet to the left or right and reassess.

Breach drywall as demonstrated in **SKILL DRILL 5-2** :

1. Complete the pre-breach sequence.
2. Drive your tool (Halligan's and axes work best) completely through to the other side, making sure it has penetrated completely. Drive this tool through the wall very carefully, probing to see if there is an obstruction on the other side. (**STEP 1**) You might encounter the back of an appliance or a shower or bathtub wall that will cause the tool to bounce back at you. You might also encounter a civilian or

a downed fire fighter against the wall. *Always* use caution!

3. If the other side of the wall seems clear, quickly enlarge the hole on your side by pulling the materials towards you. Start by making a series of holes in the approximate shape of your breach hole by "punching" the tool through the drywall. (**STEP 2**) The hole needs to be large enough so that a fire fighter can fit.
4. Open the wall on the other side by pushing, with the head of the tool, the materials away from you.
5. Once the opening on the other side is large enough, examine what is on the other side. Are you in a

---

SKILL DRILL 5-2 Breaching Drywall
NFPA 1407, 7.7(1)

1. Complete the pre-breach sequence. Drive your tool (Halligan's and axes work best) completely through to the other side, making sure it has penetrated completely. Drive this tool through the wall very carefully, probing to see if there is an obstruction on the other side.

2. If the other side of the wall seems clear, quickly enlarge the hole on your side by pulling the materials towards you. Start by making a series of holes in the approximate shape of your breach hole by "punching" the tool through the drywall.

3. Open the wall on the other side by pushing, with the head of the tool, the materials away from you. Once the opening on the other side is large enough, examine what you see and determine if you can continue or if you have to find another breach location. Before you pass through, place your tool on the other side to the side of the hole made. Pass through the breach hole.

cabinet or a closet or blocked by furniture? Is there a floor on the other side or have you opened up a wall on the second floor overlooking a cathedral-type ceiling area? Assess what you see and determine if you can continue or if you have to find another breach location. Before you pass through, place your tool on the other side to the side of the hole made. **(STEP 3)**

6 Pass through the breach hole.

Breach wood lathe and plaster as demonstrated in **SKILL DRILL 5-3** [NFPA 1407, 7.7(1)]:

1 Complete the pre-breach sequence

2 Drive your tool completely through to the other side, making sure it has penetrated completely.

3 Place the tool downward through the hole created and then pull the top of the tool towards you, breaking through the plaster and lathe. Continue to pull the materials towards you to create a large enough opening.

4 Open the wall on the other side by pushing the materials away from you with the head of the tool.

5 Once the wall on the other side is large enough, examine what is on the other side. Are you in a cabinet or a closet or blocked by furniture? Assess what you see and determine if you can continue or if you have to find another breach location.

6 Pass through the breach hole.

■ Breaching an Exterior Masonry Wall

In a commercial building setting, many times we encounter a large "box type" store (Type III, Ordinary Construction) with very limited exterior access points (doors/windows). As part of an RIC, you might be called upon to create an entry point that is in close proximity to the downed fire fighter. Many times this access point will be created by breaching a <u>masonry block wall</u>.

Breaching a masonry block wall is difficult and labor-intensive. The recommended tool of choice is a sledgehammer or heavy maul-type tool, and the minimum weight is 8 pounds (4 kg). Breach a masonry block wall as demonstrated in **SKILL DRILL 5-4**:

1 Size up the most appropriate location on the exterior wall.

2 Using a sledgehammer, the fire fighter will make a series of holes through the block's exterior in the shape of an inverted V. **(STEP 1)**

3 After the holes have outlined the opening shape, begin to knock out all the remaining masonry starting from the top and working downward. Quickly clear the debris from the area the fire fighters will be crawling on. **(STEP 2)**

Crew Notes

Breaching a masonry wall is difficult and labor-intensive. As the RIC Operations Chief, consider using a second company to complete the breach so as to keep the RIC physically fresh for the entry and search.

SKILL DRILL 5-4   Breaching a Masonry Block Wall
NFPA 1407, 7.7(1)

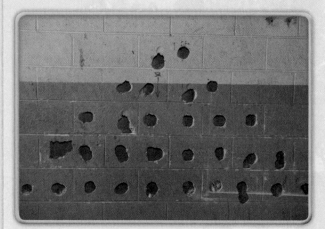

1 Size up the most appropriate location on the exterior wall. Using a sledgehammer, the fire fighter will make a series of holes through the block's exterior in the shape of an inverted V.

2 After the holes have outlined the opening shape, begin to knock out all the remaining masonry starting from the top and working downward. Quickly clear the debris from the area the fire fighters will be crawling on.

Crew Notes

The masonry joint will be the weakest point. The hole needs to be large enough for fire fighters to pass through safely.

Crew Notes

Use two fire fighters to complete the breach. While one fire fighter is creating the outline holes, the second fire fighter can begin to break through, again starting at the top and working downward.

All fire fighters, especially those assigned to the RIC, need to have a very solid understanding and foundation in all forcible entry skills.

Fire Fighter Techniques for Passing Through the Framed Wall Breach

Once a framed wall has been breached, a fire fighter needs to pass through it quickly, safely, and while maintaining their air supply. To achieve this, three different skill sets are presented here. The one that works at that given moment is the right one for you to use!

Most walls today and in recent times have been constructed with the studs placed every 16 inches (41 cm) on center. With 2 × 4 construction, this provides a pocket or opening of 14 ½ inches (37 cm) **FIGURE 5-5**. Even though this

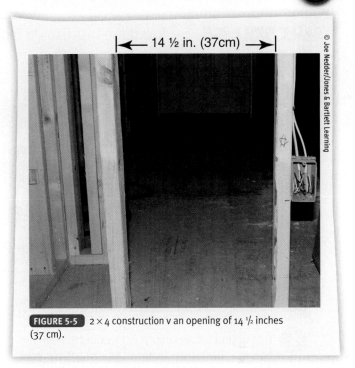

FIGURE 5-5 2 × 4 construction v an opening of 14 ½ inches (37 cm).

© Joe Nedder/Jones & Bartlett Learning

sounds very restrictive, most fire fighters, using one of the three methods presented below, will be able to fit through.

The backstroke pass-through technique allows the fire fighter to maintain the SCBA in the donned position, without any shifting or repositioning of the unit. The backstroke pass-through technique is demonstrated in **SKILL DRILL 5-5**:

1 In a seated position, place the cylinder in the "pocket" created in the wall. (**STEP 1**)

SKILL DRILL 5-5 Backstroke Pass-Through Technique
NFPA 1407, 7.7(2)

1 In a seated position, place the cylinder in the "pocket" created in the wall.

2 Extend your right arm over your head and through the hole in a "backstroke" manner.

(Continued)

SKILL DRILL 5-5   Backstroke Pass-Through Technique *(Continued)*

3 Extend your left arm over your head and through the hole in a "backstroke" manner. As your left arm goes through, begin to twist your head and shoulders through the hole.

4 Pivot your body onto your side and pull yourself through the breach.

5 Retrieve your tool.

© Joe Nedder/Jones & Bartlett Learning

2 Extend your right arm over your head and through the hole in a "backstroke" manner. (**STEP 2**)

3 Extend your left arm over your head and through the hole in a "backstroke" manner. As your left arm goes through, begin to twist your head and shoulders through the hole. (**STEP 3**)

4 Pivot your body onto your side and pull yourself through the breach. (**STEP 4**)

5 Retrieve your tool. (**STEP 5**)

The repositioned SCBA pass-through technique involves sliding the SCBA cylinder onto the left shoulder, thereby assisting in low profiling the fire fighter to fit through the breach hole. The repositioned SCBA pass-through technique is demonstrated in **SKILL DRILL 5-6**:

1 Loosen your right shoulder strap *completely* and your waist strap *slightly*.

2 Grab your shoulder strap with your left hand and grab your waist strap on the extreme left side with your right hand. (**STEP 1**)

3 At the same time, push the left shoulder strap towards the right and pull the waist strap to the right, repositioning your SCBA cylinder high on your left shoulder. (**STEP 2**)

4 Maintain pressure on the shoulder strap with your left hand and, at the same time, drop your head and right shoulder through the hole. **Note:** If you wear a traditional-style helmet, you will need to insert your head face-down first, as the helmet will not fit otherwise. (**STEP 3**)

5 As you drop your right shoulder, you must maintain pressure on the left shoulder strap and keep the cylinder high on your left shoulder.

SKILL DRILL 5-6

Repositioned SCBA Pass-Through Technique
NFPA 1407, 7.7(2)

1 Loosen your right shoulder strap *completely* and your waist strap *slightly*. Grab your shoulder strap with your left hand and grab your waist strap on the extreme left side with your right hand.

2 At the same time, push the left shoulder strap towards the right and pull the waist strap to the right, repositioning your SCBA cylinder high on your left shoulder.

3 Maintain pressure on the shoulder strap with your left hand and, at the same time, drop your head and right shoulder through the hole. **Note:** If you wear a traditional-style helmet, you will need to insert your head face-down first, as the helmet will not fit otherwise.

4 As you drop your right shoulder, you must maintain pressure on the left shoulder strap and keep the cylinder high on your left shoulder. Work your body through the breach hole. After you are through, reposition your SCBA straps to the proper location and tighten and secure. Retrieve your tool.

6 Work your body through the breach hole. (**STEP 4**)

7 After you are through, reposition your SCBA straps to the proper location and tighten and secure.

8 Retrieve your tool.

Crew Notes

On most SCBAs, the left shoulder strap is the strap that has the low-pressure hose feeding through it. By grabbing this strap, you will help to protect and maintain your air supply.

You might encounter a situation where you cannot fit through the space while the SCBA is on your back. In this case, you need to remove the SCBA while still maintaining your air supply, pass through the breach hole, and then re-don SCBA. The SCBA removal while maintaining your air supply pass-through technique is demonstrated in SKILL DRILL 5-7 :

1 Completely loosen your waist and shoulder straps. (**STEP 1**)

2 Unbuckle your waist strap. (**STEP 2**)

3 Remove your arm from your right shoulder strap and let the cylinder drop to the left (like removing a jacket) while maintaining a hold on your left shoulder strap with your left hand. From this point forward until you are ready to re-don the SCBA, you must always maintain your grip on the left shoulder strap, as it is the strap with the low-pressure hose that goes to the mask-mounted regulator (MMR). Maintaining contact with this strap will assist you in a continuous supply of air. (**STEP 3**)

4 Position the SCBA cylinder and frame on the ground, stem facing towards you, and place it through the breach hole while maintaining your grip on the left shoulder strap. (**STEP 4**)

5 Push the SCBA through the breach hole and follow with your body. (**STEP 5**)

6 After you are completely through, re-don the SCBA, tighten all straps, and retrieve your tool. (**STEP 6**)

Crew Notes

By completely loosening your shoulder and waist straps, they will be properly positioned for redonning when needed.

SKILL DRILL 5-7

SCBA Removal While Maintaining Your Air Supply Pass-Through Technique
NFPA 1407, 7.7(2)

1 Completely loosen your waist and shoulder straps.

2 Unbuckle your waist strap.

(Continued)

SKILL DRILL 5-7 SCBA Removal While Maintaining Your Air Supply Pass-Through Technique (*Continued*)

3 Remove your arm from your right shoulder strap and let the cylinder drop to the left (like removing a jacket) while maintaining a hold on your left shoulder strap with your left hand. Maintaining contact with this strap will assist you in a continuous supply of air.

4 Position the SCBA cylinder and frame on the ground, stem facing toward you, and place it through the breach hole while maintaining your grip on the left shoulder strap.

5 Push the SCBA through the breach hole and follow with your body.

6 After you are completely through, re-don the SCBA, tighten all straps, and retrieve your tool.

Using a Hoseline to Find Your Way Out to Safety

If you as a fire fighter become separated from your company or lost within a building and find or stumble upon a hoseline, you can use this hoseline to find your way out to safety. This technique is commonly known as the long lug out escape drill and works well with 1½-inch (38-mm; used on 1¾-inch hose [44-mm]) and 2 ½-inch (64-mm) threaded couplings. Almost all threaded couplings on hand lines (the primary exception is Stortz connectors) utilize similar lugs. FIGURE 5-6 illustrates the male and female coupling. When a hoseline is deployed, the water will come out of the male-coupled end. This coupling will have a longer lug than the female coupling. On the female coupling we find that the swivel coupling has a distinctly short lug. A fire fighter with gloved hands can easily distinguish the difference between the two types of lugs on a coupling. Once determined, moving in the direction of the long lug will, in essence, have you going against the flow direction and back towards the pump and safety.

The long lug out escape drill can be used with either a charged hoseline or an uncharged hoseline. The most important skill to remember is to *never let go of the line!*

The long lug out escape drill is demonstrated in SKILL DRILL 5-8:

1. If lost or separated from your company, first and foremost notify command and declare a MAYDAY!
2. Attempt to locate a hoseline if possible
3. If a hoseline is found, choose a direction and begin to follow. There is no general guideline as to which direction to choose. (**STEP 1**)
4. Once you find a coupling, examine it and determine in which direction the long lug out points. Follow the hoseline and *never lose contact with it!* There could be another line adjacent to the one you are following, and if you lose contact, you could pick up the wrong hose. (**STEP 2**)
5. Whenever you encounter another coupling, check it to make sure you are still going in the right direction. (**STEP 3**)

Crew Notes

If you are following the line and are sure of the direction, but the line begins to go up a set of stairs, continue to follow it, as it might be a bight or loop of hose that comes back down. The key is to determine the proper direction and then use your skills to follow it.

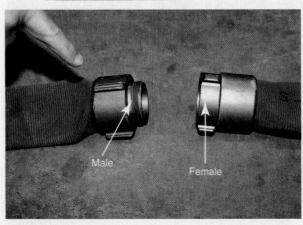
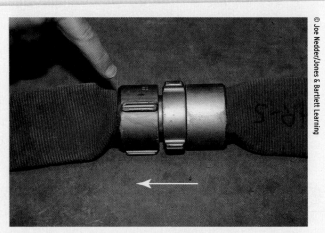

A.   B.

FIGURE 5-6 A set couplings includes one female coupling (on the right) and one male coupling (on the left), shown **A.** unconnected and **B.** connected. The female coupling has threads on the inside of the coupling and short lugs, and the male coupling has exposed threads on the outside and includes the long lug (as indicated). The white arrow below the connected coupling indicates that following the long lug will lead the way out.

© Joe Nedder/Jones & Bartlett Learning

SKILL DRILL 5-8   Long Lug Out Escape Drill
NFPA 1407, 7.13

1 If lost or separated from your company, first and foremost notify command and declare a MAYDAY! Attempt to locate a hoseline if possible. If a hose line is found, choose a direction and begin to follow. There is no general guideline as to which direction to choose.

2 Once you find a coupling, examine it and determine in which direction the long lug out points. Follow the hoseline and *never lose contact with it!* There could be another line adjacent to the one you are following, and if you lose contact, you could pick up the wrong hose.

3 Whenever you encounter another coupling, check it to make sure you are still going in the right direction.

© Joe Nedder/Jones & Bartlett Learning

Extrication of a Fire Fighter From Debris

It is a very common occurrence in building fires to have debris fall during firefighting operations and overhaul. The extrication of a fire fighter from debris can be a simple lift of the debris and extraction of the fire fighter operation, or a complex technical rescue involving a lot of personnel and equipment to lift and remove large and heavy elements off of the fire fighter. For the purpose of this text, we need to remember that the basics of RIC are more focused on the former, rather than the latter.

So, what is debris? Debris can be anything and everything contained in a building, as well as the components that make up a building. It can be fragments of interior ceilings, walls, or structural components. It can be broken masonry blocks, bricks, and other materials from a building that has fallen down or has been destroyed by the fire. It can be exterior parts, such as heavy slate or tile roofing materials or interior stone work. Many times, ceiling components, such as sheetrock, plaster and lathe, suspended ceiling tiles, and decorative objects, will fall from fire impingement and the application of water.

The contents of a room, such as bookcases, shelving units, large TVs, and other furniture can land on unsuspecting fire fighters, trapping them suddenly. A sheetrock ceiling can hold large amounts of water until a fire fighter opens up the ceiling, causing a large piece of the ceiling to suddenly fall on them. Suspended ceilings contain metal wires to hold them in place; these wires can melt or fail, easily dropping large amounts of debris (ceiling tiles and metal grid) onto crews in the affected area. In many instances, electrical wire and conduit, heating and ventilation ductwork, and other equipment may be above a ceiling. Debris can also be the contents of commercial properties, such as department stores and supermarkets, or it could be manufactured products at a factory. Debris can be in the form of hazardous materials, requiring even more specialized rescue operations. In short, debris is part of a building or the contents within a building. For the purpose of this text, we will define debris as either light debris or heavy debris.

Light debris is something that, although trapping the fire fighter, can be removed by other fire fighters by hand or with the use of the hand tools they have with them. Light debris might be a bookcase, a shelving unit, or the ceiling (suspended, wallboard, or plaster and lathe) that has come down during overhaul, trapping the fire fighter. In a commercial setting, it might be inventory, boxes full of materials, or other unsecured items impinged by the fire. Occasionally a fire is fought in the residence of a hoarder, and the opportunity to get entangled or have all sorts of "debris" fall on and entrap the fire fighter is significant.

For the most part, light debris can be removed by hand, with the hardest part being finding the buried fire fighter. Remember, if the fire fighter is almost completely covered, his PASS device might be activated, but the sound will be severely muffled. Furthermore, the debris covering the fire fighter will make searching with the thermal imaging camera (TIC) very difficult, depending on the degree of coverage. If a MAYDAY is received from a fire fighter trapped under possible light debris, the RIC Operations Chief or Incident Commander should realize and plan for the numerous teams that will be needed for the search and rescue. The first RIC deployed has the assignment of trying to find the downed fire fighter and establish an air supply. Subsequent teams will extract the fire fighter and get him out.

Heavy debris can be defined as "everything else." It is heavy, cannot be moved by hand or with the use of your hand tools, requires the use of technical expertise in collapse rescue, and will require extensive manpower and the use of equipment such as hydraulic jacks, air bags, saws, struts, and shoring to remove the fire fighter from entrapment. In some instances, structural components can fail, causing whole or partial collapse of a building and trapping fire fighters under heavy debris. In addition, heavy debris includes when the floor above has collapsed and trapped the fire fighter and when furniture and large appliances are involved in the collapse, adding extensive weight and technical challenge to the rescue. The very debris trapping a fire fighter could potentially be what is holding up a section of the structure from further collapse, making a rescue much more challenging. Heavy debris involving suspected structural elements or structural stability can and will affect the safety of all fire fighters involved in the rescue attempt. Typically, heavy debris is involved when

there is concern for what will happen when the debris entrapping the fire fighter is moved. Will it cause another event of collapse or structural failure, endangering, trapping, or injuring the rescuers? Removing a fire fighter from heavy debris can quickly become a major task, requiring a significant amount of extra personnel and technical expertise. It is not a basic RIC operation.

When a MAYDAY call is received for a fire fighter(s) trapped by heavy debris from an interior collapse or partial collapse, the situation must be quickly assessed to determine, if possible, the extent of the collapse and how many fire fighters are trapped. The IC must perform a risk analysis to determine the best course of action for the RIC and the rescue of the fire fighter(s) trapped. Additional questions include: Do we have enough personnel to effect a rescue? Do we have the equipment and training to undergo what could basically become a technical rescue situation under fire conditions? Are we potentially putting large numbers of personnel in a situation that more than likely is beyond the abilities of crews on scene? How far away are the extra help and/or the technical expertise needed to perform the rescue? The answers to these questions will have a direct impact on the action plan created by the IC. A constant risk analysis must be conducted throughout any RIC operation to keep more fire fighters from becoming victims.

Wire Entanglement Survival

The dangers of a fire fighter getting entangled in wires has increased dramatically in the past 20 years, as we have seen a tremendous increase of technology-based wires within structures, both residential and commercial. The most significant increase has been with low-voltage wiring, which includes but is not limited to:

- Cable TV
- Computer and data-based communications wire
- Telephone
- Alarm

What these wires all have in common is that:

- They are low-voltage wires, which are safe to cut.
- They are found in abundance above suspended ceilings in commercial settings.
- They usually have minimal securing to the structure and are thin in size.
- They are easy to become entangled in.
- They are difficult to disentangle from without some skill training.

Envision that you are working a fire and enter into an area in which a suspended ceiling has fallen, or worse, you are under the ceiling when it collapses. Besides the ceiling tiles and metal frame of this ceiling, you will probably find many low-voltage wires that were passing through. These small wires are now waiting to entrap you, snagging on your cylinders, helmet, the flashlight you have mounted on your helmet, and other items mounted on your equipment and turnout gear.

In addition to commercial buildings, these wires are also used in abundance in hospitals, apartment buildings, housing for the elderly, condominiums, multifamily homes, and private residences. Virtually every structure today is filled with wires that can become deadly to fire fighters. Because much of this wire is low voltage (12 volts), it is safe to use your cutters to cut yourself free, as the 12 volts possess little if any danger to you.

Other wires and entanglement dangers encountered are standard Romex-type wires; MC wire (cabled wire), typically found in overhead ceilings providing power to the lighting; and <u>HVAC</u> coiled ductwork or dryer vents. **FIGURE 5-7**. The issue with the HVAC system is that the fire may burn through the insulation, but the wire used to give the ductwork its shape is a significant entanglement hazard and if entangled it is difficult, if not impossible, to cut free with wire cutters. The Romex and MC type wire is typically 110 volts and as such it is not advisable to attempt to cut unless it is confirmed that the power is off.

If you have become entangled in wires, there are a few methods you can use to attempt to free yourself. The first three steps in any entanglement escape are:

1. Always try to avoid getting entangled.
2. If you find yourself entangled, always call a MAYDAY!
3. After declaring the MAYDAY, activate your PASS.

With these three steps in mind, we will look at some different techniques to escape an entanglement.

Once you determine you are entangled or are covered with wires and have the possibility of getting entangled,

A.

B.

FIGURE 5-7 **A.** HVAC coiled ductwork and **B.** the wire that is exposed after the insulation burn off.

Safety

All fire fighters should equip themselves with a good pair of wire cutters. The cutters need to be easily manipulated with one hand, and, when open, the jaw opening needs to be wide enough to fit over the wire easily **FIGURE 5-8**.

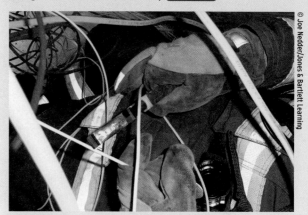

FIGURE 5-8 Wire cutters need to be easily manipulated with one hand, and, when open, be able to fit over the wire.

the entanglement escape technique should be attempted, as demonstrated in **SKILL DRILL 5-9**:

1. If you are near a wall, try and position yourself against the wall, lying on your side with the cylinder positioned to the point where the wall and floor meet. If you are not near a wall, lie on your side, if possible with the regulator line towards the floor to protect the low-pressure hose from getting snagged or caught. (**STEP 1**)
2. Sweep the high arm from below your waist up and across your body, continuing to sweep and lift up the wires until your arm is extended beyond your helmet. While still supporting the wires, move in the direction of your escape and then repeat the procedure until you have moved through the entanglement area. (**STEP 2**)
3. If needed, cut any *low-voltage wires* you identify to expedite the escape.
4. If the entanglement is severe and you cannot escape, do not further entrap yourself by trying to force yourself through the wires, as they will not give and break. Advise command of your severe entrapment, try and control your breathing, and wait for the RIC to assist you. (**STEP 3**)

Fire Fighter Rapid Emergency Egress Techniques Through Windows

Rapid emergency egress techniques are headfirst rapid exits, commonly known as "bailouts," and are survival skills that every fire fighter must have and be able to execute as a last-ditch effort. What is a last ditch? Simply put, conditions have deteriorated to a point that you either get out fast or you will die in the fire! Bailouts are a dangerous technique and should not

SKILL DRILL 5-9 Entanglement Escape

NFPA 1407, 7.7(3), 7.7(4), 7.13.1(1), 7.13.1(2)

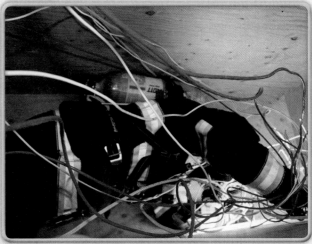

1 If you are near a wall, try and position yourself against the wall, lying on your side with the cylinder positioned to the point where the wall and floor meet. If you are not near a wall, lie on your side, if possible with the regulator line towards the floor to protect the low-pressure hose from getting snagged or caught.

2 Sweep the high arm from below your waist up and across your body, continuing to sweep and lift up the wires until your arm is extended beyond your helmet. While still supporting the wires, move in the direction of your escape and then repeat the procedure until you have moved through the entanglement area.

3 If needed, cut any *low-voltage wires* you identify to expedite the escape. If the entanglement is severe and you cannot escape, do not further entrap yourself by trying to force yourself through the wires, as they will not give and break. Advise command of your severe entrapment, try and control your breathing, and wait for the RIC to assist you.

be used unless all other avenues of escape have been exhausted or your air supply is almost empty. If there is another way to escape use it, if not—bail out!

■ The Go/No Go Decision

What conditions would warrant a quick escape?

- The area you are operating in has had a rapid change in fire conditions, and the room or area you are in is near flashover.

- You have run out of air or have had a catastrophic failure with your SCBA; are in a heavy smoke, pre-flashover condition; and there is no other quick way out.
- As you and your company advanced, you have been trapped by fire blocking your only egress. The fire is beginning to overtake you and your company, and if you do not get out quickly you face injury or death.
- A structural failure has entrapped you and/or your company, and you have no other alternative but to rapidly exit with an emergency procedure.

■ Types of Rapid Emergency Egress Escape Techniques or Bailouts

There are three rapid emergency egress escape techniques:

1. The ladder escape
2. The cylinder wrap escape
3. Using an escape descent control device (DCD) system

■ Training Fire Fighters to Execute Rapid Emergency Egress Escapes

When fire fighters are trained in emergency escape techniques, three critical elements are emphasized:

1. Understand that an emergency escape technique is a last-ditch effort, and know what constitutes a last-ditch effort.
2. Understand that learning to execute a skill today a few times and needing it 3 years from now does not guarantee you will remember how to do it. As with all firefighting skills, there is a need to drill frequently in order to keep skills sharp.
3. Whenever training, all safety precautions need to be taken to protect those training, including the use of harnesses and belay lines.

Safety

Whenever you are drilling on the bailout skills, it is imperative and absolutely necessary that the student executing the egress skill be properly harnessed and on a belay line. There is no excuse not to use a belay line. Taking common sense precautions such as this will prevent training injuries. Belays should be run by trained fire fighters qualified in setting up and managing safety belay lines. Only the proper ropes and equipment (NFPA-approved) should be used. These skills are dangerous under the best of conditions, and every precaution must be taken to protect the fire fighters during any training exercise.

Ladder Rapid Emergency Escape Techniques

The ladder escape is an excellent starting point when training fire fighters on bailouts. The ladder escape should encourage fireground commanders to be very proactive in placing ground ladders on every floor that is reachable of a burning structure where fire fighters are operating.

To execute a ladder rapid emergency escape, the fire fighter either needs to declare a MAYDAY and request a ladder be placed rapidly to a designated window, or locate a window where a proactive IC has already placed a ladder for fire fighter egress.

■ Ladder Position: Angle and Tip Placement

Ladder position for this escape is very important and is significantly different from the conventional methods taught. In basic skills training, we are taught that the ladder must be placed at a 75-degree angle as per the manufacturer's specifications. This angle provides the most benefit from the ladder during normal operations or use. However, a 75-degree angle is not safe or conducive for a ladder escape. First, the steep angle might not hold up to the energy and inertia exerted onto it by a rapidly moving fire fighter escaping, thus causing the ladder tip to lift off the building and crash to the ground with the fire fighter under it. Second, a fire fighter trying to exit, and observing the ladder at the 75-degree angle, will "see" a ladder that looks like it is almost at 90 degrees, causing the fire fighter to hesitate or retreat back into the building. Either way, the fire fighter in peril might not get out. To overcome this issue, a ladder set up for rapid emergency egress needs to be set at a reduced angle, typically about 60 degrees **FIGURE 5-9**. Caution must be observed when a ladder is placed at this angle for two reasons:

1. The steeper angle can contribute to the bottom of the ladder kicking out during the escape.
2. The angle reduces the working load of the ladder and the number of fire fighters it can support.

When a ladder is set for an emergency escape, the ladder tip should be set at or directly below the sill **FIGURE 5-10**. Keeping the tip below or flush with the sill will help prevent the escaping fire fighter from snagging gear or equipment onto the tip when going out headfirst rapidly.

FIGURE 5-9 A ladder set up for rapid emergency egress needs to be set at a reduced angle.

© Joe Nedder/Jones & Bartlett Learning

© Joe Nedder/Jones & Bartlett Learning

FIGURE 5-10 A ladder set for an emergency escape should be set at or directly below the sill.

4 Continue descending out and, with the left hand, grab the fourth rung (Grab 4) with your palm down. (**STEP ③**)

5 Your body will be completely out of the window at this point. Place your legs together, bend your knees to at least a 90-degree angle and shift your weight to the *left*. (**STEP ④**)

6 During this transition, allow your waist and hip area to slide against the ladder. (**STEP ⑤**) You will transition easily back into an upright position and land with both feet on a rung. (**STEP ⑥**)

7 If you are the only member escaping, proceed down the ladder to safety. (**STEP ⑦**)

8 If there are additional fire fighters escaping behind you, perform a ladder slide to get out of the way quickly. (**STEP ⑧**)

9 Once you are on the ground, quickly clear away so as not to impede other fire fighters escaping.

An Appeal for Change

Perhaps it is time for the fire service to reconsider the numerous ways ladders are positioned to and in windows. The methods taught today were developed when windows were very large. To place a ladder one or two rungs into today's typical house window and expect a fire fighter in full gear, on air, to *safely* climb off the ladder and into the structure is antiquated and obsolete. If we place our ladders to windows with the tip at or below the sill, it will allow for entry, emergency egress, and rescues over ladders *without* having to reposition the ladder. We need to be proactive and recognize that it is time for change and that this change is to our benefit.

As mentioned earlier, all steps to the ladder rapid emergency escape technique are preceded by the assessment indicating the need to exit rapidly with no other escape avenue. If there is no other option, then a bailout is warranted. The ladder rapid emergency escape technique is demonstrated in **SKILL DRILL 5-10**

1 Determine the window to exit and if a ladder is in place. If there is no ladder preplaced, do you have time to call a MAYDAY and request a ladder? If the answer to this is no, then a rope bailout is required to escape. (Rope bailouts will be discussed later in this chapter.) If a ladder is in place, call a MAYDAY (if time and conditions allow—if not, get out!) and indicate where you are exiting. The IC must assign one or two fire fighters to assist with the heeling of the ladder and to help the escaping fire fighter.

2 Starting from a kneeling position inside the window, extending your chest over the sill and begin to exit headfirst over the ladder. (**STEP ①**)

3 Take your right arm and hook it under the second rung (Hook 2), palm up. Do not grab the third rung! (**STEP ②**)

Crew Notes

A simple way to remember this technique is to commit to memory "HOOK 2, GRAB 4." This indicates your hooking onto rung 2 and grabbing rung 4.

Safety

When teaching the ladder rapid emergency escape skill, the student must be attached to a belay line. In addition, a *second ladder* should be placed to the side of the practice ladder so an instructor can assist with direction and instruction as the student completes the maneuver **FIGURE 5-11**.

© Joe Nedder/Jones & Bartlett Learning

FIGURE 5-11 A second ladder should be placed to the side of the practice ladder so an instructor can assist the student with direction and instruction.

SKILL DRILL 5-10 Ladder Rapid Emergency Escape Technique
NFPA 1407, 7.13.2

1 Determine the window to exit and if a ladder is in place. The IC must assign one or two fire fighters to assist with the heeling of the ladder and to help the escaping fire fighter. Starting from a kneeling position inside the window, extending your chest over the sill and begin to exit headfirst over the ladder.

Belay line

2 Take your right arm and hook it under the second rung (Hook 2), palm up. Do not grab the third rung!

3 Continue descending out and, with the left hand, grab the fourth rung (Grab 4) with your palm down.

4 Your body will be completely out of the window at this point. Place your legs together, bend your knees to at least a 90-degree angle and shift your weight to the *left*.

(Continued)

SKILL DRILL 5-10 Ladder Rapid Emergency Escape Technique (*Continued*)

5 During this transition, allow your waist and hip area to slide against the ladder.

6 You will transition easily back into an upright position and land with both feet on a rung.

7 If you are the only member escaping, proceed down the ladder to safety.

8 If there are additional fire fighters escaping behind you, perform a ladder slide to get out of the way quickly. Once you are on the ground, quickly clear away so as not to impede other fire fighters escaping.

© Joe Nedder/Jones & Bartlett Learning

The ladder slide is a technique to be used, after a ladder emergency exit, to get out of the way quickly so that other fire fighters (possibly members of your company) exiting behind you can do so quickly and safely without having to worry about landing on you. The ladder slide is demonstrated in **SKILL DRILL 5-11** :

1 Grab the outside beam of the ladder. (**STEP 1**)
2 Crouch and extend your knees to the outside of the ladder beam. (**STEP 2**)

3 Using your hands as a braking control and keeping friction on the ladder with your legs, slide down the ladder. (**STEP 3**)
4 By keeping your body in a crouched position and your arms extended, you will be in a position to assist with the descent and to prevent items such as flashlights, radios, and other equipment from snagging onto the ladder.

SKILL DRILL 5-11 Ladder Slide
NFPA 1407, 7.13.2

1 Grab the outside beam of the ladder.

2 Crouch and extend your knees to the outside of the ladder beam.

3 Using your hands as a braking control and keeping friction on the ladder with your legs, slide down the ladder. By keeping your body in a crouched position and your arms extended, you will be in a position to assist with the descent and to prevent items such as flashlights, radios, and other equipment from snagging onto the ladder.

© Joe Nedder/Jones & Bartlett Learning

■ Securing or Aggressively Heeling a Ladder During a Ladder Escape

When a ladder is used for a ladder emergency escape, the ladder *must* be heeled aggressively. The conventional way to

heel a ladder for climbing is for a fire fighter to stand under the ladder with arms extended and grab onto the beams. Because the ladder is being set at less than 75 degrees (about 60 degrees) during a ladder escape, the butt is much more prone to kick out. To prevent this from happening, we need

© Joe Nedder/Jones & Bartlett Learning

FIGURE 5-12 An aggressive escape heeling technique prevents the ladder's butt from kicking out.

© Joe Nedder/Jones & Bartlett Learning

FIGURE 5-13 The fire fighters securing the ladder are in a position to assist the rescuer by lifting the unconscious fire fighter off the rescuer's arms.

to retrain ourselves to use a more aggressive heeling technique **FIGURE 5-12**. This aggressive escape heeling technique has numerous advantages, including:

- The fire fighters securing the ladder are in a position to look up and observe the escaping fire fighter.
- The fire fighters securing the ladder are in a position to render assistance to the escaping fire fighter, including ascending the ladder to assist.
- The fire fighters securing the ladder are using both their arms and legs to aggressively heel and secure the ladder, thus preventing the butt from kicking out.
- If this technique is used when a rescue or RIC fire fighter is carrying an unconscious fire fighter down the ladder, the two fire fighters securing the ladder are in position to assist the rescuer and lift the unconscious fire fighter off the rescuer's arms **FIGURE 5-13**.

Emergency Escapes Using a Personal Rope or an Escape DCD System

Two rope emergency escapes will be discussed here: (1) the cylinder wrap escape; and (2) the rope escape with an escape DCD. The escape DCD system can be standalone, attached to your escape belt, or as part of a system integrated into your SCBA. As with the ladder emergency escape, these techniques are intended as a last-ditch effort to escape. As such, the go/no go decision matrix is the same. Three things are needed to complete these emergency escapes:

1. The proper equipment
2. A secure anchor point
3. Knowledge and ongoing practice of the skill

■ Proper Equipment

Personal Rope for an Emergency Egress Rope Escape

The NFPA designates 7.5 to 9.5 mm for personal escape rope, which must be static kernmantle and in compliance with NFPA 1983, *Standard on Life Safety Rope and Equipment for Emergency Services*. These types of personal escape ropes are available in either nylon or Aramid yarns. The primary advantage of the Aramid is that it is heat-resistant up to 932°F (500°C) and is more abrasion- and cut-resistant than nylon. It is suggested that the rope be 50 feet (15 m) long, which provides enough length to escape from a third-story window, while anchored inside the room, with some extra length if the grade below the window is a walkout cellar. The 50-foot (15-m) length will also fit in your bunker pant's bellows pocket.

How Do I Carry My Personal Rope Without an Escape DCD System: Bag or Pocket?

There are three methods to carry your personal rope. The first is to place the rope in your bellows pocket on your bunker pants, the second is to place the rope in a bag that is attached to your SCBA, and the third is to place the rope in a bag that is designed to be placed in your bellows pocket. Many departments will purchase one escape rope in a bag per SCBA. The primary advantage of this is cost, but the disadvantages are:

(1) it is not *your personal* rope, so you have no idea if it has been taken care of or inspected; and (2) the bag attached and hanging from a SCBA waist strap adds bulk and size to the fire fighter, which can affect the ability to move through tight spaces or cause entanglements.

Putting your personal escape rope in your bellows pocket works fine and is easy to use. It can be loaded like any rope bag: just feed it in and lay the carabiner on top. A rope being put into a bag that fits easily into your bellows pocket has the added advantage of, once the rope is anchored and you are at the window, removing the bag from the pocket and thrown out the window to the ground. This will ensure that the rope does not tangle up as it is coming out of your bellows pocket.

■ Emergency Rope Escape Anchoring

In addition to the rope, an anchoring device is needed. Traditionally, these have been carabiners. Many new securing hooks have been introduced to the market, including the Crosby hook and the Lightning hook **FIGURE 5-14** .

Carabiners

Many times when an escape rope is purchased, it has been prepackaged with a medium carabiner that is self-locking. Remember, NFPA requirements for self-locking carabiners are for *technical rescue*. Rapid emergency escapes are not technical rescues, but rather a lifesaving emergency technique. With that said, consider these two questions:

1. When trying to get out rapidly to survive, do you really want to have to manipulate an autolock on the carabiner?
2. With a gloved hand and with no visibility, how easy do you think it is to handle a medium-size carabiner? Can you open it, quickly thread the rope through, and get out rapidly?

Using an XL modified D-type carabiner with a *screw lock* is the best alternative. The extra-large size makes it easy to handle and operate with a glove on, and, most importantly, easy to thread the rope through for your escape **FIGURE 5-15** .

Anchor Hooks (Escape Anchor Device)

Anchor hooks or escape anchor devices (EADs) are designed to be used in conjunction with a mechanical escape device.

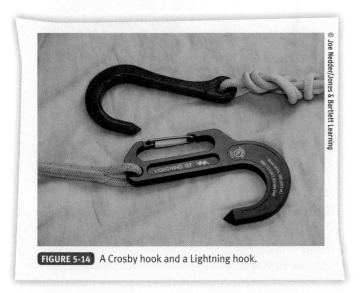

FIGURE 5-14 A Crosby hook and a Lightning hook.

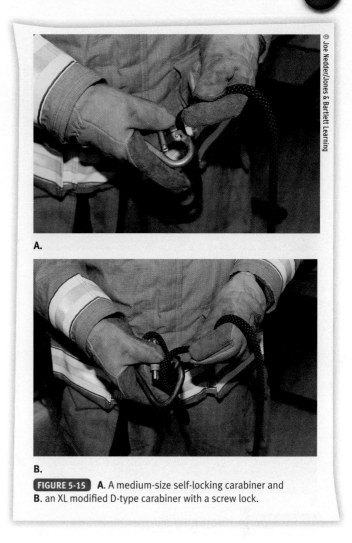

FIGURE 5-15 **A.** A medium-size self-locking carabiner and **B.** an XL modified D-type carabiner with a screw lock.

They are not intended and never should be used for a cylinder wrap descent! The two most popular anchor hooks currently on the market are the Crosby hook and the Lightning hook, which differ in weight, size, and ease of use. These hooks are designed to grab onto a windowsill or another right angle at or near the window. The advantage of this type of hook is that it enables you to hook at the exit point versus an anchor point in the room and thus use less rope. A significant caution with these types of devices that must be remembered is that they need to be "hand managed" until you are completely out the window, exerting downward pressure on the rope and to the hook, "loading it" to prevent it from slipping loose. With the cylinder wrap escape this is nearly impossible, as both hands are needed to manage the rope. If you have a hook and are executing a cylinder wrap escape, you need to tie off the rope at a hard anchor point or it will not work. With most hooks this can be accomplished by wrapping the rope around a secure object and then securing the rope to the hook with a clove hitch knot. Some hooks such as the Lightning hook have a built-in ability to pass through and wrap the rope around the hook itself, thus getting a safe and secure anchor.

Anchor Points

When a cylinder wrap escape is required, you must have a hard anchor point to which you secure the rope. Anchor points are usually created by the escaping fire fighter. There is only one

FIGURE 5-16 The rope is anchored by securing one end with a **A.** carabiner or **B.** a hook around something solid and not movable within the room.

safe way to properly anchor the rope for a cylinder wrap escape: secure the end of the rope, with a carabiner or hook, around something solid and not movable within the room, such as a stud in the wall near the exiting window **FIGURE 5-16**.

What Are Good Anchor Points?

The most common method to create an anchor point is to punch through the drywall or plaster and wrap the rope around an interior wall stud, which should be as close to the

Crew Notes

The personal rope issued or carried, and in service, should never be used for training and practice. Once an escape rope has been used for a real-time escape, it should be removed from service. When training, additional ropes should be purchased and used specifically for that purpose. Consult with your rope's manufacturer to determine how many "descents" the rope can be subject to in training before being replaced. Trainers should keep a rope log, as per NFPA requirements, to document the use the rope has been subjected to.

escape window as possible to utilize the least amount of rope. If you have a door opening with in a reasonable reach to the window, the door framing is substantial and only requires breaking out the wall sheathing. Other anchor points might be a substantial piece of furniture that will not move, such as a large bureau or chest of drawers on the wall near the window you are exiting. In older homes you might find cast iron radiators under the windows that can offer a solid anchor point. However, great caution must be taken if you wrap a nylon rope to this type of radiator, as the nylon will begin to melt and possibly degrade the strength of the rope. Whatever you do, it must be a solid enough anchor point that will not let go during your descent!

Safety

Hand placement when executing the cylinder wrap emergency escape is critical! If you fail to ensure your hands are placed outside the sill, when you exit, both your hands will become trapped under the rope with your entire weight bearing down on them. You will not be able to free them and will be in extreme pain. **FIGURE 5-17** shows the correct and incorrect hand position for successfully and safely completing this emergency escape.

FIGURE 5-17 **A.** Correct hand placement when executing the cylinder wrap emergency egress escape, and **B.** incorrect hand placement when executing the cylinder wrap emergency egress escape.

■ Cylinder Wrap Emergency Egress Escape

Of all the rapid emergency egress methods to bailout, the cylinder wrap is clearly the most dangerous. However, if it is all you have, and your choice is to bail out or die, it works! The cylinder wrap emergency escape technique is simply that. With one end of the rope secured to a good anchor point, the escaping fire fighter wraps the rope around his body and SCBA cylinder, and, using his hands as a descending device, drops out of the window and descends to safety. Even though it sounds rather simple, not using the proper equipment or technique can lead to serious injury or death. The two most important things to remember with this technique are: (1) the way you position your hands with the rope outside the window; and (2) to never let go of the rope as you descend or you will fall to injury or death.

The cylinder wrap emergency egress escape technique is demonstrated in **SKILL DRILL 5-12** :

1. Secure your personal escape rope to the anchor point, wrapping the rope around one or two times and then, if using a carabiner, lock the rope through the carabiner (**STEP ❶**) or, if using a hook, tie off the rope to the hook securely. In most situations a clove hitch will work. (**STEP ❷**) Once the anchor point has been established and the rope secured, you are ready to execute the emergency cylinder wrap escape.

2. With the rope properly anchored and staying low, advance to the window you will exit from, keeping the working end tight or snug to the anchor.

3. While kneeling, place the rope around your body and bring the rope together with one hand in front of yourself. (**STEP ❸**)

4. Position yourself on the windowsill with one leg out. The working end of the rope should be kept taut. (**STEP ❹**) Make sure you have enough slack in the rope to allow both your hands, while gripping the rope, to be placed outside the windowsill. (**STEP ❺**)

5. With your hands gripping the rope tightly, drop your shoulder out of the window and allow your body to follow. (**STEP ❻**) As you drop out, you will self-right yourself, feel the rope dig in around your body and find yourself with your hands gripping the rope somewhere around your face level. (**STEP ❼**)

6. To descend, slightly release your grip on the rope and you will go down. The more you release the faster you will go. (**STEP ❽**)

7. When you reach the ground, quickly move out of the way if other fire fighters are escaping behind you.

Safety

The cylinder wrap descent is not rappelling. As such, you should not be bouncing off the building as you descend. The descent should be straight down without wall contact.

SKILL DRILL 5-12   Cylinder Wrap Emergency Egress Escape Technique
NFPA 1407, 7.13.2

❶ Secure your personal escape rope to the anchor point, wrapping the rope around one or two times and then, if using a carabiner, lock the rope through the carabiner.

❷ If using a hook, tie off the rope to the hook securely. In most situations a clove hitch will work.

(Continued)

SKILL DRILL 5-12   Cylinder Wrap Emergency Egress Escape Technique (*Continued*)

3   With the rope properly anchored and staying low, advance to the window you will exit from, keeping the working end tight or snug to the anchor. While kneeling, place the rope around your body and bring the rope together with one hand in front of yourself.

4   Position yourself on the windowsill with one leg out. The working end of the rope should be kept taut.

5   Make sure you have enough slack in the rope to allow both your hands, while gripping the rope, to be placed outside the windowsill.

6   With your hands gripping the rope tightly, drop your shoulder out of the window and allow your body to follow.

7   As you drop out, you will self-right yourself, feel the rope dig in around your body and find yourself with your hands gripping the rope somewhere around your face level.

8   To descend, slightly release your grip on the rope and you will go down. The more you release the faster you will go. When you reach the ground, quickly move out of the way if other fire fighters are escaping behind you.

■ Rope Emergency Egress Escape Using an Escape DCD

A third option for an emergency escape technique is the use of an escape DCD. A DCD is part of a *complete system* that meets the requirements of NFPA 1983, 2012 edition **FIGURE 5-18**. There are currently numerous systems available on the market, and the advantages of this type of system are:

1. The system is complete and in full compliance with NFPA 1983.
2. The system is rigged and secured to the harness during the donning of your PPE; therefore, it is immediately ready for deployment and use.
3. Once out the window, the DCD will hold the fire fighter in place if he lets go of the rope or loses consciousness.
4. The system provides for more control over the descent.

The use of any escape DCD system requires the use of a life safety <u>harness</u>. Currently, there are three options: (1) a Class II harness attached or built into your bunker pants by the PPE manufacturer; (2) an emergency escape belt worn over your PPE; and (3) an emergency escape belt built into your SCBA waist strap **FIGURE 5-19**. The important considerations when choosing what system to purchase are simplicity of use, ease of anchoring, and ease to deploy.

Like the personal rope, there is a case to be made for issuing an escape system to every fire fighter versus one per SCBA. As previously indicated, many departments will purchase one escape system, in a bag, per SCBA, with the primary advantage of this being cost. The disadvantages are, first, it is not *your personal* rope, so you do not know how it has been cared for or if it has been inspected, and second, the bag attached and hanging from a SCBA waist strap adds bulk and size, affecting the ability to move through tight spaces and increasing the potential for entanglement. If you do carry your personal escape rope in this manner, the rope should be inspected and maintained regularly, in accordance with the NFPA requirements. The rope should never be used to practice emergency escapes or bailouts. Once used in an emergency, it should be removed from service. SCBAs with built-in escape systems must also be inspected regularly. This includes completely removing the rope from the storage pocket, inspecting it, and then repacking as per the manufacturer's instructions.

If your bunker pants have a built-in Class II harness, putting your bagged personal escape system in your bellows pocket works fine and is easy to use. The key is to make sure the system is secured onto the harness when donning, so if needed, it is ready to go to work. Many gear manufacturers now make special bellows pockets to accommodate and work well with different brand mechanical escape systems **FIGURE 5-20**. The rope in a bag that fits easily into your bellows pocket has an increased advantage in that, once the rope is anchored and you are at the window, the bag can be taken out of the pocket and thrown out the window to the ground. This will ensure that the rope does not tangle up as it is coming out of your bellows pocket. Some SCBA manufacturers have developed special waist straps that are fully NFPA-compliant emergency escape belt and many have designed ways to fit the escape rope into the belt without adding a lot of bulk or entanglement hazards.

Use of and Training with the System

Whatever system you choose, you must consult with the manufacturer regarding training requirements. Each manufacturer has different requirements that must be met before the fire fighter is considered qualified and capable in the safe and proper use of their system. Once met, there must be continued and on-going training to maintain your skills and competence in the use of the system. This is critical, because if the need to execute this lifesaving skill under untenable conditions becomes apparent, you will be able to perform and escape.

The Basic Steps for Using a DCD With an EAD Hook

When using a DCD, there are basic steps that must be taken regardless of the manufacturer. The steps for using a DCD with an EAD hook are demonstrated in **SKILL DRILL 5-13**:

1. Ensure that *during the donning of your PPE* you always engage the DCD as required for use.
2. If an emergency arises and you must execute an emergency escape, deploy the device from its storage. **(STEP 1)**

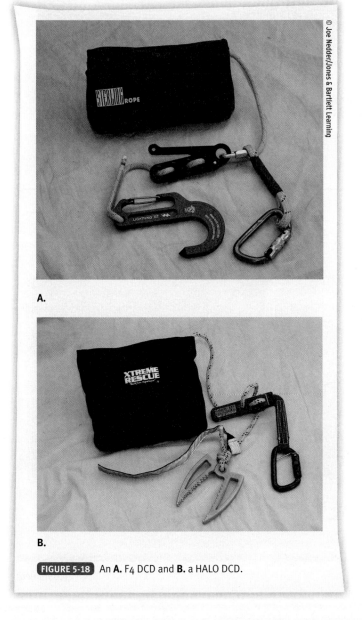

© Joe Nedder/Jones & Bartlett Learning

A.

B.

FIGURE 5-18 An **A.** F4 DCD and **B.** a HALO DCD.

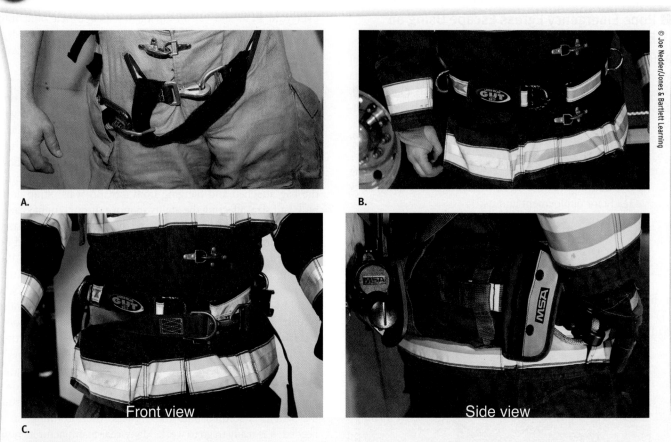

A.

B.

Front view

Side view

C.

FIGURE 5-19 There are three options for a life safety harness: **A.** a Class II harness attached or built into the bunker pants by the PPE manufacturer; **B.** an emergency escape belt worn over PPE; and **C.** an emergency escape belt built into the SCBA (front and side views).

© Joe Nedder/Jones & Bartlett Learning

SKILL DRILL 5-13

Using a DCD With an EAD Hook
NFPA 1407, 7.13.2

1 If an emergency arises and you must execute an emergency escape, deploy the device from its storage.

2 Anchor the hook as trained.

(Continued)

SKILL DRILL 5-13 Using a DCD With an EAD Hook (*Continued*)

3 Position yourself at the window.

4 Roll out the window, maintaining tension on the hook.

5 After the emergency egress, engage the DCD to lower yourself to the ground. Then move away from the landing point so other fire fighters can execute the emergency egress if needed.

© Joe Nedder/Jones & Bartlett Learning

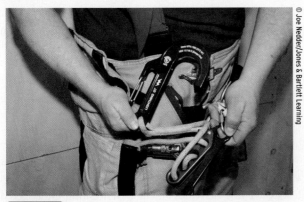

© Joe Nedder/Jones & Bartlett Learning

FIGURE 5-20 A bellows pocket that accommodates different brands of mechanical escape systems.

3 Anchor the hook as trained. (**STEP** **2**)
4 Position yourself at the window. (**STEP** **3**)

5 Roll out the window, maintaining tension on the hook. (**STEP** **4**)
6 After the emergency egress, engage the DCD to lower yourself to the groun. (**STEP** **5**)
7 Disengage from the rope and DCD in whatever manner is prescribed by the manufacturer and move away from the landing point so that other fire fighters can execute the emergency egress if needed.

Conclusion

Survival skills are a very important part of any fire fighter's training and abilities. They can assist you if you are lost or trapped, allow you to participate in your own rescue, or, as part of a RIC, provide a personal foundation to escape if things go bad during the RIC operation. Regardless of what your level of skills might be, your ability to help in your own survival is critical.

Wrap-Up

Chief Concepts

- The RIC is not just saving another fire fighter; it is also having the survival skills to save yourself.
- Skills and techniques to save one of our own are necessary basic skills.
- Rapid intervention can be broken into two parts: survival and rescue.
- Fire fighter survival skills can be used to not only save yourself but also to rapidly allow a company to egress to safety.
- The first step to any survival training is to always be prepared just in case of an emergency.
- All fire fighters need to recognize and understand that when they think they are lost, trapped, or stuck, they are.
- The low air alarm contained on a 30-minute SCBA means you have 25 percent of your cylinder's capacity left for the pre-2013 NFPA 1983 standard, and 33 percent for SCBAs manufactured after the implementation of the 2013 edition of the NFPA 1983 standard.

Hot Terms

Accountability system A method of accounting for all personnel at an emergency incident and ensuring that only personnel with specific assignments are permitted to work within various zones.

Anchor hook Hook used in conjunction with a personal escape system that is placed on a window sill and used as an anchor prior to bailing out of a window.

Anchor point A single secure connection point at which a rope can be secured.

Belay To protect against falling by managing an uploaded rope (the belay line) in a way that secures one or more individuals in case the main line rope or support fails, or to protect against falling while practicing dangerous skills at a height greater than head level.

Carabiners Metal snap links used connect elements of rope rescue equipment.

Crew integrity The ability to maintain control of an organized group of fire fighters under the leadership of a company officer, crew leader, or other designated official.

Descent control device (DCD) A device used in conjunction with a personal escape system that creates friction by means of a rope running through it.

Drywall Also known as plasterboard, wallboard, or gypsum board; drywall is a panel made of gypsum plaster pressed between two thick sheets of paper. It is used to make interior walls and ceilings.

Escape DCD An emergency rope escape system that can be standalone, attached to the escape harness, or part of a system integrated into the SCBA.

Fire and smoke behavior The science of understanding fire, it stages and how it grows and spreads along with understanding what he smokes volume, velocity, density and color indicate.

Harness Class II or class III: a device used to secure the body to an anchor point, found either as part of the PPE or it can be separate.

Heavy debris Debris that cannot be removed by hand or with the use of hand tools, but requires the use of technical expertise in collapse rescue and extensive manpower and the use of equipment such as hydraulic jacks, air bags, saws, struts, and shoring to remove the fire fighter from entrapment.

HVAC Heating, ventilation, and air conditioning; a system designed to heat and cool a building.

Immediately dangerous to life and health (IDLH) Used to describe an environment that could cause harm and/or death if not immediately escaped.

Incident Commander (IC) The person in charge of the incident site and responsible for all decisions relating to the management of the incident.

Light debris Debris that, although trapping the fire fighter, can be removed by other fire fighters by hand or with the use of the hand tools they have with them.

LIP Location, identification, problem; an acronym used when calling a MAYDAY to advise the IC of the situation being encountered by the fire fighter in peril.

Long lug out Technique used for self-rescue by which a fire fighter identifies the longer lug on the female hose coupling, which will lead out of the structure to the water source.

LUNAR Location, unit number, name, assignment, resources needed; an acronym used when calling a MAYDAY to advise the IC of the situation being encountered by the fire fighter in peril.

Masonry block wall The building of structures from individual units laid in and bound together by mortar; the term *masonry* can also refer to the units themselves. The common materials of masonry construction are brick, stone, marble, granite, travertine, limestone, cast stone, concrete block, glass block, stucco, and tile.

Personal alert safety system (PASS) A device worn by fire fighters that sounds an alarm if the fire fighter is motionless for a period of time.

Rescue Those activities directed at locating endangered persons (fire fighters) at an emergency incident, removing those persons from danger, treating injured victims (fire fighters), and providing for transport to an appropriate healthcare facility.

<u>Self-rescue</u> A fire fighter's use of techniques and tools to remove him or herself from a hazardous situation.

<u>Survival skills</u> Learned skills and techniques used by fire fighters to increase chances of surviving an unexpected event while performing fire suppression/rescue operations.

<u>Wall breaching</u> Survival skill technique where the use of mechanical force is applied to open or penetrate a wall for the purpose of escape from and IDLH atmosphere.

<u>Wood lathe and plaster</u> A building process used mainly for interior walls until the late 1950s. Typically made of strips of wood covered by plaster-type cement.

References

National Fire Protection Association (NFPA) 1407, *Standard for Fire Service Rapid Intervention Crews*. 2015. http://www. nfpa.org/codes-and-standards/document-information-pages?mode=code&code=1407. Accessed February 11, 2014.

National Fire Protection Association (NFPA) 1981, *Standard on Open-Circuit Self-Contained Breathing Apparatus (SCBA) for Emergency Services*. 2013. http://www.nfpa.org/codes-and-standards/document-information-pages?mode=code&code=1981. Accessed October 29, 2013.

National Fire Protection Association (NFPA) 1983, *Standard on Life Safety Rope and Equipment for Emergency Services*. 2012. http://www.nfpa.org/codes-and-standards/document-information-pages?mode=code&code=1983. Accessed October 14, 2013.

National Institute for Occupational Safety and Health (NIOSH), *ALERT Preventing Deaths and Injuries of Fire fighters Using Risk Management Principles at Structure Fires*. Publication Number 2010-153. http://www.cdc.gov/niosh/docs/2010-153/pdfs/2010-153.pdf. Accessed October 29, 2013.

You are an experienced fire fighter at your department. You are working as part of an attack crew at a building fire. You become separated from your crew, get disoriented, and subsequently partially entangled in wires. Conditions within the structure are deteriorating rapidly.

1. What steps could you have taken that would have prevented you from becoming in trouble?

   A. Avoid disorientation. Know where you are in the building, what floor, your approximate location (i.e., A/B corner, etc) at all times, and your company's assignment.

   B. Maintain crew integrity, stay together, work together, and exit together.

   C. Do not overextend your air supply and do not overestimate your skills and abilities.

   D. All of the above.

2. Why is it so important to declare a MAYDAY the very instant you think you are in trouble?

   A. Because you have a limited supply of air and delaying the call for could allow the fire fighter to run out of air prior to the team reaching the fire fighter.

   B. Because that is what your department standard operating guidelines state to do.

   C. The RIC company may be performing other fire-fighting functions and are not immediately ready to be deployed.

   D. All of the above.

3. All fire fighters need to recognize and understand that when they think they are lost, trapped, or stuck, they are!

   A. True

   B. False

4. How much air is left in a pre-2013 standard SCBA when the low-air alarm sounds?

   A. 25 percent of the cylinder.

   B. 30 percent of the cylinder.

   C. 5 minutes of air.

   D. 10 minutes of air.

5. Typically, what type of wires can be found in a residential structure that could potentially trap or entangle a fire fighter?

   A. Electrical wires.

   B. Phone, Internet, and cable television wires.

   C. Suspended ceiling support.

   D. All of these can be found in the typical residential structure.

CHAPTER 6

Skills and Techniques of the Rapid Intervention Search

© Photos.com

© Joe Nedder/Jones & Bartlett Learning

Knowledge Objectives

After studying this chapter you will be able to:

- Understand the importance of thermal imaging in a rapid intervention crew (RIC) search and why it is critical to understand and interpret the images correctly. (NFPA 1407, 7.6(3)), p 94–97, 106–109)
- Know the two types of search techniques used in rapid intervention. (p 97)
- Understand which search technique is best suited to each situation. (NFPA 1407, 7.4(2)), p 97–99)
- Understand the importance of using a search rope and when it needs to be used. (NFPA 1407, 7.4(5), 7.6(1)), p 99)
- Know the four positions of RASP and the tool assignments for each position. (p 100)
- Understand the process of the rope-assisted search procedure (RASP). (NFPA 1407, 7.6(1), 7.6(2), 7.6(3)), p 100–113)
- Understand the importance of using the proper equipment for a RASP search. (p 100–101)
- Understand the 12 key methods, skills, and techniques of RASP. (p 102)
- Know what the best options are for a proper rope anchor position. (NFPA 1407, 7.6(1)), p 102)
- Understand that safety is paramount during an RIC operation and what is the role of the RASP Control Officer. (p 103)
- Understand the importance of having distance indicator knots. (NFPA 1407, 7.6(1), 7.6(2), 7.12.1), p 104–105)
- Understand the importance of securing a change in direction. (p 105–106)
- Understand what a rapid room search is as it relates to a RASP search. (p 97–98, 107–110)
- Understand the dangers and concerns of tethering during a RASP search. (NFPA 1407, 7.6(1), 7.13.1(3)), p 108)
- Recognize the importance of good communications and understand the person(s) with whom you should be communicating during an RIC search. (p 109)
- Understand the importance of individual air management. (NFPA 1407, 7.4(4), 7.13.1(4)), p 109)

Skills Objectives

After studying this chapter and with hands-on training, you will be able to perform the following skills:

- Use a thermal imaging camera (TIC) during an RIC search, and discern how you must interpret the images in a working fire environment versus training in ambient temperatures. (NFPA 1407, 7.6(3)), p 94–97)
- As the Company Officer, use the TIC for best results. (NFPA 1407, 7.6(3)), p 95–96)
- Conduct an oriented search. (NFPA 1407, 7.6(1), 7.6(2), 7.6(3)), p 97–99)
- As the Company Officer, use the TIC in the scan, target and release technique. (NFPA 1407, 7.6(3)), p 106–107)
- Conduct a RASP search using the tools, skills, and techniques as presented in this chapter. (NFPA 1407, 7.6(1), 7.6(2)), p 100–113)

Additional NFPA Standards

- NFPA 1801, *Standard on Thermal Imagers for the Fire Service*
- NFPA 1983, *Standard on Life Safety Rope and Equipment for Emergency Services*

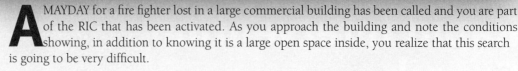

A MAYDAY for a fire fighter lost in a large commercial building has been called and you are part of the RIC that has been activated. As you approach the building and note the conditions showing, in addition to knowing it is a large open space inside, you realize that this search is going to be very difficult.

1. What search techniques, including both skills and equipment, do you as the RIC have available?

2. Has your department ever conducted training on using an RIC search off a rope line, and if so, have you and every member of the company retained the skills?

Introduction

As fire fighters, we have learned to conduct a search for civilians. However, the search for trapped or missing fire fighters will be much more intense and emotional, because we are looking for one of our brothers or sisters. In order to be able to conduct a rapid and skillful search that can make a difference to the survival of one of our own, we must utilize not only our basic search skills but also additional techniques and tools that will assist and enhance our abilities. Here we will present skills, tools, and concepts that will give you this ability, but only *with continued training*. Although considered a basic skill, searching is a tough job that requires sharpened skills that should be instinct. When we enter a structure to look for a lost, trapped, or missing fire fighter, it is not the time to review search skills and abilities, but a time for action requiring an aggressive organized search.

Search Cautions and Reminders

Before the RIC search begins, the company members must:

- Realize and understand the dangers they are going into and exposing themselves to, and not blindly enter a building without basic information, such as the type of construction, where the fire is and if it has been contained, where the target search is, if all the equipment is ready and complete, and, most importantly, if the crew is capable physically and mentally to do the job.
- Remember to enter, work, and exit as a team. Crew integrity is a must, and the Company Officer must keep his crew intact and working as a team.
- Focus not only on the search, but also on air management. It is not in anyone's best interest to have the RIC get themselves into an "out of air" emergency.
- Have the correct tools staged and ready for the given situation.
- Be properly trained in the techniques of fire fighter rescue and the use of the equipment.
- Most importantly, know their way out.

Thermal Imaging and the RIC

One tool that greatly enhances an RIC's abilities is a thermal imaging camera (TIC). In order to be effective using the TIC, you must understand what the camera is telling you, what the camera's limitations are, and how to properly use it. The technology has increased significantly over the years, and the abilities of the cameras being sold are impressive. All fire fighters are encouraged to learn as much about thermal imaging and what the technology can do. Basically, thermal imaging uses infrared technology and identifies the different temperatures by displaying light and dark images on the camera screen; many of the cameras today operate and present a colored image screen. Regardless of what model you have, it is very important to:

- Be properly trained on the use of the camera.
- Be trained to clearly understand what you are looking at when you interpret or translate the screen images.
- Be trained to operate the camera in extremely limited visibility.
- Have the training and skills to change out the unit's battery in blacked out conditions with your gloves on.

■ Understanding What the Image Is Showing You

As fire fighters, the biggest challenge encountered when learning to use the TIC is that the image colors viewed in the classroom or apparatus floor are the *reverse* of what they will look like under real fire conditions. A basic TIC will show "heat" in white and "cool" as dark. A human being, normally about 98.6°F (37°C), is usually the warmer object in the classroom and will show on the screen as white. However, an injured or missing fire fighter inside a building that is on fire and hot will appear as the darker (cooler) object, along with water in the hoselines and self-contained breathing apparatus (SCBA) cylinders, while the heat and fire will appear as the white or lighter objects. This thermal inversion or reverse polarity experienced in training versus reality is critical to understand when using a TIC during any search **FIGURE 6-1**. With the publication of National Fire

Fire fighters must train and learn how to use and interpret the TIC under realistic conditions. The TIC will also show walls, doors, windows, furniture, office partitions, and victims. It will also show hot radiators, thermal currents, and hidden fire in partitions. Each and every one of these will be seen as a variation of white, grey, or black or color, if equipped. It is very important that you are fully capable of recognizing the thermal image of a downed fire fighter or civilian. Remember that what is in front of the camera will give you a good indication of the room or area layout if you know how to interpret what the camera is displaying.

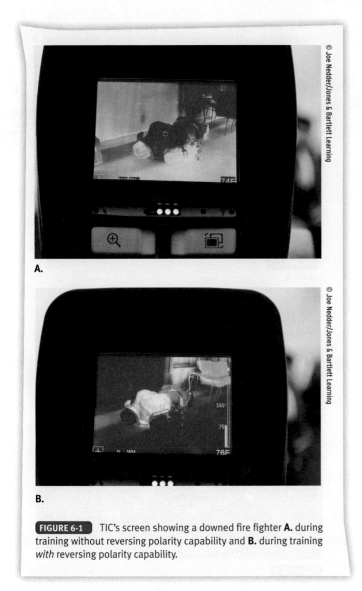

© Joe Nedder/Jones & Bartlett Learning

A.

© Joe Nedder/Jones & Bartlett Learning

B.

FIGURE 6-1 TIC's screen showing a downed fire fighter **A.** during training without reversing polarity capability and **B.** during training *with* reversing polarity capability.

Crew Notes

Learning to use a TIC properly is involved and demanding. Many of the current cameras include a digital readout of temperatures. It is a great tool, but do not rely on looking at temperatures showing on the screen; instead, let your body tell you when it is too hot. Also, due to insulating properties, bad conditions on the other side of walls, partitions, and ceilings might be masked and not be detectable. TIC training is a key component for fire fighter safety today.

Sweep Slowly When Using the TIC

Another issue that can affect a search team's ability to see, process, and interpret a <u>TIC sweep</u> of the floor is moving the camera too quickly. When looking for a downed fire fighter, you need to be very diligent and move the camera slowly in your sweep (look) of an area. You have to interpret what you see on the screen. Moving too quickly might cause you to overlook the fire fighter and pass him by! Make your sweeps slow regardless of what generation TIC you have or what you think or know the camera's capabilities to be. The speed and ability of the camera will not overcome your moving too fast and missing who you are looking for. Your interpreting capabilities are what matter most in the end.

■ Using the TIC for Best Results

Whenever possible, and hopefully without exception, an RIC will always have a TIC to use. How the camera is used will greatly influence your ability to locate the fire fighter. Having a camera does not mean you can stand or walk around. Remember, you are in a limited visibility environment; therefore, you should stay off your feet, stay low, and crawl.

It is also not intended for the officer to put the TIC up to his face upon entry and use it as they move and search throughout the building without ever putting it down. Using the TIC in this manner will give the user "tunnel vision," which greatly inhibits the ability to move rapidly and get the most assistance the camera can provide. <u>Tunnel vision</u> tends to cause the user to focus and concentrate on just one thing and may also cause the user to disregard what else is happening in

Protection Association (NFPA) 1801, *Standard on Thermal Imagers for the Fire Service*, many changes were brought to thermal imaging. As a part of advancing technology, a few manufacturers have allowed the user to reverse polarity during training. This means that when using this camera while training, the images will show exactly how they would appear in a fire—with the fire showing as white and other objects that are cooler than the fire, including people, appearing as black.

Safety

A word of caution if you are using a camera with this reverse polarity capability during training. After training, the camera must be reset to the proper setting or it will not operate as intended. Failure to do this might mean missing a victim or misinterpreting a dangerous heat situation.

his surroundings and other information he is receiving from his senses.

When Technology Fails

Another concern with using and relying on the TIC is if the camera fails to work or malfunctions. A TIC is battery-powered—batteries can run out of power or fail, and electronic devices have been known to malfunction. Equally important to consider is whether or not the fire fighters have been trained to change out batteries (and bring a spare battery) in blacked out conditions and with their gloves on.

A competent RIC will not solely rely on a TIC to find their way in and out. They will be searching on a rope line and will maintain their orientation within the structure, even if they are using a TIC. They will use basic search techniques in conjunction with the TIC. This ensures that, in the event of camera failure or malfunction, the crew is operating safely. It is important to be proactive and know where you are and how to get out alive if the TIC fails.

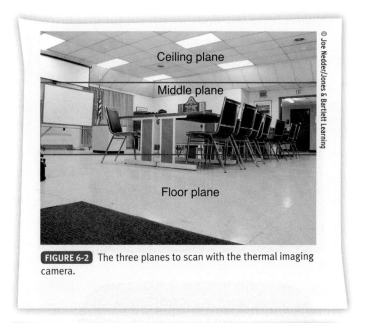

FIGURE 6-2 The three planes to scan with the thermal imaging camera.

Safety

The use of a TIC greatly increases your abilities to rapidly search for the down fire fighter. As technology changes and newer equipment is purchased, it is your responsibility to stay current with your training and know how to use your department's available equipment.

How RIC Can Utilize the TIC

The best way an RIC can use the TIC is for the Company Officer to lead out and use the camera for two purposes: (1) to target the path the crew will travel; and (2) to scan the area for the missing fire fighter. When the RIC makes entry into the structure or the target search area, the officer should use the TIC to scan and see the area up ahead, looking for targets such as doorways to rooms and offices. Look at the size of the hallway or area you are heading into and look for exterior windows. Create an orientation and try and maintain it. Stop and hold the camera up in front of you and scan side-to-side in three planes: (1) the floor in front of you; (2) the area ahead of you; and (3) the area above and in front of you **FIGURE 6-2**. This third scan will keep you aware of fire and heat conditions.

It is also advisable, in poor visibility, to look behind you to check on your crew. This technique is called <u>back scanning</u> **FIGURE 6-3** and enables the Company Officer to keep track of his crew (accountability) and keep them together (crew integrity). Once he has back scanned, the Company Officer can then take another forward look and pick the next target (door, room, or area) that he will move the crew up to. This is commonly referred to as a scan and release technique, and its number one advantage is speed.

After the Company Officer has performed a back scan and has determined the next target, the camera should be suspended on its strap or harness as the crew is moved forward. You can move ahead a lot faster without looking through the

FIGURE 6-3 The Company Officer will back scan to keep track of the RIC.

TIC, and the TIC will not be dragged along the floor in your hand as you crawl forward. When you get to the next target area, the Company Officer should stop and repeat his scans, always communicating with the crew and letting them know when to stop and move. Nothing should be left to chance. If something is seen in the TIC and you are not sure what it is, have a company member move up and investigate if it is safe to do so. You can observe, direct, and communicate with this fire fighter while watching through the camera.

Keep in mind that you are looking for a downed fire fighter. If the missing fire fighter reported being trapped under or entangled in something, you might be looking for a hand, a leg, SCBA, or some other portion of his body not covered. Debris piles will show on the camera, but how they will be defined (what shade of white/black or color) is the question. The camera cannot see through anything, so if an area that needs to be searched is blocked by an object, relocate to view the other side or send a member of the team to check it out.

Crew Notes

Different TICs have unique lens-to-screen alignments, including some that have lens-to-screen alignments that are angled **FIGURE 6-4**. It is important to know your equipment and train frequently to know how to use it properly.

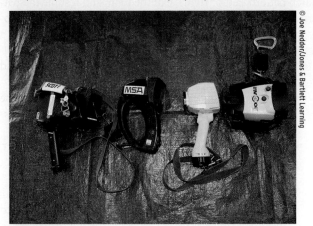

© Joe Nedder/Jones & Bartlett Learning

FIGURE 6-4 TICs with different lens-to-screen alignments.

Search Techniques

There are two types of search techniques that can be used in RIC:

1. The <u>oriented search</u> technique (using a wall as the landmark).
2. The rope search using the <u>rope-assisted search procedure (RASP) search</u> technique.

Each type of search technique has it applications and advantages. It is strongly recommended that all RIC searches utilize a search rope (RASP bag), even if the intent is to do an oriented search, because if things go wrong or conditions rapidly deteriorate, the RASP line will lead you out safely. With either technique, it is important that the Incident Commander (IC) provide as much information as possible regarding the possible whereabouts of the fire fighter in peril. An RIC search is a targeted search, and this information can be used to focus on the most likely location, thereby reducing the time spent searching areas where it is highly improbable that the fire fighter will be found.

Imagine, for example, a MAYDAY is received from a fire fighter on Ladder 1 caught under a ceiling collapse. From the MAYDAY call, we know who the fire fighter is and what company he is with. Knowing the company will tell us what the assignment was and where that crew was working. In this case, if we know that Ladder 1 was assisting Engine 3 with overhaul on the second floor in the B/C corner, we can direct and target the RIC to the second floor B/C corner.

With either technique, remember your basic skills, air management, and crew integrity. Also remember that time is critical: you are searching for someone who is in trouble in a toxic environment with a very limited air supply.

Safety

Depending on fire conditions, the IC might choose to commit a dedicated handline for the protection of the RIC. Keep in mind that a hoseline cannot move and advance as quickly as the RIC can and needs to. If the area the RIC is entering is so heavily involved in fire that the company cannot advance, then the fire must be contained and controlled so entry can be made, or another, safer entry point must be found.

■ Oriented Search Technique

The oriented search technique is based on the basic search techniques learned during Fire Fighter I/II (basic) training. The oriented search technique is best used in smaller structures, when a TIC is not available, or in structures where a searcher can safely and rapidly search a room or area using the walls to maintain orientation and a way out. The fire fighters, of which there should be a minimum of two, should maintain contact with each other at all times. This can be done by touch, communicating (talking to each other), or by sight, which is unlikely in a smoke-filled environment. The oriented search technique has two search patterns: the left-hand search and the right-hand search. Some refer to the patterns as clockwise and counter-clockwise, but the important thing is to keep instructions as simple as possible, considering that it can be confusing under duress to remember which way is clockwise or counter-clockwise, but chances are we all know our left from our right. It is imperative that the Company Officer communicate the search pattern (left or right) that the RIC will be using with the company members before beginning the search and get feedback that everyone clearly understands. A lack of communication can result in additional RIC companies looking for you!

The search is conducted as you would a primary search: rapidly and skillfully. To believe that we as fire fighters can conduct a rapid search under adverse conditions, with the added burden that we are looking for one of our own, and take our time to be extra thorough is *not realistic*. We need to learn to conduct our search quickly and with great skill. Working in teams of two, one fire fighter can maintain the doorway/exit of the room while the other quickly enters and, using the walls to maintain orientation, sweeps the room. It is important to remember when using this technique that keeping yourself "glued" with your hand or shoulder to a wall **FIGURE 6-5** and just following it around is not the intended idea and might not give you a positive result.

Even with a tool to reach out with, you might not reach the middle of the room. Furthermore, one fire fighter following another, while holding onto his ankle, around the four walls of a small room is counterproductive. A fire fighter using his foot to keep in contact with the wall can extend his body outward into the center of the room to get the necessary coverage **FIGURE 6-6**.

FIGURE 6-5 Fire fighter performing the search oriented technique incorrectly by just following the wall.

FIGURE 6-6 Fire fighter performing the search oriented technique correctly by anchoring his foot to the wall and extending his body outward into the center of the room.

In residential bedrooms, especially children's rooms, floors are often covered with toys and other items. These types of situations make it very difficult to stay on the wall and use your tool to probe for the missing fire fighter. By reaching out and using your hands, one skilled fire fighter can quickly and skillfully search a typical room. The second fire fighter can maintain a position at the doorway, monitoring the progress and safety of the other fire fighter. He is also available to assist if the downed fire fighter is found. In addition, most rooms are filled with furnishings. The RIC fire fighters conducting the search, without a TIC, must always have a tool. The tool can be used to extend your reach to probe and can also be used to assist you if your exit needs to be forced. Use caution when sweeping for a victim with a tool. Swinging it too aggressively

might injure the person you are looking for. Also, the tool can be inhibited and obstructed by the various items found on the floor in any given room.

In a smaller structure or residential home, an RIC of four fire fighters can split into two groups, with each pair searching a specific room, and then, while maintaining the assigned search pattern (left or right), exit the room and resume the pattern to the next room. In this type of circumstance, by "leap-frogging" over each other the teams can search two rooms at a time **FIGURE 6-7**.

Some additional tips that will help with the search and tactical assignment include:

- Before entering and while staged as the RIC, look at the building and observe doors, windows, and other egress points in the area you will be searching.
- If the search is targeted to the second floor or higher, check that ladders are in place on the target floor for rapid egress or rescue.
- Be fully aware of the fire conditions you are entering into. Are you searching ahead or above an uncontrolled fire?
- As you search, periodically stop, hold your breath, and listen for the victim. Do you hear a tool being banged as a signal for you? Do you hear a SCBA regulator or a cry for help? Do you hear a PASS device?
- If the search area is heavily charged with smoke, work with the IC for additional ventilation, or, if allowed by the IC, vent as you go with due regard to the possibility that venting might draw the fire upon you.
- Always keep at least one member in touch with the wall you are searching off of; in larger spaces, loss of the wall might mean loss of your way out.
- Keep the RIC Chief informed of your progress.
- It is critical to remember your and the crew's air management.
- Know, use, and drill on your basic search skills. Skills not practiced and drilled on are lost!

The two keys to success with the oriented search are to keep oriented and keep moving as fast as possible.

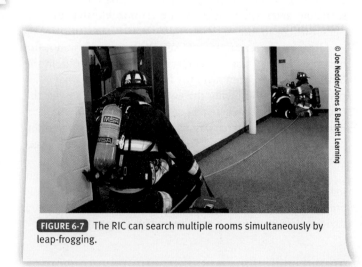

FIGURE 6-7 The RIC can search multiple rooms simultaneously by leap-frogging.

■ Conducting a Rope Search

The use of a rope to assist in conducting a search is not a new concept to the fire service. Over the years, however, some techniques and improvements, including technology, updated equipment, and new search techniques, have been added to assist the fire fighters conducting the search. Technology has been greatly improved with the introduction of the TIC. In fact, one could argue that this single item has had the most significant impact on our ability to search with more success for missing civilians and fire fighters.

Updated equipment includes the development of the 7.5-mm and 9-mm (1/3-inch and 2/5-inch) search rope. In the past, fire fighters have used 1/2-inch (13-mm) rope for conducting a search off a rope line. Our comfort with the 1/2-inch (13-mm) rope is because we have been using it for water rescue and utility rope; however, the 1/2-inch (13-mm) rope is bulky and heavy. It has been argued that the larger rope gives you more to hold on to, but it also has less flexibility, adds weight to be carried, and is manufactured in nylon only. The development of 7.5- and 9-mm (1/3- and 2/5-inch) rope that is both heat-resistant and certified to NFPA 1983, *Standard on Life Safety Rope and Equipment for Emergency Services*, as an escape rope was a significant leap forward. Developed for the fire service by The Sterling Rope Company, the concept behind the rope's development was to have a rope that is large enough for a fire fighter to locate and feel with a gloved hand, be high-heat resistant, be easy to pack, and have a secondary purpose as a personal escape rope. The use of 9-mm (2/5-inch) rope allows us to utilize a smaller rope bag that is significantly lighter.

New search techniques are always being developed, but not all are making the job easier or less complex. Any technique that is to be used must be:

- Simple to learn and understand.
- Simple and fast to deploy.
- Cost-efficient to the point that most fire departments can afford the required equipment.

The most significant improvement in rope search skills was the development of the RASP technique.

When to Use a Rope for a Search

The decision on whether or not to use a search rope is usually guided by two things:

1. The type of building you will be entering.
2. The RIC situation and the possibility of a prolonged search and extended operation.

The type of buildings can be broken into two basic categories: industrial/commercial or other large buildings, including residential apartments and single and duplex houses.

■ Industrial, Commercial, or Other Large Buildings

The rope search is designed primarily to be used in larger buildings, such as industrial or commercial buildings, large retail (big box) stores, malls, large homes (mansions), multi-family homes, and apartment complexes, and offices buildings, where the use of the open plan concept is used in conjunction with paneled office furniture. The rope search should always be used in these types of buildings, regardless of what the environment and fire conditions are upon making entry. What if your company chooses to conduct a basic left or right hand search, using the walls for orientation, and without warning the conditions rapidly deteriorate and the company needs to quickly exit? A search rope line will give a predetermined and safe way out. Remember, the search rope serves three purposes:

1. It allows you to move forward in a more rapid and organized manner.
2. It provides you and your company a way out.
3. It provides additional companies, sent to assist or locate you, a definite "path" to follow.

In rapid intervention, this is critical to the rescue! The most important reason an RIC should always use a search rope in large buildings is that if things go drastically wrong, the line will give you a direct way out. As the RIC, we must be aggressive in searching for the lost, trapped, or missing fire fighter, but sometimes in our aggressiveness, fire fighters will increase the risks taken to save their brother or sister. The use of a search rope, therefore, helps to control the company, better organizes the efforts of the company, allows the crew to move more rapidly, and provides a positive indicator of the safe way out. Life safety for ourselves and other fire fighters on scene is the number one priority on the fireground.

Crew Notes

Even if searching a smaller building or residence, consider always using the search rope. Smaller buildings might not at first seem to require it, but if conditions deteriorate quickly and the crew needs to exit rapidly, the use of the search rope will greatly assist in getting out safely. Be proactive, and have a plan if things go wrong.

■ Average and Smaller-Size Homes

Using a search line on a traditional or smaller home is not always required. In most homes we can count on interior walls, numerous exterior windows, and multiple doors for egress to assist us in maintaining our orientation and location. The most important words here are "in maintaining." Remember, if you fail to maintain your orientation to a wall and things go terribly wrong quickly, there is a good chance your company might become disoriented, lost, or require an additional RIC to rescue them. In most average homes, if you become disoriented you stand a good chance of locating a window or other means of egress quickly and escaping, but you cannot always count on this. Be proactive when searching average or smaller residences and *maintain your orientation at all times*. If, for whatever reason, you feel that a search rope is needed to search what might be considered a smaller facility or home, use it! It is always better to err on the side of safety than to wish you had as you call a MAYDAY!

Conducting a Rapid Intervention Rope Search

When the RIC is activated and a search off a rope line is needed, a proven, simple, and safe search technique system is the rope-assisted search procedure (RASP).

■ Rope-Assisted Search Procedures

RASP is based upon three key elements: (1) a TIC; (2) a RASP bag; and (3) a four-person company trained in the technique of RASP. The concept is simple: a search team is provided with a TIC and the training to truly utilize and understand it; the officer is provided with a hands-free rope bag that deploys easily and contains a rope that is small enough to manage but large enough to be easily felt and held onto by company members; and, finally, the company is trained to utilize the simple but aggressive techniques of the RASP search. This simple and coordinated technique has given the fire service an effective way to conduct rope-assisted searches. It is an effective and exceptional tool for searching office spaces, large-area commercial spaces, large homes with confusing layouts, big box stores, or virtually any space that fire fighters need to be able to search quickly, while maintaining their orientation and the way out. Although RASP was originally developed for searching for civilians, it has become a very powerful tool for rapid intervention.

■ The History of RASP

The concept and techniques of the RASP search program were conceived and developed by the Chicago Fire Department. After 2 years of extensive fire fighter rescue and rapid intervention team training with 5,100 Chicago fire fighters, the instructors from the Chicago Fire Academy and the Illinois Fire Service Institute, led by Lt. Will Trezek and Chief Rick Kolomay from Carol Stream, IL, had evaluated the results of the extensive training program. It was determined that there was an additional need to train the fire fighters on search procedures during a MAYDAY incident. Although the department as a whole had been trained on many new self-survival and MAYDAY rescue procedures, the task of locating the fire fighter victim(s) before being able to use any of the newly learned life-saving procedures required improved methods of using the search rope and also thermal imaging and communication techniques.

The instructors worked hard for 2 years in formulating the fire fighter rescue training program, and in 2003, they began to work on a basic, simplistic, and effective method of conducting searches using search rope during MAYDAY incidents. Such incidents involved collapse scenarios and smoke-filled buildings with complex and/or wide-area floor plans. The acronym RASP was coined, and the instructors developed search procedures based on actual MAYDAY incidents, many of which had involved the instructors themselves. One of the predominant improvements is with the rope bag. It was found that a "hands-free" rope bag had to be developed for the officer in charge of the search team. That rope bag was exclusively developed by Chicago fire fighters

Brian Herli and Bryan Valez. As the group of instructors continued to refine and record the various procedures for training purposes, Herli and Valez also refined the construction of the rope bag. Once again, the 5,100 members of the Chicago Fire Department were provided citywide training on RASP, to compliment their recent rapid intervention training. It was found that RASP had become very "street worthy," rapid, and safe, especially in light of some of the complex searches and need for SCBA air management. After the 5 months of training was completed, the RASP rope bags were distributed to the fire companies and they were put to use quickly in numerous fire incidents in warehouses, high rises, and commercial buildings. The landmark fire where RASP searches were used extensively was at the December 29, 2004 fire in the six-story Chicago Sun-Times Building. During its demolition, to provide land for the newly planned 90-story Trump Tower, a fire broke out in the late morning hours, with workers trapped and some reported missing. As the building had been riddled with large holes on multiple floors, searches had to be conducted under heavy smoke conditions. Rescues were made, and no fire fighters were disorientated or lost during the searches, although three fire fighters were injured due to heat exhaustion.

■ RASP Search Team Positions

A RASP team typically is made up of three or four members; the ideal is four, as the team can split into two teams of two, allowing for a search off the line into small spaces while still maintaining a fire fighter on the search line to monitor and or assist the other fire fighter. For the purpose of this book we will focus on the suggested team of four **FIGURE 6-8**. The four team positions, and the tools and equipment they are responsible for, are:

Position #1: Company Officer with a TIC, a portable radio, a RASP bag, and a hand light.

Position #2: Fire fighter #2 or tool man with irons (Halligan bar and flathead axe), a portable radio, and a hand light.

Position #3: Fire fighter #3 or tool man with an additional set of irons or, if warranted by the type of occupancy, a hydraulic rabbit tool, a radio, and a hand light.

Position #4: Fire fighter #4 or airman with emergency SCBA or RIC air, a radio, and a hand light.

For an RIC search, if the team is made up of only three fire fighters, the second tool man position should be dropped. For a non-RIC search in a large, confusing building, consider having the fourth position still bring the RIC air. If a team member overextends him- or herself, the RIC air can be used to supply emergency air for the team to make a rapid emergency egress.

■ The RASP Bag and Why It Needs to Be Specific in Its Abilities

The chapter *Planning for a Prepared Rapid Intervention Crew* discussed tools, both personal (what every member should have) and team tools, that are problem- or technique-specific. The RASP bag is a technique-specific tool and needs to be

FIGURE 6-8 The RIC staged with tools and wearing SCBA masks.

© Joe Nedder/Jones & Bartlett Learning

Safety

The type of building and its construction should play a role in what tools the RIC team uses. For example, if you are searching an apartment building, it might be in your best interest to use a hydraulic rabbit tool for forcing numerous doors versus solely relying on the irons.

specific in its abilities. Most RASP bags tend to be rectangular in design and should:

1. Be able to carry 150 to 200 feet (46 to 61 m) of 9-mm (²/₅-inch) rope, designed for this purpose. The ideal length is 200 feet (61 m), but some departments have gone to 150-foot (46-m) lengths. Under no circumstances should more than 200 feet (61 m) of rope be used in a RASP bag. The rope is simply stuffed in, as one would a rope rescue bag.
2. Have an adjustable shoulder strap with a quick release buckle on each end. This allows the bag to be used on the left or right shoulder without making any adjustments.
3. Have a flap that is self-closing, allowing the rope to deploy from either side, making it usable from the left or right shoulders without any adjustments.
4. Have a small tool pouch or pocket that can accommodate a set of cutters, some webbing, and some door chocks. In addition, a few extra carabiners should be carried to assist with attachments (i.e., the RIC air bag if it is a regulator swap), additional anchor points, or change of direction.
5. Have a ring or anchor point inside at bottom of the bag where the end of the rope should be secured, indicating the end of the deployable rope.
6. Have the company identification clearly tagged on the bag and the rope end, which is first deployed as an anchor. This is key when numerous lines are deployed from the same anchor point, because it will ensure that additional companies sent in to assist will follow the correct line.
7. The anchor end of the rope should have a large carabiner attached to it for easy securing.

8. Have one knot on the 9-mm (²/₅-inch) rope at 50 feet (15 m), two knots at 100 feet (30 m), and, if, it is a 200-foot (61-m) rope, three knots at 150 feet (46 m). The knots should be 6 to 12 inches (15 to 30 cm) apart. The bag itself indicates the end of the line, so no knots are needed.

The primary purpose of using a RASP bag is that it is designed to be carried and the rope deployed and managed with the greatest possible ease. *It is suggested that duffle-type bags, rope stuff bags, and other rope bags used in technical rescue never be used.* They do not fit the requirements stated above, and with the non-RASP bag, the officer typically would not carry the bag when deployed but rather place the bag on the floor ahead of himself and push it as they move forward. This type of deployment can put you at risk of losing the bag if you break hand contact for any reason or the bag getting caught or snagged onto something on the floor. As with many tools in the fire service, the RASP bag is designed to do its job in the best manner possible FIGURE 6-9 .

Crew Notes

Conducting a RASP search is not an easy task. The company has much to focus on and deal with, which is why fire fighters are cautioned to keep their tools and techniques as simple as possible. There are many *off the shelf* search rope systems that can be purchased, but what most have in common is high cost, complexity, and the illusion that a more technical system is better. Knots are not needed every 5 or 10 feet (2 or 3 m). No one could keep track of the count. Our focus should not be counting knots, but finding the missing or downed fire fighter. Using a spacing of 50 feet (15 m) is consistent with the hose lengths we use and know. Complex metal devices to hold onto that will click over balls to measure distance in or LED lighting to indicate where the rope is are not needed, nor would they be visible in heavy smoke conditions. In order to maintain orientation and know where the rope is, it is necessary to keep holding on and never lose contact.

FIGURE 6-9 RASP bag showing key points and features.

© Joe Nedder/Jones & Bartlett Learning

RASP Deployment and Search

When the RIC is deployed and the RASP technique is to be used, the team needs to stick to the skills, methods, and techniques of RASP. The RASP deployment and search technique include:

1. Activation of the RIC company with a target area of the building that they are to deploy into and search
2. Preparing to make entry and anchoring the RASP line
3. Assigning the RASP Control Officer
4. Making entry and having team members take their designated positions on line
5. Advancing the RIC on the RASP line
6. Understanding the distance indicator knots and how to use them
7. Changing direction while on the RASP line
8. The scan, target, and release technique
9. The rapid room search
10. Communications
11. The RASP exit
12. Managing two companies on the RASP line, one exiting and one entering

Each one of these subjects and skills are critical to the success of the RASP operation and to the safety and survival of the team.

Skills, Methods, and Techniques of RASP

Activation of the RIC Company

When an RIC is activated, it is important that all known information about the possible location of the downed fire fighter is shared with the team. This will help establish the search target area. Based upon this information, the RIC will select the best entry point, which should be the fastest and most direct route to the target area. From the 360-RECON conducted as part of their size-up, the RIC should know the building and access points. All of this will help to expedite the search.

Preparing to Make Entry and Anchoring the RASP Line

Every time a company is conducting a rope search, the leading end of the search rope must be secured and anchored to a nonmoving object **FIGURE 6-10**. If the search is being conducted in a larger building with a fire on a different floor, the RIC might find themselves entering the main entrance and proceeding in a non-immediately dangerous to life or health (non-IDLH) environment a great distance. In this case, to deploy and anchor the line at the front door might use all of the 200 feet (61 m) of rope in the bag even before you get to the area you need to enter to begin the search. Using common sense, secure the rope in an area that will allow for a safe and rapid egress. Remember: this is your way out. If it is a high-rise or multi-story building, the best anchor point might be in the stairwell a floor or half-flight below the target area entry point or fire floor. It is not advisable to secure your anchor point to

a door or door hinge, because if the door closes, it might not clearly indicate the way out if conditions rapidly worsen and drive the company to back out. *Never* use a door knob as the anchor point. Ropes tend to slip or fall off, and it can easily be dislodged by another person unintentionally. Remember, the anchor point will secure the line in place and serves two purposes: (1) it provides a way out with the line secured by an exit point to a safe area, and (2) it provides an entry point for another company entering to relieve or assist you. In this situation, the anchor must have an identification tag denoting the company. In multiple search line situations, this is critical to ensure that the relieving companies enter on the correct search line **FIGURE 6-11**.

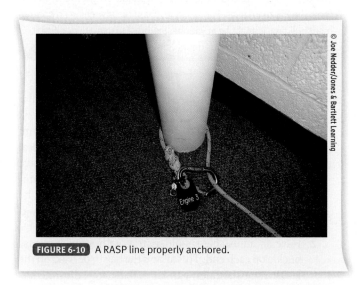

FIGURE 6-10 A RASP line properly anchored.

FIGURE 6-11 Multiple search lines at one anchor point, showing the company ID tags.

■ RASP Control Officer

If manpower allows or conditions warrant, it is important to utilize a RASP Control Officer. This position follows the crew to the target entry point where they will tie off and anchor. There are situations where the control position must be manned, including building size, number of RASP lines being deployed, construction type, present conditions, and anticipated conditions, based upon what is observed and the experience of the search team. When only one RASP line has been deployed and the search is in a small structure, the decision to utilize the Control Officer should be made by the IC and the RIC Company Officer, or if staffed, the RIC Operations Chief and the RIC Company Officer. If staffing allows, it never hurts to have this position filled. Because of the responsibilities this position holds, it is advised that it be staffed by a Chief Officer or a senior Company Officer. In either case, the Officer staffing the position must be skilled in the RASP search so as to understand what is happening.

This position's primary responsibility is for the safety of the search team(s). Because of this, the Control Officer must be in close proximity to the target entry point. This allows him or her to visually monitor conditions inside the building where entry was made and to also monitor conditions outside in the general area they are operating. At a single family residential, the position might be staffed at the front door, and at a multistory commercial building, the Control Officer might be in the stairwell adjacent to where the team tied off (anchored). When multiple RASP lines have been deployed from the same entry point, the Control Officer needs to monitor who is where and how long they have been inside. If multiple lines have been deployed, the Control Officer can help make sure that a relieving company enters on the correct line. The Control Officer should also monitor the team's time in as it regards SCBA. It should always be noted if the crews are wearing 30-, 45-, or 60-minute cylinders, as this will help determine the time in and when they should be exiting safely.

If RASP lines are being deployed from different positions, each entry point should have its own Control Officer for the safety of the crews operating FIGURE 6-12 . The Control Officer should report any change in conditions that might affect the safety of the team(s) operating to the RIC Operations Chief. If conditions are rapidly worsening it is advisable to work with the RIC Operations Chief and terminate the search, backing all the crews out.

FIGURE 6-12   The RIC Control Officer in place.

■ Making Entry: Team Members Take Their Designated Positions on the Line

Once the target entry point has been reached, team members must take their designated positions on the line. The members should be approximately 3 to 5 feet (1 to 2 m) apart. As with anything, how you start a process has a direct effect on its outcome. It is very important that the team members be placed properly. This will help the Company Officer to maintain crew integrity, by knowing who is where, and it will also help maintain the equipment flow. The tools are directly behind the officer, while the RIC airman is last. Crew members need to all be on the same side of the line as the Company Officer, and which side is best depends on the Company Officer and how he deploys the line out of the bag. As the crew advances into the building, they will most likely turn left or right, so beginning the search thinking that the rope should be near or away from the wall has little bearing on reality as search conditions change FIGURE 6-13 .

■ Advancing on the RASP Line

When a company advances on a RASP line, it must be done with skill, coordination, and proven techniques. Remember, you are an RIC and are entering a hostile environment searching for a fire fighter in distress—there is not a lot of extra time to discuss what techniques to use. The primary mission is to find the distressed fire fighter as quickly as possible, deliver him air, and get him out rapidly. All this has to be done without causing injury or death to the crews doing the search.

FIGURE 6-13 An RIC on the rope line advancing.

FIGURE 6-14 RASP Company Officer managing the rope line from the back of the bag in an area with visibility.

FIGURE 6-15 RASP Company Officer deploying the rope from the front of the bag, while crawling forward.

With the crew on the line and ready to move up, the Company Officer will order the advance. It is advisable that all orders or commands given by the Company Officer be passed down the line from one fire fighter to the next for clarity and good communications. Ensuring crew integrity is not only keeping them together under tremendous stress and duress, but also making sure they all are fully aware of what they are doing, and how they are progressing.

The rope is deployed out the bag and controlled by the Company Officer, and it is strongly suggested that the Company Officer always maintain hand contact with the rope. Advancement will be in two types of environments—with or without visibility—and each requires a different rope deployment method.

Area With Visibility
If the area we are entering has light, minimal visibility (examples include an enclosed stairwell or an area where the crew is moving from an unburned or non-involved section of the structure into a more involved area), or no smoke conditions where the company can safely stand and walk, the rope deploys best from the back of the bag with the officer managing and controlling the deployment. The Company Officer needs to maintain some tension on the line with his hand or the crew behind him might freely pull additional rope out the bag, defeating the purpose of the RASP line and shortening the available rope needed to advance further **FIGURE 6-14**.

Areas With Limited or No Visibility
In areas of poor or no visibility, the company will be crawling. In this case, the rope will deploy best from the front of the bag. The Company Officer will thread the rope through his hand in-between his thumb and index finger. As he crawls, and his hand pushes forward, he will pull additional rope out of the bag. By using this technique, he will maintain contact with the line, provide tension on the line so the crew behind cannot pull out additional rope, and be able to feel the knots that indicate 50 feet (15 m), 100 feet (30 m), and 150 feet (46 m) **FIGURE 6-15**.

The crew members maintain hand contact with the rope at all times and advance as a team **FIGURE 6-16**. If at any time something hinders a member from advancing with the company, he or she must shout out for the crew to stop. If he is last in line, this call out should be heard, at minimum, by the fire fighter in front of him, who can pass the message along so that the members can stop as a group. The Company Officer can also use the TIC to back scan and see if he can visually determine what the issue is **FIGURE 6-17**.

FIGURE 6-16 The RIC advancing on the line.

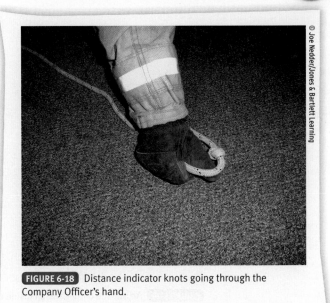

FIGURE 6-18 Distance indicator knots going through the Company Officer's hand.

FIGURE 6-17 The RASP Company Officer back scanning the RIC company to maintain crew integrity.

This information is critical when deciding when to back the company out before they run out of air. The knots will assist the team in determining, if the downed fire fighter is found, if assistance is needed to move him out. Going a long distance to find the downed fire fighter taxes the air supply and the physical capabilities of the RIC. An increased distance in should be a leading indicator that another RIC is urgently needed to assist in getting the downed fire fighter out.

> **Safety**
>
> It cannot be stressed enough that the warning device on your SCBA is *not* a "time to get out" indicator, but rather a low air situation alarm. Every RASP team and their officers must utilize good air management assessments to ensure that the crew will have enough air to make a safe exit.

■ Distance Indicator Knots

As the RASP rope line is knotted to indicate the amount of rope deployed, it is a smart and safe practice that whenever a knot distance indicator comes out, the Company Officer call the distance **FIGURE 6-18**. This information is then communicated down the line from one fire fighter to the one behind him, making sure that everyone knows how deep in the company has advanced. If the Company Officer fails to feel the knot or call the distance, the first fire fighter down the line who feels the knot(s) should immediately call out the distance and make sure that the information is communicated up and down the line. Knowing how far along you are on the line is important. If the RIC calls the IC for assistance, they can clearly indicate how far in or deep they are. This knowledge is also helpful to another company making entry to assist, and helps them to anticipate and stay proactive. Knowing the distance in also helps the company know how far they will need to go to exit.

■ Changing Direction While on the RASP Line

As the crew advances, it will most likely encounter a situation where a right or left turn needs to be made. This change in direction might involve an inside corner or an outside corner **FIGURE 6-19**. If unsecured, the rope will pull away from the corner as tension is added to the line by the advancing RIC. This places your rope out over an unsearched area and increases the possibility of encountering hazards and obstructions, such as a hole in the floor or entanglement problems, during your exit, or if another team is entering to assist and the rope is unsecured, the rope will pull away from the unsecured inside corner.

It is very important to always tie off and secure the RASP line whenever a change in direction with an inside corner is needed. It most likely will be in a large open space, such

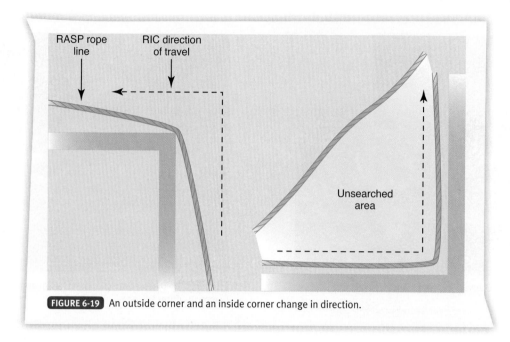

FIGURE 6-19 An outside corner and an inside corner change in direction.

as in a factory, warehouse, interior loading dock, school, or church hall. There are a few options for tying off the line in this situation. The first is to tie off to a heavy object in the immediate area. Another option is to breach the wall and wrap the RASP line around the wall stud or a door jamb **FIGURE 6-20**. An extreme measure would be to drive the point end of the Halligan into the floor with the flat head axe or other striking tool and secure the rope to this, but remember that this will mean your company has just given up a very valuable tool. What if you need it? Consider this a last-ditch effort. Another last-ditch effort is to leave a company member at the turning point to become a human anchor. Be cautious—leaving a company member can have a serious impact on the effectiveness and safety of the crew.

You might think that if you have a TIC, the outside corner does not have to be secured and it does not matter if the RASP line follows you around the unsearched area, because you will be able to spot any hazards as you exit with the camera. Great fire fighters never rely on just technology to get them out; they practice common sense, using their basic training and planning for the worst situation. The TIC is electronic and could fail, or the batteries could go dead. Your life and the life of the fire fighter you are searching for depend on you doing the right thing always. In the event the TIC fails, a secured RASP line will ensure the company has a clear path of egress without having to exit over an area with unknown hazards. Securing the RASP line quickly is being proactive and smart.

■ Scan, Target, and Release Technique

The Company Officer using the TIC will scan the area they are entering or moving into. The scan is to determine where they are going (and selecting the target area to move toward). The scan should be of the floor before the crew and then the

FIGURE 6-20 A breached wall with the RASP line tied off to a wall stud.

surrounding area. Depending on fire and smoke conditions, the Company Officer can also scan above, looking for thermal currents, rollover, or fire. Remember, if the fire fighter is lost and disoriented, one of the floor scans might discover him in an area that was not considered, so always be vigilant. The Company Officer completes the scan and then selects a target **FIGURE 6-21**. This target can be a doorway, window, corner, or hallway. It is also advisable to stop on occasion and back scan the company to ensure crew integrity and accountability. After the target is

FIGURE 6-21 The Company Officer identifying the next target to move towards.

selected, the Company Officer should release (put down) the TIC, tell the crew they are moving up, and start to advance.

As part of the <u>scan, target, and release technique</u>, the Company Officer and crew members should also be listening. You might hear a PASS device, or the distressed fire fighter banging their tool to draw attention, or even radio communications from the downed fire fighter's portable radio.

Safety

From preplans and from the 360-RECON and size-up, the RIC will have clues as to the type of building and possible layout they are entering. It cannot be stressed enough that an RIC needs to be proactive in its assessments as soon as they are staged.

When you get to the target area, if the downed fire fighter is not found, the TIC scan is repeated, a new target area to move toward is selected, and the TIC is released. This technique is repeated over and over as the crew moves up the line. If you encounter a room or space that is adjacent to the search line (e.g., an office or small room off the hallway) and you are in the area where the missing fire fighter might be, the Company Officer must scan it. As a room is scanned, obstructions such as a desk, furniture, office partitions, or closet space may be encountered that block the TIC from seeing the hidden area behind it. In this case, the area must be searched by a fire fighter using a technique called the rapid room search.

■ Rapid Room Search

The term <u>rapid room search</u> as applied to RASP is defined as a left- or right-hand primary search, off the rope line, that is done very quickly. Typically, the Company Officer will use the TIC to monitor the fire fighter doing the searching. The room could

Crew Notes

For ease of use, it is suggested that a retractable holder or strap be used to secure the TIC **FIGURE 6-22**. This allows the Company Officer to extend the camera out when needed and return it to a secure place on his gear when not in use. Carrying the TIC in this or a similar manner allows the Company Officer the use of both hands when moving and will also protect the camera from being dragged on the floor. Also, with this type of system the Company Officer will not have to put the camera down on the floor, avoiding the potential difficulty of retrieving it.

A.

B.

FIGURE 6-22 A TIC with a retractable holder **A.** deployed and **B.** in storage mode.

be a small bedroom or an office in a commercial building, but it needs to be small and manageable for a rapid room search to be conducted off the line. If the fire fighter searching is being watched by the Company Officer with the TIC, the fire fighter does not necessarily have to maintain contact with the walls during the search.

In RASP, once an obstruction is identified, the Company Officer will call up the #2 fire fighter (tool man with irons) to do the search off the line. *Do not deploy tether lines or anything else to secure the searching fire fighter to the line.* The Company Officer can observe his progress through the TIC. Also, the fire fighter has not been asked to search a large, confusing area or room off the line. He has been asked to do a rapid room search, and, without a tether, the fire fighter's search will be unhampered and executed rapidly.

Crew Notes

The use of a tether to attach a fire fighter to a main line for a rapid room search has been rendered all but obsolete with the introduction and use of TICs. When a fire fighter searches a small or cluttered space, the tether rope is prone to getting snagged and caught on different things and obstructions. After a four-wall sweep around a room on a tether, a fire fighter would return back from where he started with 25 to 50 feet (8 to 15 m) of rope trailing out and around the four walls. Even if the rope could be gathered up, the fire fighter now has a pile of tangled rope in his hands. This is not smart firefighting and is not safe. Using a tether line in a small space is not necessary with TICs.

Before the #2 fire fighter enters the room, the Company Officer will quickly share a TIC view with him **FIGURE 6-23**. This TIC view is intended to give the searcher an idea of the type of room, its size, configuration, and the target areas to search (what the camera cannot see behind). When the fire fighter enters the room to be searched, he must maintain a basic search pattern, left- or right-hand search. One tip suggested by the Chicago Fire Department is that if the door he is entering swings to the right, do a right-hand search, and if it swings to the left, do a left-hand search; the thought being that you will always search behind the door first for the victim. The Company Officer should monitor the #2 fire fighter performing the search using the TIC and provide direction if needed. The fire fighter is not searching the entire room, but rather the

areas that the TIC could not see. Once the area is searched, and with negative results (all clear), the fire fighter returns to the RASP line with assistance from the Company Officer, and after ensuring he has the rope securely in his grasp, the company moves up.

Safety

If you encounter an interior door (office, hallway, storage area) that must be opened, use the TIC camera to scan the door for heat before you open the door.

During any rapid room search off the main line, a door safety position must be maintained **FIGURE 6-24**. Never leave a fire fighter to search a room without having another fire fighter at the door to maintain the safety position. If the searching fire fighter gets disoriented, the safety position can assist in guiding him back to the door way by calling out to him, using the TIC to guide him back, or even by banging a tool to identify the direction he must travel to get out.

A major safety issue with this type of search is if the searcher discovers other doorways and connecting rooms in the room he is searching untethered. Continuing the search into this type of floor plan might put the searcher at risk of getting lost in a complex maze of walls and additional rooms. If this type of layout or floor plan is discovered, the searcher should shout out to the Company Officer and report what has been discovered. It is the Company Officer's decision to determine what needs to happen. If the search is extended into the area, it might require moving the RASP line within and continuing the search. Whatever the decision, the Company Officer and the searcher need to always maintain contact either verbally (not by radio) or visually with or without the TIC. Remember: you are the RIC, and the objective is to find the distressed fire fighter. All decisions should be based upon that objective and what is known at that given moment.

FIGURE 6-23 The Company Officer sharing the TIC image with the fire fighter who is about to conduct the rapid room search.

FIGURE 6-24 A fire fighter in the door safety position.

As with any search operation, all members of the team should have gathered some of the basic size-up information initially from the 360-RECON, during the ongoing size-up of the situation preactivation, and finally as they enter and progressed into the structure. While in the structure, they should not only be looking for the downed fire fighter, but also staying proactive in monitoring the conditions they are in. Heat, smoke, and fire conditions will keep a crew aware and safer as they operate within the structure. Keep track of the interior layout as you progress and look for emergency egress points, windows, exit doors, and stairways. What you observe and remember might save your life or provide a faster exit for the downed fire fighter if found.

Searching an Area with Multiple Offices or Rooms

If the RIC encounters an area with multiple rooms or offices that need to be scanned or searched, the Company Officer can utilize all the members of his company to execute a more rapid search. To achieve this, the Company Officer sends in the #2 fire fighter to search the room. While the search is in progress, the Company Officer can continue to move forward to the next room, bringing the #3 fire fighter with him. The #4 fire fighter (airman) will assume the door safety position and communicate with fire fighter #2 that he is in place. When fire fighter #2 does exit the room, fire fighters #2 and #4 then move up the line to where the officer and fire fighter #3 are currently searching and reunite. Again, the Company Officer might decide to move up, leaving a fire fighter behind as the door safety for the fire fighter searching the current room. A skilled and practiced company can cover a lot of area in a minimal amount of time by using this technique **FIGURE 6-25**.

■ Communications

During a RASP search, the Company Officer must communicate important information to the RIC Operations Chief. When a MAYDAY is declared and a full rapid intervention operation is commenced, the Company Officer of the RIC operating in the building must communicate important information in a timely manner. This information will include:

- The progress or lack thereof that the RIC is making. How far have you advanced? If it is a lack of progress, why? What physical obstacles or company issues have prevented the search to move rapidly?
- Conditions within the structure, especially if they worsen. What is your current situation with the environment surrounding your crew? Are we getting into a situation we need not be in for the safety of the company?
- Requests for additional help. Do you need another company? Is additional venting required? Is a hoseline needed immediately to protect your egress?
- You have found the fire fighter.
- You are moving up a floor (i.e., from floor 1 to 2).
- You hear a PASS. This type of feedback is important because an IC can quickly determine if there is an accidental PASS activation from another company operating within the structure or if it is from a PASS left on and outside.

Remember that providing timely feedback to your boss will assist him in assessing the progress of the rescue operation. All too often the RIC Operation Chief might be heard calling the company for a progress report. This causes the company to stop what they are doing and answer the request. By being proactive with important and timely reports that are *initiated by the RIC*, will keep those who need to be informed up-to-date with timely information of the operation without their needing to request feedback or a progress report.

Crew Notes

Good communication is vital with any search team, whether it is for a downed fire fighter or a civilian. If it is not an RIC search, the search crew must communicate important information to the Rescue Sector Officer, the IC, or the Operations Chief. Staying proactive and providing benchmarks and updates are important.

The Company Officer and all members of the RIC must also maintain good communication with each other. In addition to providing information up and down the line, other things that need to be communicated include:

- **One of the RIC members thinks he hears the downed fire fighter.** He might be hearing the fire fighter's PASS, a tool banging, the fire fighter's regulator drawing air, or even a call for help. If you think you hear something, tell the other members of the company. We are all working together for a common goal: the rescue of a brother or sister fire fighter in dire trouble.
- **The air status of every member of the RIC.** It is important that the Company Officer and crew members are very cognizant of exactly how much air they and every member of the RIC has left in their cylinder. It is advisable that every so often the Company Officer stop and call back to the team for an air status report. Each team member needs to look at their gauge, not the heads up display (HUD), and call out at what level they are at in the cylinder. The response should be "three-quarters" or "about one-half." It should never be reading your gauge and calling out 3850 or 2053, or whatever the digital readout says. Answering in numbers might sound good in theory, but in reality, answering that you have one-half a cylinder left is easily understood—especially if three members callout one-half and the fourth member calls out "about one-quarter" just before the low air alarm activates.

■ RASP Exit

When a company needs to exit the building, they must continue to maintain the skills of RASP for their own safety. There are three types of exits that we can experience:

1. The company is still searching, but the available air indicates that it is time to exit.

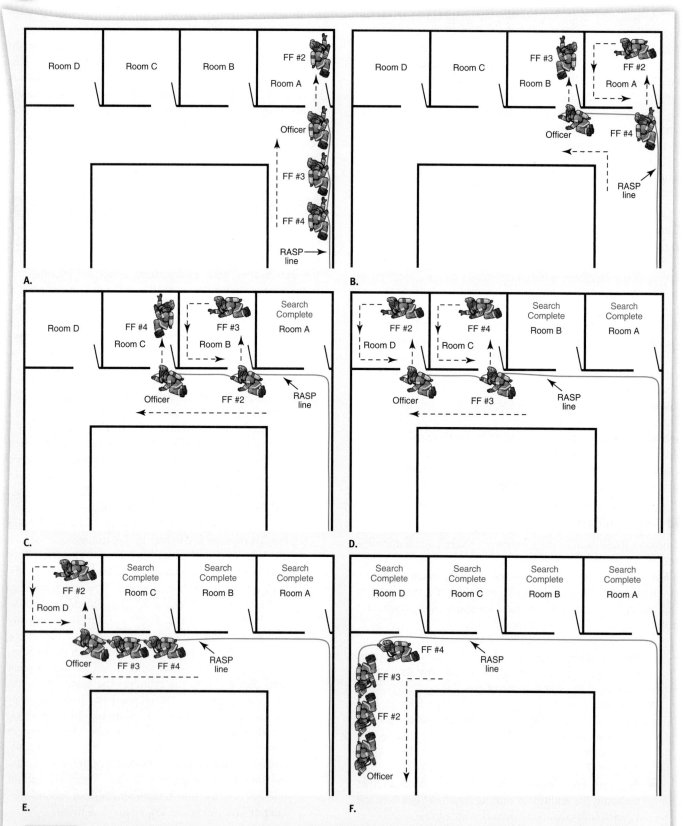

FIGURE 6-25 A rapid room search of multiple rooms. **A.** With the Officer at the door, fire fighter #2 conducts a rapid room search (room A). **B.** Fire fighter #4 assumes the door safety position and the Officer and fire fighter #3 move up to the next room (room B). Fire fighter #3 begins the rapid room search (room B) with the Officer as the door safety. **C.** With room A's search complete, fire fighters #2 and #4 move up to the Officer. The Officer takes fire fighter #4 and moves up to the next room (room C) and begins to search, while fire fighter #2 assumes the door safety position for room B. **D.** With room B's search completed, fire fighters #2 and #3 move up to the Officer. The Officer takes fire fighter #2 and moves up to the next room (room D), where fire fighter #2 searches room D. Fire fighter #3 assumes the door safety position for room C. **E.** With room C's search completed, fire fighters #3 and #4 move up and rejoin the RIC Officer while fire fighter #2 conducts the search of room D. **F.** With the search complete in that area, the RIC continues on to the next area to be searched.

2. An emergency has arisen (team member having difficulty, equipment failure, rapidly deteriorating conditions, structural concerns), and the company needs to get out as soon as possible.
3. The downed fire fighter has been found and the company is exiting with him.

In every case presented above, the RIC will utilize the RASP line to guide them back to safety. There are two RASP exit techniques: the routine exit and the emergency exit.

The Routine Exit

This exit would be used for a normal exit, not under emergency conditions. Examples of a routine exit include:

- The search might have been called off as the fire fighter self-extricated himself and is outside safe and sound.
- The Company Officer has determined that the crew has advanced as far as they should and all team members have enough air to exit without worry.

To make a routine exit, the Company Officer must first inform RIC Chief that they are exiting. It is important to inform command of your current situation with the search. Is another company needed to continue the search or are other resources required besides additional crews? After communicating with RIC Operations, the Company Officer must inform the crew that they are making a routine exit. When the order to back out is given, the Company Officer must be the last person to exit. He should first scan the crew with the TIC to confirm accountability and then he will initially keep the RASP rope line taunt **FIGURE 6-26**. The RIC members will change direction and begin to crawl, duck walk, or walk out, depending on the conditions encountered **FIGURE 6-27**. As the RIC follows the line out, the officer will follow.

The Company Officer must also make a decision to either leave the RASP line or take it with him. The decision to take the rope or leave it should be based on the following considerations:

- Is another company entering on your line to continue the search where you left off?
- Has the rescue been called off because the fire fighter has been located, accounted for, or has self-extricated, and if so, are conditions safe enough to gather the rope with you as you exit?
- Does the company have enough air to complete this task of removing the rope? Gathering the rope will take additional time, and if you are in an environment that requires the use of SCBAs, this must also be taken into consideration.

The primary advantage of gathering the rope as you exit is that once outside, the RASP bag can be repacked immediately and available for another deployment if needed on scene. If you do decide to gather the rope, the Company Officer will simply collect it into loops as he exits, not worrying about it tangling. Once outside, the rope line can be tended to.

FIGURE 6-26 The RIC preparing to exit and the Company Officer back scanning the crew to determine accountability.

FIGURE 6-27 The RIC exiting by duck walking.

As the company members exit, they must maintain crew integrity and stay together. Since they have reversed direction and the #4 fire fighter, the airman, is leading the way out without a TIC, he must maintain contact with the RASP line at all times. As he comes across knots, or a change in direction, he should call it out so that the entire company maintains a working knowledge of their approximate location on the line.

Crew Notes

Any time a RASP line is deployed, after the operation, the rope should be thoroughly inspected, dried if required, and repacked for the next use. Damaged ropes should be taken out of service immediately.

Safety

It is important that as an RIC makes entry and moves in on the RASP line, every member of the team maintain an awareness of where they are on the line (what is the distance in, what are the number of corners made) and where they are in the building (second floor at about the B/C corner). This awareness is important to recall when making any exit. *Know where you are at all times!*

As they exit, if the RIC is in very poor visibility conditions, the Company Officer should maintain contact with the team. This can be done by voice, calling out periodically, "How's everyone is doing?" and by using the TIC to scan the crew. Crew integrity is not only the Company Officer's responsibility, but the responsibility of every crew member on the RASP line. Talk to each other, and if you advance upon the man in front of you, ask who it is. You might think it was the #3 fire fighter, but find out it is #4. What happened to the #3 fire fighter? Is he off the line, lost and disoriented? Always work as a team and stick together, especially during an exit.

The Emergency Exit

This exit is used when the company needs to get out fast! It might be because of rapidly deteriorating conditions, a report of a collapse, a SCBA emergency, overextending your air supply and needing to get out right now, or any other dangerous situation. The two RASP exits have similarities, except in the emergency exit, the Company Officer will call an urgent message to the RIC Chief that the company is making an emergency exit, and he will leave the RASP bag behind. The crew is ordered to reverse direction and get out. They can either crawl or duck walk—whatever will get them out the fastest that conditions will allow. A duck walk is faster than a crawl, but always take precautions with dangers such as high heat, falling debris, and weakening floors. It is these types of dangers that might dictate your exit mode. *Most importantly, if you are in limited visibility, never let go of the rope line!* All members should bring their tools if possible. The officer will, by using the seat belt-type connectors, drop the RASP bag in place and leave it FIGURE 6-28, put some tension on the line to assist the crew in getting going, and then follow them out. Crew integrity is critical. All members need to maintain voice and/or touch contact. The Company Officer can use the TIC to check on the crew periodically, but remember this is an emergency exit, so get out fast!

■ Working With Two Companies on the Same RASP Line: One Exiting and One Entering

Whenever an RIC operation becomes an extended operation and rope search lines are used, we will encounter the issue of companies trying to pass each other on the same RASP line, with one team entering and one exiting. The exiting company might be doing an emergency exit as they are critically low on air. Their Company Officer notified the RIC Chief of the situation and another RIC has been sent in to take up the search where they left off or, if the downed fire fighter has been found, to commence the rescue extraction. Regardless the circumstances, the two companies will "collide" on the rope line and possible chaos, anger, and frustration might arise. This highly possible situation should be planned for. Both crews have an urgent mission, so this transition needs to be made workable. There are many solutions, but be careful not to select just one idea or procedure. There is a need to plan for different circumstances and conditions. For example, say that the RIC will always enter and exit with the rope in their right hand; however, if you as the Company Officer are right-handed, you would want to

FIGURE 6-28 The RIC Company Officer using the seat belt-type connectors to rapidly disconnect from the RASP bag for an emergency exit.

use the TIC with your right hand. This is just one example of why there is a need to plan for different variables.

One thing is certain, the exiting company always has the right of way. Incoming companies must yield and let them pass. If you are an incoming company and collide with an exiting RIC, they will be calling for you to yield. This is not the time for discussions and disagreements—it is important to get out of the way! One safe method to achieve this is for the incoming crew's Company Officer to have his company cross over the line. Once on the other side, the officer maintains contact with the rope and the other members hold onto the fire fighter's ankle or boot in front of them in a fashion similar to how search was once taught. This will clear the right of way and the rope line, except for one of the incoming member's hands, and will allow the exiting crew to keep moving quickly. If this method is used, it is *urgent* that the incoming member who is maintaining a hold on the line *never* lose control or contact with the rope and that the other incoming members maintain contact by touch with each other so as not to lose crew integrity FIGURE 6-29.

Officers Passing on the RASP Line

As two companies pass on the RASP line, the two Company Officers will pass each other. We know that the incoming crew is led by the Company Officer and that the Company Officer of the exiting crew is last on the line. When these two officers meet on the line, if conditions and circumstances allow, they should communicate. The exiting officer can say how far up the line the incoming crew still needs to go, if they found the fire fighter and what his status is (converted the harness, provided air, etc.), if the RASP line is secured to the downed fire fighter, any hazards they may have encountered, or any other pertinent information FIGURE 6-30. This update needs to be brief and concise; your crew needs to get out and the other crew needs to keep moving in.

A.

B.

C.

FIGURE 6-29 Two companies on the RASP line, **A.** one exiting and one entering. **B.** The entering company yields to the exiting company, with the entering RIC's Company Officer maintaining contact with the rope line and the other three members off the rope line, but maintaining physical contact with the member of his crew in front of him. **C.** Once the incoming company fields to the exiting company, the exiting company proceeds out.

FIGURE 6-30 The incoming and outgoing RIC Company Officers communicating on the RASP line.

Wrap-Up

Chief Concepts

- A search for trapped or missing fire fighters will be very intense and emotional; we are looking for one of our brothers or sisters.
- Entering a structure looking for a lost, trapped, or missing fire fighter is not the time to review your search skills and abilities. It is a time of action that requires an aggressive organized search.
- One of the greatest tools an RIC should have that will greatly enhance its abilities is a TIC.
- It is important to understand how images appear on a TIC. Understand the difference between what images are under real fire conditions and what images are when training in the fire house.
- When sweeping with a TIC, you must sweep *slowly* in order to get an accurate image on the screen.
- Learn to use the TIC as another tool. *Do not* totally rely on the camera; remember your basic search skills. TICs have batteries, which can run out!
- When searching for a downed fire fighter, remember that time is critical; we are searching for someone who is in trouble in a toxic environment and has a very limited air supply.
- The rope search is designed primarily to be used in larger buildings.
- The most important reason an RIC should always use a search rope in large buildings is because if things go drastically wrong, the line will give you a direct way out!
- Always anchor the RASP line to an area of safety prior to deployment.
- It is best to use a crew of four when conducting a RASP search.
- When a company advances on a RASP line it must be done with skill, coordination, and proven techniques.
- It is very important to always tie off and secure the RASP line whenever a change in direction with an outside corner is needed.
- *Never* lose control or contact with the rope or other members of the RIC team.

Hot Terms

Back scanning Using the TIC to look behind you to monitor heat and fire conditions while maintaining a visual contact with the crew.

Infrared technology Used in thermal imaging, identifies the different temperatures viewed by displaying light and dark images on the camera screen.

Oriented search Searching an area while keeping your basic search skills in mind, such as keeping a wall to your left or right as a landmark, maintaining contact with the rest of the search crew, and utilizing tools to search wide areas.

Rapid room search Left- or right-hand primary search, off the main line, that is done very quickly.

RASP Control Officer Officer in charge of monitoring the RASP search. He or she needs to be close to where the RIC has entered to conduct the search and have direct communication with the RIC Operations Chief and/or Incident Command.

Rope-assisted search procedure (RASP) search Rope techniques typically used in open areas.

Scan, target, and release technique A technique used while advancing forward with a TIC: looking at the area, looking at the target area checking for fire and heat conditions, and then proceeding.

TIC sweep Slow, deliberate side-to-side view of the conditions in front of you at the floor, mid-level, and ceiling level.

Tunnel vision To focus and concentrate on just one thing.

References

National Fire Protection Association (NFPA) 1407, *Standard for Fire Service Rapid Intervention Crews.* 2015. http://www.nfpa.org/codes-and-standards/document-information-pages?mode=code&code=1407. Accessed February 11, 2014.

National Fire Protection Association (NFPA) 1801, *Standard on Thermal Imagers for the Fire Service.* 2013. http://www.nfpa.org/codes-and-standards/document-information-pages?mode=code&code=1801&tab=committee. Accessed October 14, 2013.

National Fire Protection Association (NFPA) 1983, *Standard on Life Safety Rope and Equipment for Emergency Services.* 2012. http://www.nfpa.org/codes-and-standards/document-information-pages?mode=code&code=1983. Accessed October 14, 2013.

An RIC deployment has been requested by the IC for a missing fire fighter in a large commercial warehouse structure. The last known point of the company was closest to the A/B corner of the structure. There is a heavy smoke condition in the building and attack crews have not yet controlled the fire. Once you establish your target area and enter the building to search, you are faced with moderate heat and heavy smoke.

1. The best technique for searching a large open structure is to use a RASP bag.
   A. True
   B. False

2. A RASP Control Officer has been staged at the entry point of the search area. His/her responsibilities include:
   A. Maintaining the safety of the search team(s).
   B. Monitoring multiple RASP lines and tracking who is where and how long they have been in.
   C. Monitoring the teams' time in as it regards SCBA.
   D. All of the above.

3. A search rope line will give a predetermined and safe way out of a structure while a search is in progress. What other purposes does a search rope provide?
   A. It allows you to move forward in a more rapid organized manner.
   B. It can be used to haul equipment up to a search area.

C. It is an NFPA requirement for searching all commercial properties.
D. There is no other reason to use a search rope.

4. It is acceptable to split a four-man search team into two groups of two in order to search small areas, provided the TIC can monitor the crew.
   A. True
   B. False

5. In a high-rise fire, the RASP bag can be anchored in a secure stairwell at the entry point to the search area.
   A. True
   B. False

CHAPTER 7

RIC Team Actions Once the Downed Fire Fighter Is Found

© Photos.com

© Joe Nedder/Jones & Bartlett Learning

Knowledge Objectives

After studying this chapter you will be able to:

- Understand why we must protect, assess, and prepare the downed fire fighter. (**NFPA 1407, 7.4(6), 7.8(2)(b), 7.8(2)(c), 7.11(2), 7.11(4)**, p 117)
- Know the actions that must be taken to protect, assess, and prepare the downed fire fighter to be moved to safety. (**NFPA 1407, 7.11(1), 7.11(2), 7.11(4)**, (p 117–127)
- Know the steps needed to evaluate the downed fire fighter's air situation, physical condition, and rescue needs. (**NFPA 1407, 7.8(1), 7.8(2)(a), 7.8(2)(b), 7.8(2)(c), 7.8(3)(a), 7.8(3)(b), 7.8(3)(c), 7.8(3)(d), 7.8(3)(e), 7.8(3)(f)**, p 117–127)
- Understand the importance of knowing the RIC operational abilities of both your self-contained breathing apparatus (SCBA) and those departments you mutual aid with. (**NFPA 1407, 7.8(1)**, p 119–121)
- Realize the importance of good communication and coordination among the members of the RIC while it assesses and prepares the downed fire fighter. (p 123–125, 127–129)
- Understand the importance of delivering emergency air and the difficulties in doing so. (**NFPA 1407, 7.8(2)(a), 7.8(3)(e), 7.8(3)(f), 7.11(1)**, p 120–127)
- Know the strengths and weaknesses of the four ways to provide a downed fire fighter with emergency air. (**NFPA 1407, 7.8(3)(a), 7.8(3)(b), 7.8(3)(c), 7.8(3)(d)**, p 120–127)
- Understand what the rapid intervention crew (RIC) Company Officer must monitor and communicate to Command when the fire fighter is found. (p 128–129)
- Describe how an incoming RIC can communicate with an outgoing RIC and the key components of the communication. (p 128–129)

Skills Objectives

After studying this chapter and with hands-on training, you will be able to perform the following skills:

- RIC Officer's assessment and communication of pertinent information to the RIC Operations Chief or the Incident Commander (IC). (p 127)
- RIC member's assessment of the downed fire fighter's situation, condition, and needs. (**NFPA 1407, 7.8(1)**, p 117–119)
- Four ways to supply emergency air.
  1. Using the universal air connection (UAC) (**NFPA 1407, 7.8(3)(c)**, p 120–122)
  2. Swapping out the mask-mounted regulator (MMR) (**NFPA 1407, 7.8(3)(b)**, p 123–125)
  3. Swapping out the SCBA face piece (**NFPA 1407, 7.8(3)(a)**, p 125–126)
  4. Utilizing the SCBA emergency buddy breathing system (EBSS) if so equipped (**NFPA 1407, 7.8(3)(d)**, p 125–127)

Additional NFPA Standards

- NFPA 1500, *Standard on Fire Department Occupational Safety and Health Program*

You Are the Rapid Intervention Crew Member

The search has been under extreme conditions, but as you reach out you feel a boot, and moving in, you discover the missing fire fighter! As the Company Officer transmits a radio message to RIC Operations Chief, you and the crew must prepare the downed fire fighter to be moved.

1. What assessment needs to happen?
2. How will you deliver emergency air?
3. What must we do to prepare him to be moved?

Introduction

This chapter covers what is, in many ways, the most important part of an RIC rescue operation. Extraordinary efforts have been made, under poor and dangerous conditions, to find the downed fire fighter, and he has been located! We know how to rapidly rescue and move him in numerous situations, but what actions must be taken to prepare for the move? Think of the RIC activation as three steps:

1. Locate the downed fire fighter.
2. Protect, assess, and prepare the downed fire fighter.
3. Move the downed fire fighter to safety.

The search for the downed fire fighter was discussed in the chapter *Skills and Techniques of the Rapid Intervention Search*, and here, the different skills and techniques to remove the fire fighter to safety will be presented. It is the second step, however—to protect, assess, and prepare the downed fire fighter—that can make all the difference to the success of the operation. The actions that must be performed to prepare the fire fighter before his removal from the scene include:

- Protecting the fire fighter.
- Assessing his air situation.
- Assessing his level of consciousness and any obvious injuries.
- Supplying emergency air.
- Communicating with the Incident Commander (IC) or RIC Command.
- Coordinating the exit rescue.

You Have Located the Downed Fire Fighter: Now What?

Your crew has worked hard and rapidly with success. You have located the downed fire fighter, but now what actions need to be taken? It is important to realize that your mission is only one-third complete: the search was successful, but you now need to protect, assess, and prepare the downed fire fighter for the rapid removal to safety and medical care. These actions are critical to the success of your mission. Imagine finding the fire fighter, moving him out at record speed, but then finding out that he ran out of air and suffocated in his own mask while you were moving him out. Many actions need to happen simultaneously, but they can easily be achieved because the RIC is composed of more than one rescuer.

■ Protecting, Assessing, and Preparing the Downed Fire Fighter

Protecting the Fire Fighter in Place

When the fire fighter is located, the very first consideration of the RIC Officer must be the level of danger the fire fighter is in. Do we need to request a hoseline to protect the path of egress? What else can we do to protect the downed fire fighter in place? Is the fire gaining and do we need a line or lines in place right away to contain or control the fire in the area? What are the smoke and heat conditions, would additional venting assist? The RIC officer's assessment of the situation is critical at this time. Also, as part of this assessment, you must also begin to plan the way out and whether the egress will need to be protected. Do we need to request another RIC to assist in the rescue or rapid exit, or are we faced with an extended intervention operation because the downed fire fighter is seriously entrapped? In this case, a quick size-up of the situation and the equipment and skills that are required need to be communicated to the RIC Operations Chief, who will coordinate additional resource needs with the IC. While the Company Officer is quickly evaluating the situation, the other RIC members must work together to accomplish the other actions that need to be completed rapidly.

Assessing and Preparing the Downed Fire Fighter: The Crew's Actions

When the fire fighter is found, the crew must take simultaneous actions to help assess and prepare for the rescue exit. These actions will include evaluation of the downed fire fighter's air situation, his level of consciousness and obvious injuries, and supplying emergency air. One way to help coordinate these actions is to divide the work up by the different positions utilized with the rope-assisted search procedure (RASP) procedures.

1. The **Company Officer** will assess the situation, control the crew, communicate with RIC command that they have located the downed fire fighter, provide updates on the downed fire fighter's status and condition, request additional help or resources if required, and constantly size-up not only the rescue but also the environment—watching the smoke conditions, fire conditions, and heat levels to ensure that the RIC is not trapped or overrun by extending fire.
2. **RIC fire fighter #2,** as a <u>tool man</u>, will be responsible for the air evaluation, possible identification, and level

of consciousness evaluation. It will also be his job to convert the SCBA waist strap into a harness if needed.

3. **RIC fire fighter #3,** also a tool or specialty tool man, will stay with the Company Officer and will assist in the evaluation and preparation of the downed fire fighter if needed. This position will assist in the rapid exit, leading out and helping the crew to exit.

4. **RIC fire fighter #4** is the <u>airman</u> and it will be his job to supply air the downed fire fighter.

Evaluation and Preparing to Move the Downed Fire Fighter

RIC fire fighter #2 is responsible for the evaluation of the downed fire fighter and needs to move quickly. The RIC is not functioning as a medic performing a complete evaluation; when a fire fighter is down, it is the RIC's job to get him out of the burning building. To assist in this job we offer the following steps:

1. As you approach the downed fire fighter, listen—do you hear the low air alarm? Can you hear his regulator as he is breathing? Lack of either sound could be an indication that there is no air or that the fire fighter is in respiratory or cardiac arrest.

Crew Notes

When you get to the downed fire fighter, attempt to turn off his PASS device. This will allow you to listen for regulator breathing and also to better communicate with your team. The PASS will begin to alarm approximately every 30 seconds, so be diligent in trying to silence it as best you can. The odds are that when you begin to move the downed fire fighter, the PASS will reactivate. Keep moving, however, and do not waste valuable time trying to prevent the PASS from activating each time it goes off.

2. Approach the fire fighter and give him a shake and call out to him **FIGURE 7-1**. Is he responsive? If the fire fighter is responsive, assure him the RIC is there and they are going to get him out. This type of assurance is important. Look at his helmet shield and attempt facial recognition. Is this who you are looking for, or is it someone else? Are there any obvious traumatic injuries? Shout out your findings to the Company Officer: "Got him, he's responsive. It's [name] from Engine 5!"

3. Assess the downed fire fighter's air situation. If he is responsive, look at the air gauge on his chest. If he is unresponsive, the fastest way to assess if he has any air is to grasp his mask and roll it slightly to the side to break the seal **FIGURE 7-2**. If there is air in the cylinder, you will hear air flow out. If you do not hear anything, the cylinder is empty or there is a SCBA problem or failure. Try and avoid having to roll the fire fighter onto his chest so as to check his air supply level by looking at the cylinder gauge. It is possible to find a fire fighter who is completely out of air with his mask-mounted regulator (MMR) still inserted. If the fire fighter had a medical condition or was struck by something and became unresponsive, he would continue to breathe,

FIGURE 7-1 An RIC fire fighter determining the responsiveness of the downed fire fighter.

FIGURE 7-2 Roll the mask of the downed fire fighter to slightly break the seal and determine if he has air in his cylinder.

but when his air ran out, he would suffocate in his own mask. Is the MMR still in place or did he remove it? If you breathe when your cylinder empty, the final breath will pull the mask into a negative-pressure position; you will not be able to breathe, and you will instinctively remove the regulator. If you find a fire fighter without his MMR in place, it is a good indication that he is out of air and the additional air assessment actions do not need to take place. Again shout out your findings to the Company Officer: "No air, his regulator is out," then call for the airman (RIC fire fighter #4) to move up. The role of the airman will be discussed in detail later in this chapter.

4. As the airman takes position, the RIC fire fighter #2 provides him with updates. If the visibility is zero or near-zero, RIC fire fighter #2 should take one of the airman's hands and place it on the downed fire fighter's mask. This will save him from having to feel and stumble around looking for body orientation **FIGURE 7-3**.

FIGURE 7-3 RIC fire fighter #2 guides the airman's hand to the downed fire fighter's mask.

Crew Notes

Always plan for and expect that when you find the downed fire fighter, you will be supplying him with air.

5. The number one priority at this point is to supply the downed fire fighter with air.

6. Once the airman is in place, fire fighter #2 needs to evaluate if a SCBA harness conversion will be needed. If the downed fire fighter will need to be dragged any distance or if stairs are involved in the exit, it is highly advisable that the harness conversion be done now. Position yourself between the downed fire fighter's legs and make the conversion as taught in the chapter *Rapidly Moving a Downed Fire Fighter*.

7. Once the harness conversion is made, communicate with the airman, and if he is ready to go, consider what skills that will be used to begin the exit—will you need to spin the downed fire fighter or is he ready to be dragged? Shout out to the Company Officer that you are ready to move and how you are going to be doing it. At this point the Company Officer will control the rescue.

It is the role or job of the airman to bring and supply the emergency air to the downed fire fighter and, when initially staging at the incident, to make sure that the RIC rescue air unit has a full cylinder, is in working order, and is ready to deploy **FIGURE 7-4** . Once the downed fire fighter is located, the airman will supply emergency air as demonstrated in **SKILL DRILL 7-1** [NFPA 1407, 7.8(1)]:

1. When fire fighter #2 confirms that a downed fighter has been found, the airman will move up on the rope line, past fire fighter #3, and position himself with the Company Officer, who will be using his thermal imaging camera (TIC) to show what is happening so as to better prepare the airman. At this point the airman should turn the RIC air unit's cylinder open all the way.

2. When the airman moves up to the downed fire fighter, he will converse quickly with fire fighter #2 and make the determination if the air is going to be supplied via the universal air connection (UAC), by doing an MMR swap, or by using another manufacturer-specific device, such as a dual emergency breathing safety system (EBSS) shared air device. The situation and the equipment might dictate the mode of air delivery. For example:

   - Air packs manufactured prior to 2002 did not come with a UAC. Does the downed fire fighter have a UAC?
   - Your RIC rescue air unit is a different SCBA manufacturer than the one worn by the downed fire fighter (who is from a mutual aid company) and the MMRs are not compatible.

   Whatever mode of air supply selected, it must be rapid and manageable for the extraction.

3. Supply the downed fire fighter with air. In the chapter *RIC On Scene: Preactivation Considerations and Actions While Staged*, the situation of the downed fire fighter wearing a different brand SCBA than the RIC air unit your department possesses was discussed. In departments that use mutual aid frequently, SCBA incompatibility can become an issue. When you need to supply the downed fire fighter with air is not the time to learn about SCBA compatibility and how to handle it.

4. Once the air supply has been established or completed, "shout out" to the Company Officer so he is kept informed.

■ Supplying Emergency Air: Considerations and Methods

Considerations

The importance of supplying the downed fire fighter with air cannot be overemphasized. However, there will be times

FIGURE 7-4 The airman turns the RIC air units' cylinder open all the way.

Crew Notes

If you are not able to supply the fire fighter with air and the exit is far away and will take time, consider moving him to a window and placing him hanging out so that he is in a breathable air situation **FIGURE 7-5**. A ladder can then be called for and the rescue made.

FIGURE 7-5 The downed fire fighter is placed outside a window and over the sill, so as to attain breathable air.

when the decision is made not to supply emergency air. Some examples of this type of situation include:

- The fire fighter is located, still has air, and is close to a rapid point of exit. In this case, move him right out to safety.
- The unresponsive fire fighter has removed his mask and is in respiratory arrest. He is not breathing, and supplying emergency air will do nothing but extend the time until you can get him to medical care. Every effort needs to be made to get him *rapidly* out to medical care.
- The downed fire fighter is wearing a brand of SCBA that is not compatible with your RIC air. In this situation, you need to evaluate if it will be faster to do a mask swap, which will take several minutes, or to move the downed fire fighter immediately and rapidly to the closest exit or window.

Methods

When the situation dictates that emergency air must be delivered, there are four methods **FIGURE 7-6** that can be utilized:

1. Using the UAC if so equipped.
2. Swapping out the MMR, if compatible.

A.

B.

C.

D.

FIGURE 7-6 The four methods to deliver emergency air are: **A.** using the UAC; **B.** swapping out the MMR; **C.** swapping out the SCBA face piece; and **D.** using the EBSS.

3. SCBA face piece swap.
4. Using the SCBA manufacturer's "buddy breathing system," or EBSS, if the air packs are so equipped.

Every situation will be different. The equipment you have in hand, your team's level of competence and training, and the SCBA being worn by the downed fire fighter will dictate which technique will work best. It is necessary, therefore, to have a variety of different skills and techniques available to choose from during an RIC operation. Changing out an entire SCBA or swapping cylinders is not discussed in this text, and common sense will dictate if the length of time it will take to accomplish this action in zero visibility is prudent. There may, however, be a specific situation where this extreme action might be warranted, but always remember that getting the downed fire fighter out *rapidly* is always your top priority.

Crew Notes

Fire fighters have a time-honored tradition of making every effort to save their own. However, if you do not frequently drill and practice these techniques to rapidly deliver air to a downed fire fighter, simulating realistic conditions of stress and limited or zero visibility, you are not honoring your commitment to each other. The only way these techniques will work is if you continue to train on how to execute these skills rapidly.

Method #1: Using the UAC

In 2002, the UAC was introduced to the National Fire Protection Agency (NFPA) standard that required all SCBAs, regardless of brand, to have an air attachment point that was universal to all brands, allowing the downed fire fighter to receive air. The primary advantages of the UAC are:

- If the SCBA is so equipped, the RIC can connect any brand RIC rescue air unit to the downed fire fighter's SCBA via the UAC connections and quickly provide air into the cylinder. RIC air units that are made by the fire department by placing an existing SCBA into a bag *will not* be able to do this except, as of this writing, for one manufacturer, MSA. In this case, the MSA Firehawk is equipped with a UAC that allows their SCBAs to be a donor and provide air to any other SCBA, regardless of brand, as long as it is equipped with a UAC fitting. Consult your SCBA's owner's manual and the manufacturer's representative to learn more about how this system works or if your SCBA brand is so equipped.

- By using the UAC, the cylinder can be supplied with air to approximately one-half the cylinder's volume, and the supply hose can then be disconnected. By disconnecting the air line, the downed fire fighter is not tethered to a cumbersome unit that can weigh 25 to 40 pounds (11 to 18 kg) and impede the rescuers' attempts to move him, which can be especially difficult up or down stairs, through holes, and out windows.

Some of the disadvantages of the UAC are:

- Currently, many of the SCBAs in use today still do not have UAC connections. Although many geographic locations will find that SCBAs in their area or mutual aid group will have 2002 or newer compliant SCBAs, there are still areas without the UAC. Unless your department has purchased all new SCBA units or has 100 percent converted over since 2002, you will need to check if the downed fire fighter's SCBA is UAC equipped. This takes time, which is something you really do not have.

- The UAC connections must be kept clean if they are to work properly. Dirt, grit, and other foreign matter will affect the UAC's ability to function properly.

- The UAC will function as a cascade system "equalizing" the pressure between the two cylinders. This means that the most a 4,500 psi (316 kg/cm^2) RIC supply unit can fill another cylinder is to approximately one-half, or 2,100 to 2,200 psi (148 to 152 kg/cm^2). The downed fire fighter now has a half-empty cylinder. If he is breathing and you have a long exit, he will probably go into a low air warning again. If this happens, the team will have to stop and repeat the air transfer, but this time the cylinder will only fill to 1,250 to 1,400 psi (88 to 98 kg/cm^2). In this instance, you will shortly be in low-air-warning mode again.

Crew Notes

The use of the UAC, if so equipped, will allow the RIC to move the downed fire fighter without having to deal with a tethered air supply unit that will weigh 25 to 40 pounds (11 to 18 kg). However, the RIC must remember that when using the UAC, regardless of manufacturer, it functions as a cascade system and will "equalize" the pressure between the full supplying cylinder and the empty or near-empty cylinder. An initial transfer will usually supply a 4,500-psi (316-kg/cm^2) cylinder at or near empty to about 2,200 psi (152 kg/cm^2), or one-half its full volume. Transfers after that will continue to equalize, but remember at a greatly reduced volume (one-half of the donor units pressure), making a rapid exit imperative.

Once the decision has been made to deliver air using the UAC, perform the following steps as demonstrated in **SKILL DRILL 7-2** :

❶ Before moving up to the downed fire fighter, the airman will open the cylinder stem valve on the unit. (**STEP ❶**)

❷ If the airman is going to supply air via the UAC located by the stem of the cylinder, he might need assistance rolling or repositioning the downed fire fighter to quickly gain access. (**STEP ❷**) Put the downed fire fighter into a position that exposes the UAC connection located near the base of the cylinder opposite the open–close valve on his SCBA and remove the rubber protective cover.

❸ Pull the UAC line from the RIC rescue air unit and insert the male coupling on the hose into the female coupling on the downed fire fighter's SCBA. Once locked in, you will immediately hear air transferring. (**STEP ❸**)

SKILL DRILL 7-2   Delivering Air Using the UAC
NFPA 1407, 7.8(3)(c)

1 Before moving up to the downed fire fighter, the airman will open the cylinder stem valve on the unit.

2 If the airman is going to supply air via the UAC located by the stem of the cylinder, he might need assistance rolling or repositioning the downed fire fighter to quickly gain access.

3 Pull the UAC line from the RIC rescue air unit and insert the male coupling on the hose into the female coupling on the downed fire fighter's SCBA. Once locked in, you will immediately hear air transferring.

4 Once the transfer is complete, if possible and if it can be done rapidly, check the downed fire fighter's air level gauge to see how much air is available. Disconnect the UAC hoseline from his SCBA and place it back into the RIC air unit to keep it clean and protected, so that it is in serviceable condition if needed again during the rescue. Inform the RIC officer that air has been transferred.

© Joe Nedder/Jones & Bartlett Learning

Crew Notes

Some SCBA manufacturers place a UAC on both the rear and the front of their SCBAs, which provides a significant advantage because you do not have to move the fire fighter around trying to locate the connection. Make a point to know your department's SCBAs, how they are equipped, and what they can do.

4 Once the transfer is complete, if possible and if it can be done rapidly, check the downed fire fighter's air level gauge to see how much air is available. Disconnect the UAC hoseline from his SCBA and place it back into the RIC air unit to keep it clean and protected, so that it is in serviceable condition if needed again during the rescue. (**STEP 4**)

5 Inform the RIC officer that air has been transferred.

Method #2: Swapping the MMR

If the downed fire fighter's SCBA or your air supply unit does not have a UAC, another technique to provide air is to change out the MMR. In this scenario, the downed fire fighter's MMR is removed and replaced with a regulator that is attached to the RIC air supply unit. The advantages of this technique are:

- Air can be supplied to a fire fighter whose SCBA does not have a UAC.
- Departments that do not have a factory-manufactured RIC rescue air unit can utilize a spare SCBA for their RIC air bag.
- By using the MMR, the RIC air unit's entire cylinder's air supply can be used, meaning that the RIC will not have to stop and refill via the UAC.
- The regulator swap is fast and, with practice, easy to perform.
  The disadvantages are:
- If the downed fire fighter is wearing a different brand SCBA, the MMRs will not be compatible, and the swap cannot be done.
- By swapping the MMR, the fire fighter is now tethered to a unit weighing 25 to 40 pounds (11 to 18 kg), which makes it more difficult for the rescuers to move him, especially for stair rescues and ladder carries. The RIC rescue air unit or the department-made RIC air bag will have to be managed and dragged out with the downed fire fighter, further complicating the rescue and extraction.
- If the downed fire fighter is not able to take a strong or deep enough breath to "open" the system, it will not work.

If the Fire Fighter's Breathing Is Not Strong

The biggest disadvantage of the MMR swap technique is if the downed fire fighter is unresponsive or unable to take a strong enough breath, he will not be able to open the regulator. Your options are very limited. Basically, if the UAC is not available, you can move the downed fire fighter without any breathing air. This decision is difficult, and if you need to make this choice, consider the following:

- How far are you from an exit point or an area of safe haven? Is there a window nearby where the fire fighter can be moved into breathable air while a ladder is being placed?
- Is the downed fire fighter in respiratory arrest (i.e., not breathing)? If this is the case, there is no need to supply air, as he is already not breathing. Get him out to medical care rapidly!

If the decision to move the fire fighter without supplying air is made, make sure that the MMR is removed when the cylinder is out of air or the fire fighter will suffocate within his own mask while you are moving him.

Swapping out the MMR is performed most effectively and rapidly when the swap is done with two rescuers as demonstrated in **SKILL DRILL 7-3** :

1 Fire fighter #4 (airman), with the RIC rescue air unit, will turn the cylinder on before he is adjacent to the downed fire fighter. (**STEP 1**)

2 Fire fighter #2 will position himself either at the head or to one side of the downed fire fighter. Fire fighter #4 will position himself on the opposite side. (**STEP 2**)

3 Fire fighter #4, the airman, will prepare the RIC air bag for the regulator swap by extending the low-pressure line with the MMR. (**STEP 3**) If visibility is poor or nonexistent, fire fighter #2 will locate the mask and MMR of the downed fire fighter and will then take the hand of the airman and "show him" where it is.

4 To execute the regulator swap, the two rescuers will communicate and indicate they are good to go. Fire fighter #4 will use the "Ready, Ready, Go!" commands. At that time fire fighter #2 will release and remove the regulator from the downed fire fighter's mask, and fire fighter #4 will then insert the regulator into the mask and ensure that the downed fire fighter has taken a strong enough breath to "open" the flow of air. (**STEP 4**)

Crew Notes

Executing an MMR swap is a skill that requires practice in order to be proficient with the technique. As with all firefighting skills, you retain when you train!

Crew Notes

It has been suggested in the past that one option with an MMR or mask swap, when the fire fighter's breathing is not strong enough to open the regulator, is to open the purge valve about one-half. This will allow air to flow unrestricted. It will also, however, deplete the air in the cylinder in a matter of a few minutes; because of this, it is no longer suggested or recommended.

RIC Rescue Command and Control Techniques

When two or more rescuers are involved in any rescue, coordination, which can be accomplished with voice commands and teamwork, is a key requirement. The person in charge of the voice commands will vary.

- For moving the downed fire fighter, whoever is in the rescue position facing the escape route will be in charge of the voice commands. In most instances it will be the fire fighter grabbing the SCBA shoulder strap, because he is facing forward and leading out, so he can see or feel what is in front of him. This rescuer can also hold onto the rope or hose if one or the other is being used to identify the escape route.
- For other RIC techniques, the person in charge of the voice commands can vary. For example, with an MMR swap, the airman should be in charge.

Regardless of who is in charge, the use of these simple voice commands and control techniques will ensure coordination and speed.

SKILL DRILL 7-3   Swapping Out the MMR
NFPA 1407, 7.8(3)(b)

1 Fire fighter #4 (airman), with the RIC rescue air unit, will turn the cylinder on before he is adjacent to the downed fire fighter.

2 Fire fighter #2 will position himself either at the head or to one side of the downed fire fighter. Fire fighter #4 will position himself on the opposite side.

3 Fire fighter #4, the airman, will prepare the RIC air bag for the regulator swap by extending the low-pressure line with the MMR.

4 To execute the regulator swap, the two rescuers will communicate and indicate they are good to go. Fire fighter #2 will release and remove the regulator from the downed fire fighter's mask, and fire fighter #4 will then insert the regulator into the mask and ensure that the downed fire fighter has taken a strong enough breath to "open" the flow of air.

© Joe Nedder/Jones & Bartlett Learning

When using voice commands, perform the steps in **SKILL DRILL 7-4**:

1 The lead fire fighter will shout "Ready."

2 The second fire fighter, if ready, will then shout "Ready." If he is not ready, the answer should be "Stop." Never say "No," as No rhymes with Go, and there could be confusion. Once you are ready, shout "All Set," and then the lead fire fighter will go back to step 1.

3 With this acknowledgement, the lead fire fighter will shout, "Go."

4 Simultaneously both fire fighters will then execute their action.

It has been suggested that if the second rescuer is not ready, he or she should not answer at all. Although this sounds reasonable, in an emergency situation with smoke, heat, lack of visibility, and an increase in adrenalin, not getting an answer from your partner might cause you to stop what you are doing to find out what is wrong and if your partner is okay. Good communication skills include providing an audible answer in restricted visibility.

Method #3: SCBA Mask Swap Out

Of the four methods to deliver emergency air, the mask or face piece swap is the least desirable, as it is difficult to do in near-zero-visibility conditions and is extremely time-consuming. However, it is a skill that the RIC needs to know, understand, and have proficiency with. Some examples of when a mask swap out is warranted are:

- The downed fire fighter is trapped, and it will be an extended rescue operation. The SCBAs are not compatible, and there are no other means to provide air to the downed fire fighter.
- The SCBAs are not compatible, and it is a great distance to the nearest exit.

Safety

Any mask swap will expose the downed fire fighter, if he is still breathing, to the heat, smoke, and toxins in the air for the entire period you are changing the masks. This is a very serious situation, and a decision must be made quickly.

Once the decision has been made to swap out the SCBA face piece or mask, perform the following steps as demonstrated in **SKILL DRILL 7-5**:

1. Fire fighter #4 (airman) will prepare by deploying the mask, which is preattached to the MMR. (**STEP 1**)
2. Fire fighter #2 will remove the downed fire fighters helmet, roll back the hood, and loosen the bottom mask straps. (**STEP 2**)
3. In a coordinated manner, fire fighter #2 will remove the mask and fire fighter #4 will place the new mask net or straps over the head of the downed fire fighter, pushing the head net down and smooth. (**STEP 3**) He will then grab the bottom mask straps and pull back to tighten them, and then tighten the other straps. The straps need to be secure and snug in order to seal the mask to the face, otherwise the air will blow by the mask unrestricted, and the total air supply will be lost in a matter of a few minutes.

If the downed fire fighter is in a position, both physically and mentally, that he can assist you with the mask swap, allow

Safety

Trying to use the bypass or purge valve to deliver air to the fire fighter when his breathing will not open the regulator is not advised. Opening this valve will provide an unrestrictive, nonstop flow of air. A typical 30-minute cylinder will bleed down in approximately 2 to 3 minutes. The question you need to ask is: Can we get the downed fire fighter out to safety within that time? If you are that close to an exit, do not expend the time to swap the mask out.

him to do so. If the fire fighter's depth of breathing is so shallow that he cannot open the regulator, the situation is the same as discussed in the section on swapping out the MMR and needs to be assessed wisely and difficult decisions made.

Method #4: Utilizing the Manufacturer's Buddy Breathing Systems to Supply RIC Air

The term "buddy breathing" has been around the fire service since the 1950s. Buddy breathing refers to the old skill of one fire fighter sharing his mask or air supply with another fire fighter, such that one low pressure regulator is shared by two fire fighters who are swapping the mask back and forth. This was possible because the old SCBAs were demand-type regulators versus the positive-pressure masks of today. With the demand-type regulator, you were able to take your mask off (break the seal) without the loss of any air, because only when you took a breath did air flow. Fire fighters were able to hold the mask to their face, take a breath, and then pass the mask to the other fire fighter who would then take a breath. Much has changed since those early SCBA times, and the fire service in general does not recommend buddy breathing in which masks or MMRs are shared.

NFPA 1500, *Standard on Fire Department Occupational Safety and Health Program*, in Section A.7.11.1.2 states:

> The danger of compromising the integrity of the SCBA by removing the face-piece in atmospheres where the quality of the air is unknown should be reinforced throughout the SCBA training program. It is natural that this same philosophy be adopted when dealing with the subject of "buddy breathing." The buddy breathing addressed herein is a procedure that requires compromising the rescuer's SCBA by either removal of the face-piece or disconnection of the breathing tube, as these actions place the rescuer in grave danger.

For two fire fighters to utilize buddy breathing is extremely dangerous. The biggest issue and concern, besides the toxic smoke and searing heat, is that two people will now be sharing one person's air supply, effectively cutting in half the available air supply the "donor" has. If you have a remaining air supply of one-half a cylinder and commence buddy breathing with another fire fighter, your actual air supply is one-fourth a cylinder, which will put you in a low-air situation with just minutes to exit to safety.

There is some confusion today because some of the SCBA manufacturers supply what they are calling an optional "buddy breathing system." In all instances, these systems do not require fire fighters to compromise the integrity of the SCBA, and they are closed systems that are not affected by the immediately dangerous to health and life (IDHL) environment. As an example, the Scott system is called the Emergency Breathing Support System, or EBSS, and the MSA system is called the ExtendAire Emergency Breathing System. In both cases, it allows two SCBAs to connect to each other via a hose and male/female connections specific to the brand SCBA and share air. The systems are designed to utilize the air in both cylinders. Although buddy breathing is not suggested or recommended, this type of setup will work very well in an RIC situation where a rescuer can connect, with the proper fittings, the downed fire fighter's SCBA to a RIC rescue air supply unit or to a spare SCBA equipped with the optional buddy breathing system.

SKILL DRILL 7-5   Swapping Out SCBA Face Pieces
NFPA 1407, 7.8(3)(a)

1 Fire fighter #4 (airman) will prepare by deploying the mask, which is pre-attached to the MMR.

2 Fire fighter #2 will remove the downed fire fighter's helmet, roll back the hood, and loosen the bottom mask straps.

3 In a coordinated manner, fire fighter #2 will remove the mask and fire fighter #4 will place the new mask net or straps over the head of the downed fire fighter, pushing the head net down and smooth.

© Joe Nedder/Jones & Bartlett Learning

Crew Notes

If you are in a non-RIC situation in which a member of your company experiences a SCBA emergency and wants to deploy and utilize the SCBA's buddy breathing system, always consider the following: In the same time it takes a fire fighter to communicate with another fire fighter that he has a SCBA emergency and wants to "buddy breathe," deploy the system, make the connections, and then attempt to exit while tethered together on a very short and limited length of hose, you might have been able to assist the fire fighter in distress in rapidly exiting or moving him to a window where he could exit or have his body positioned outside the window into breathable air while assistance is sent.

The advantages of these systems include:
- Like brand systems that are equipped with these units, they can be connected rapidly.
- The system gives the RIC members another option to utilize to deliver air to the downed fire fighter.
- The downed fire fighter will not be exposed to toxins, heat, or smoke as his mask or regulator will not be removed.
- The system operates on the low-pressure side of the SCBA.
- If deployed, the SCBAs connected together will utilize the air in both cylinders.

The disadvantages of these systems include:
- In an RIC situation, the downed fire fighter will be tethered to another unit adding an additional 25 to 40 pounds (11 to 18 kg) to move.

- The length of the hoses that connect the units will limit where the RIC rescue air unit can be positioned.
- The downed fire fighter must have a compatible system in order for this technique to work.
- Attempting to make the connection in blacked-out conditions can be difficult.

It is strongly recommended that a fire fighter will never utilize his personal air to buddy breathe with a downed fire fighter. Tethering yourself and the downed fire fighter together with a few feet of connecting hose does not allow for quick movement, and is especially difficult if you are dragging the downed fire fighter out.

Each manufacturer's system has its own specific deployment and operating procedures. Therefore, in order to best understand the system you have, refer to the operator's manuals supplied with the SCBAs and consult with the manufacturer's representatives to understand the limitations and operating procedures for your specific brand unit. The procedure outlined below is generic and is intended to give you a guideline, not to address brand-specific operating steps and protocols. When the decision is made to utilize the manufacturer's emergency buddy breathing system (EBBS), and it is confirmed that the downed fire fighter's unit is compatible, perform the following steps as demonstrated in **SKILL DRILL 7-6** :

1. Fire fighter #4 (airman) will deploy the proper hose connections from the RIC air bag and ensure that the RIC cylinder is open. (**STEP ①**)
2. Fire fighter #2 will locate and deploy the hose connection on the downed fire fighter. With the hoselines fully extended, the rescuer who is in the best location to make the connection will do so. (**STEP ②**)

Communications

Throughout the entire RIC activation, superior communication is needed. After the fire fighter has been located, it is the RIC Officer's responsibility to provide feedback, status reports, and updates to the RIC Operations Chief, which include:

- The downed fire fighter has been found.
  - Where he was found (location)
  - His condition (obvious injuries, trapped, breathing)
  - If he is responsive or unresponsive
  - If more than one fire fighter is missing, identify additional missing fire fighters if this can be done quickly.
  - If any other fire fighters have also been found adjacent
- This information will be used to be prepared with the proper level of medical care. Knowing who has been located will assist in determining accountability.
- What other resources are needed?
  - Is the fire fighter trapped?
  - What is entrapping him—a collapse, wires, or furnishings?
  - What specialty tools are needed to free him?
  - Will it be a prolonged rescue, and if so, do we have adequate personnel?
  - What about an adequate air supply for the entrapped fire fighter?
  - If other missing or downed fire fighters are found adjacent, what additional resources and assistance is needed?
- Manner and direction of exit.

Once the direction and manner (how it will be accomplished) of exit is communicated, the RIC Operations Chief can initiate a

---

SKILL DRILL 7-6   Utilizing the Manufacturer's Emergency Buddy Breathing System (EBBS) to Supply RIC Air
NFPA 1407, 7.8(3)(d)

1 Fire fighter #4 (airman) will deploy the proper hose connections from the RIC air bag and ensure that the RIC cylinder is open.

2 Fire fighter #2 will locate and deploy the hose connection on the downed fire fighter. With the hoselines fully extended, the rescuer who is in the best location to make the connection will do so.

plan to assist in the extraction. For example, if it is a ladder rescue, crews can be committed to secure and place the ladder. Is the RIC that found the fire fighter getting low on air and needs to exit quickly? If so, another team will need to make entry quickly to take over the extraction. What are the interior conditions? The officer must also monitor the conditions that the crew is operating in and provide feedback to RIC operations. This feedback will help to determine if additional suppression resources or actions are needed to support the rescue.

Coordination of the Exit Rescue

As the RIC is evaluating the downed fire fighter and preparing him to be moved, the RIC Officer must be in communication with RIC operations. Once it has been determined how the downed fire fighter will be extracted, it is extremely important that it be coordinated with the RIC Operations Chief. This coordination will ensure that the required resources are where they need to be and that the team works as one unit to make the rescue as rapid as possible.

One of the most overlooked requirements is the fact that multiple teams may be needed to make the rescue. This fact is often missed because in RIC training, one crew is used to find and "save" the downed fire fighter. The reality of the situation is more likely to be that numerous teams will be needed if a rescue is to take place. There is no definitive answer as to how many crews it will take, so it is important to know if the rescue will be a rapid grab and go or an extended intervention search.

The goal of the first team is to find the downed fire fighter and supply him with air. If they find him, they will probably not have enough air supply left to prepare him and do the extraction. Most fire departments use 30-minute cylinders. This effectively provides approximately 14 to 16 minutes of air working under stress and duress. Having other teams prepared and ready to go when needed is critical to the speed of the RIC operation. Be proactive, plan, and stage.

How Incoming RICs Can Communicate Under Combat Conditions

It is highly likely that more than one or two teams will be needed to make an extended intervention rescue, so good relevant communications need to happen between the exiting and incoming RICs. It is suggested that the most effective way is *not* to use the radio, but rather as the two teams pass each other, the officers communicate face-to-face. This type of communication will help eliminate unnecessary radio traffic, ensure that the information being given is understood, and allow for any quick questions.

When an RIC needs to exit because of low air or other urgent reason, they need to take the following actions:

1. Inform the RIC Operations Chief that the company needs to exit, where they are in the rescue process, and what still needs to happen.
2. Tie off the RASP line to the downed fire fighter **FIGURE 7-7**. It is advised that you use a carabiner and

secure it to the SCBA harness. This will ensure that the incoming team following the RASP line will go directly to the fire fighter. It also means that if the downed fire fighter regains consciousness and starts to wander away, the rope will stay attached to him and the incoming crew should still be able to locate him.

3. The outgoing RIC exits on the RASP line, with the officer the last in line as discussed in the chapter *Skills and Techniques of the Rapid Intervention Search*.

The incoming RIC will follow the designated RASP line in and the two crews will need to pass each other. If both the incoming and exiting RIC are on the same side of the rope line, the exiting crew stays in place and the incoming crew will cross over the rope to allow the other company to pass **FIGURE 7-8**. The last fire fighter exiting should be the officer who will stop when he encounters the incoming crew's officer (who will be first in line), and the two will communicate quickly and briefly **FIGURE 7-9**. The information communicated should include:

- How far up the line the incoming crew still needs to go
- What still needs to be done to prepare the downed fire fighter for the exit

FIGURE 7-7 The rescue fire fighter securing the RASP line to the downed fire fighter's SCBA using a carabiner.

© Joe Nedder/Jones & Bartlett Learning

FIGURE 7-8 The outgoing and incoming RICs passing on the RASP line.

© Joe Nedder/Jones & Bartlett Learning

FIGURE 7-9 The two RIC Officers communicating with each other as the two RICs pass on the RASP line.

- Any conditions or concerns that need to be shared
- If the downed fire fighter is combative
- The downed fire fighter's air supply situation
- The downed fire fighter's level of entrapment
- Fire and smoke conditions
- Other relative information

This type of information is important to the ongoing operation and will better prepare the officer of the inbound RIC to know what to expect, what has to be done, and the conditions they will be operating in.

The actions we take to assess and prepare the downed fire fighter, along with good communication, will greatly increase the effectiveness of the operation and provide the downed fire fighter with an increased chance of rescue. As members of the RIC we must understand that ongoing training and practice is the only way we will retain these important skills.

Wrap-Up

Chief Concepts

- The most important part of an RIC operation is getting the downed fire fighter out and to medical care.
- Think of the RIC activation as three steps: (1) locate; (2) protect, assess, and prepare to move; and (3) move.
- When the fire fighter is located, the very first consideration the RIC Officer must assess is the level of danger the fire fighter is in.
- Always plan and expect that when you find the downed fire fighter, you will be supplying him with air.
- Upon reaching the downed fire fighter, the number one priority is to supply him with air.
- Consider the pros and cons of giving the downed fire fighter air via the MMR or the UAC—both have advantages and disadvantages.
- If you are not able to supply the fire fighter with air and the distance to the exit is great and will take time, consider moving him to a window and placing him hanging outside so that he is in a breathable-air situation.
- It is strongly recommended that a fire fighter never utilize his personal air to buddy breathe with a downed fire fighter.
- Throughout the entire RIC activation, superior communications are needed.

Hot Terms

<u>Airman</u>   RIC member #4 is the airman, responsible for maintaining the RIC emergency air system.

<u>Tool man</u>   RIC member #2 or #3, responsible for carrying the irons and/or specialty tools. These two positions do the majority of searching.

References

National Fire Protection Agency (NFPA) 1407, *Standard for Fire Service Rapid Intervention Crews*. 2015. http://www.nfpa.org/codes-and-standards/document-information-pages?mode=code&code=1407. Accessed February 11, 2014.

National Fire Protection Agency (NFPA) 1500, *Standard on Fire Department Occupational Safety and Health Program*. 2013. http://www.nfpa.org/codes-and-standards/document-information-pages?mode=code&code=1500. Accessed October 11, 2013.

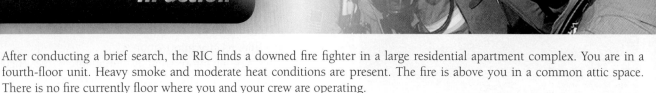

After conducting a brief search, the RIC finds a downed fire fighter in a large residential apartment complex. You are in a fourth-floor unit. Heavy smoke and moderate heat conditions are present. The fire is above you in a common attic space. There is no fire currently floor where you and your crew are operating.

1. When a downed fire fighter is located, the very first consideration the RIC Officer *must* assess is the level of endangerment the fire fighter is in.
   A. True
   B. False

2. While the RIC Officer assesses the situation, what is the number one priority the crew must undertake to assist the downed fire fighter?
   A. Assess the fire fighter's level of consciousness and deliver emergency air.
   B. Perform the harness conversion.
   C. Radio for more help and tie off the fire fighter.
   D. Remove the SCBA and prepare to remove the fire fighter from the closet window.

3. What are some disadvantages to using the UAC?
   A. You can only equalize the pressure between the emergency air and the fire fighter's SCBA.
   B. Not all SCBAs have UAC capabilities.
   C. The UAC connections must be kept clean if they are to work properly.
   D. All of these are disadvantages.

4. When performing an MMR swap on a downed fire fighter who has very shallow respirations, what difficulties may occur?
   A. Do not take the time for an MMR swap on a fire fighter with shallow respirations—just drag him out.
   B. It is always best no matter what to use the UAC.
   C. The MMR might get contaminated with smoke particulates.
   D. The fire fighter may not be able to breathe deeply enough to start the flow of air.

5. How can emergency air be delivered to a downed fire fighter?
   A. Using the UAC if so equipped.
   B. Swapping out the MMR if compatible.
   C. The SCBA face piece swap or using the SCBA manufacturer's "buddy breathing system" if the air packs are so equipped.
   D. All of the above.

RIC Skills: Rapidly Moving a Downed Fire Fighter

© Photos.com

© Joe Nedder/Jones & Bartlett Learning

Knowledge Objectives

After studying this chapter you will be able to:

- Understand the importance of having coordinated and common skill techniques to remove a downed fire fighter. (NFPA 1407, 7.4(3), 7.4(7), 7.12(1), 7.12.1 , p 132, 143, 150)
- Understand the importance of drilling and practicing these skills frequently in order to keep your skills sharp. (p 132)
- Know the basic grab points for a rapid intervention crew (RIC) removal. (NFPA 1407, 7.12(1) , p 132–133)
- Understand the steps that need to be taken in training to protect the fire fighters from injury and to keep from damaging your equipment. (p 134, 137)
- Understand and recognize the strengths and limitations of various rescue devices. (p 138–142)
- Understand the communications procedures specific to rapid intervention. (NFPA 1407, 7.1(3) , p 135–138, 144)

Skills Objectives

After studying this chapter and with hands-on training, you will be able to perform the following skills:

- One-rescuer drag (NFPA 1407, 7.12(1) , p 133–135)
- Two-rescuer drag (NFPA 1407, 7.12(1) , p 135–136)
- Two-rescuer drag/push (NFPA 1407, 7.12(1) , p 137–138)
- Three-rescuer drag/push (NFPA 1407, 7.12(1) , p 137–138)
- Use the drag rescue device (DRD) in the appropriate situation (NFPA 1407, 7.12(1) , p 138–139)
- Convert the self-contained breathing apparatus (SCBA) to a RIC rescue drag harness (NFPA 1407, 7.12(1) , p 139–142)
- Change the downed fire fighter's body orientation for a rescue (NFPA 1407, 7.12(1) , p 143–145)
- Use 1-inch (25-mm) webbing to create a rescue harness to drag the downed fire fighter (NFPA 1407, 7.12(1) , p 145–150)

Additional NFPA Standards

- NFPA 1971, *Standard on Protective Ensembles for Structural Firefighting and Proximity Firefighting*

You Are the Rapid Intervention Crew Member

As a member of an activated RIC, you have moved into the structure and located the downed fire fighter. He is unresponsive and you and the crew need to move him out to immediate medical attention.

1. How will an RIC, using two or three fire fighters, rapidly move the downed fire fighter?
2. If the downed fire fighter has removed his SCBA, how will you grab onto him?
3. If the downed fire fighter has removed his SCBA, how will you move him to safety?

Introduction

To this point, the need for rapid intervention, the mechanics of preincident and the RIC activation, self-survival skills, how to conduct a search using a search rope, and the actions that need to be taken once the downed fire fighter is located have all been discussed. We are now ready to begin training on how to rapidly move a downed fire fighter from danger and extricate him from the building. The techniques and skill sets include:

- Basic grab points that can be utilized for moving the downed fire fighter
- One-rescuer drag
- Team command and control voice commands
- Two-rescuer drag
- Two-rescuer drag/push
- Three-rescuer drag/push
- Using the drag rescue device (DRD) in the turnout coat
- Converting the self-contained breathing apparatus (SCBA) to an RIC rescue drag harness
- Additional removal skills
- Changing the downed fire fighter's body orientation for a rescue
- The use of webbing in fire fighter rescues

Moving a Downed Fire Fighter

There are two ways you will encounter a downed fire fighter: (1) witnessing a fire fighter who is part of your team going down, or (2) locating a downed fire fighter after he became missing or a MAYDAY has been called. In either case, you have to be *prepared and ready* to take immediate action to *rapidly* remove the fire fighter to safety and medical care, if needed. In the first scenario, you might witness your partner collapse on the hoseline with a medical emergency. Knowing what to do and how to do it will have a great impact on your partner's survival. Moving a downed fire fighter who is unresponsive or immobilized is not an easy feat! It takes skill and coordination among the rescuers to make the rescue happen rapidly and efficiently. The following skill sets will, with practice, give you those abilities.

There are four ways to move someone: drag (pull), lift, push, and carry. These techniques might sound simple, but they need to be taught in a coordinated manner, with practiced skill and developed techniques. Oftentimes, even today, some

still believe in the old adage "Don't worry about it! If you're in trouble I'll get you out!" When it comes to RIC, however, always remember it takes skill, courage, and common sense.

■ Basic Grab Points to Simplify and Expedite Rescue

A fire fighter in full personal protective equipment (PPE) has some basic grab points that we utilize in a RIC rescue. They are the SCBA shoulder strap, the cuff of the bunker pants, and the bunker coat collar.

SCBA Shoulder Strap

The shoulder strap, which is the most obvious, gives you a good handle on which to grab hold of and pull or lift **FIGURE 8-1**. The major issue with the SCBA shoulder straps is that they tend to loosen when pulled. It has been suggested that if the RIC ties the tail of the strap with an overhand knot, it will prevent the strap from loosening; however, as much as this might make sense, consider the following: you have found a downed fire fighter in near-zero visibility and you have begun to move him rapidly to safety. How long will it take you, in total blacked out conditions, to locate the tail of the shoulder strap and tie an overhand knot? The emphasis continues to be to move the fire fighter to safety *rapidly*. The one or two minutes it may take to try to tie an overhand knot might not be in

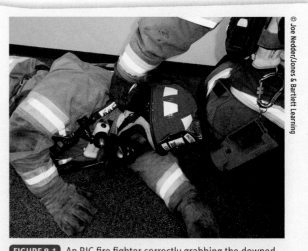

FIGURE 8-1 An RIC fire fighter correctly grabbing the downed fire fighter's SCBA shoulder strap.

© Joe Nedder/Jones & Bartlett Learning

the survivability profile of the downed fire fighter. The decision is not black and white, but it will be up to you to choose wisely and make the correct decision at that given moment.

Cuff of the Bunker Pants

The inside cuffs of the bunker pants make excellent grab handles when having to move or lift the downed fire fighter's legs **FIGURE 8-2**. By using the inside of the cuff versus just grabbing a handful of fabric, you will have a better grip and therefore better ability to move the legs as needed.

Bunker Coat Collar

If you encounter a downed fire fighter who has removed his or her SCBA and you need to perform a redirection maneuver, such as move him or her into a seated position, using the inside of the coat collar is a good grab point **FIGURE 8-3**. This is an area where a lot of material and strong stitching comes together, which makes for an excellent grab point.

■ One-Rescuer Drag

Every fire fighter needs to be prepared to move a downed fire fighter to safety by himself. The situation might be that you and your partner are on a hoseline in the fire room, and he collapses with a medical emergency. Do you know what to do and how to do it? Your first priority is to get your partner out of the fire room and to a safer location. This might mean dragging him out of the room and closing the door, or placing him in the hallway with the wall separating him from the fire room. Remember, the room is involved in fire, so you do not have any extra time. You have to move him *now*. The one-rescuer drag is a good technique to accomplish this. Please note that we are saying that your *first* action is to move the downed fire fighter, not take time to call a MAYDAY or do an evaluation. This is because the situation is that you are in a room with an *uncontrolled fire*. Move your partner to safety first. Then, once you are clear of the fire room, declare a MAYDAY! If the fire conditions do not warrant an emergency move, declare the MAYDAY first and then begin the one-rescuer drag.

When a fire fighter is injured or becomes unresponsive, he will typically go into one of two positions: (1) lying on his side, as the SCBA cylinder prevents him from being on his back **FIGURE 8-4**, or (2) lying face down **FIGURE 8-5**. The most effective position to execute the one-rescuer drag is for the downed fire fighter to be on his side. The one-rescuer drag can also work if the fire fighter is face down, although it is more difficult

FIGURE 8-2 An RIC fire fighter correctly grabbing the cuff of the downed fire fighter's pants.

FIGURE 8-4 The downed fire fighter might be lying on his side.

FIGURE 8-3 An RIC fire fighter using the collar for a grab point.

FIGURE 8-5 The downed fire fighter might be lying face down.

to perform if the fire fighter has a lot of equipment hanging on his coat (i.e., flashlights, radio), as the equipment will inhibit the dragging. An additional issue and of greater concern is the possibility of dislodging the downed fire fighter's mask. To perform the one-rescuer drag, follow the steps as demonstrated in **SKILL DRILL 8-1** :

1 Grab the SCBA strap on the high side as close to the SCBA back frame as possible, making sure you are grabbing it with your hand inserted palm down. It is critical that you grab the shoulder strap as shown. (**STEP** 1) Grabbing it using other methods has resulted in permanent injuries to fire fighters, specifically damage and separation of tendons.

2 Extend your arm and position your front knee on the floor as a pivot point and your back leg in a position such that your foot is planted firmly on the floor, ready and cocked like a piston to exert energy by pushing off. (**STEP** 2)

3 Face in the direction you are going, keeping your body low and placing your free hand in front of you for balance and leverage. (**STEP** 3)

4 When properly positioned, push off with your rear leg and at the same time pull on the downed fire fighter's SCBA shoulder strap. (**STEP** 4)

5 Reposition and continue this technique until you have moved the fire fighter to safety.

Safety

When using your SCBAs for training on these techniques, it is important that you protect your equipment. If you are using anything but an aluminum cylinder, great caution and care must be taken to not damage the cylinder. Dragging a fiberglass-wrapped cylinder on an abrasive surface will cause the coating to scuff off. If the coating is scuffed off and you "see" black, the integrity of the cylinder for filling has probably been destroyed. The cylinder must be taken out of service and removed from use immediately. There is no repair for this type damage. When practicing these techniques, either use an old, out-of-service cylinder for the drags (commonly referred to as a drag cylinder) or place Masonite, plywood, or a scrap piece of carpet on the ground where the dragging will take place. Failure to take these precautions will result in permanent and irreparable damage to your SCBA cylinder.

Safety

One-Rescuer Drag Safety Tips:

- Always face and look forward.
- Always stay as low as possible.
- Always use your leading hand to ensure there is a floor ahead of you.
- Always be fully aware of the fire conditions around you.

SKILL DRILL 8-1   One-Rescuer Drag
NFPA 1407, 7.12(1)

1 Grab the SCBA strap on the high side as close to the SCBA back frame as possible, making sure you are grabbing it with your hand inserted palm down. It is critical that you grab the shoulder strap as shown.

2 Extend your arm and position your front knee on the floor as a pivot point and your back leg in a position such that your foot is planted firmly on the floor, ready and cocked like a piston to exert energy by pushing off.

(Continued)

SKILL DRILL 8-1 One-Rescuer Drag (*Continued*)

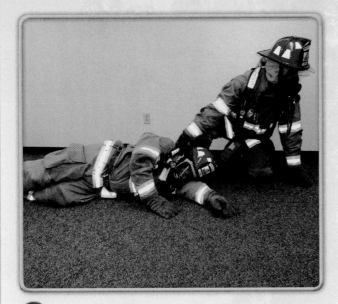

3 Face in the direction you are going, keeping your body low and placing your free hand in front of you for balance and leverage.

4 When properly positioned, push off with your rear leg and at the same time pull on the downed fire fighter's SCBA shoulder strap. Reposition and continue this technique until you have moved the fire fighter to safety.

© Joe Nedder/Jones & Bartlett Learning

■ Two-or-More-Rescuer Techniques

Moving a downed fire fighter to safety is a difficult task for one fire fighter to perform. Even if the downed fire fighter is small in size, trying to move him on a wet rug, around common household obstacles, with his gear and SCBA adding to the difficulty, is a significant physical challenge. To overcome this, it is suggested that whenever possible, two or more rescuers move a downed fire fighter. There are a few options to the <u>two-rescuer drag</u> technique, including low and high versions of the <u>two-rescuer side-by-side drag</u>.

RIC Rescue Command and Control Techniques

The RIC rescue command and control techniques utilizing the "Ready, Ready, Go" commands, discussed in the chapter, *RIC Team Actions Once the Downed Fire Fighter Is Found,* must be used with all RIC techniques employing two or more rescuers. Using these simple commands and techniques enables the RIC to move the downed fire fighter to safety in a rapid and coordinated manner.

The two-rescuer side-by-side low drag technique can be used if the area you need to move through is wide enough to fit two fire fighters side-by-side and conditions allow for two fire fighters to position themselves on both sides of the head of the downed fire fighter. Two rescuers can then both drag the downed fire fighter by his SCBA shoulder strap. To perform the two-rescuer side-by-side low drag, follow the steps as demonstrated in **SKILL DRILL 8-2**:

1 Fire fighter 1 positions him as in the one-rescuer drag. (**STEP 1**)

2 Fire fighter 2 positions himself on the opposite side and grabs the low shoulder strap using the same technique as in the one-rescuer drag. (**STEP 2**)

3 One fire fighter then initiates the "Ready, Ready, Go!" command and both fire fighters, at the same time, initiate the drag.

4 The two fire fighters continue step 3 until they reach safety or cannot physically continue.

SKILL DRILL 8-2 — Two-Rescuer Side-by-Side Low Drag
NFPA 1407, 7.12(1)

1 Fire fighter 1 positions him as in the one-rescuer drag.

2 Fire fighter 2 positions himself on the opposite side and grabs the low shoulder strap using the same technique as in the one-rescuer drag. One fire fighter then initiates the "Ready, Ready, Go!" command and both fire fighters, at the same time, initiate the drag. The two fire fighters continue step 3 until they reach safety or cannot physically continue.

© Joe Nedder/Jones & Bartlett Learning

If the downed fire fighter is not wearing his SCBA, this technique can also be done by using the downed fire fighter's collar as a drag point **FIGURE 8-6**.

Two-Rescuer Side-by-Side High Drag
This technique can be used when visibility and heat are at a reasonable level, allowing the two RIC fire fighters to safely stand up. The skills and commands are similar to the low drag technique, except that the two rescuers are going to stand/crouch on both feet and drag the downed fire fighter out **FIGURE 8-7**.

Heat conditions will dictate if the RIC rescuers can stand or not, as will the visibility in the smoke. If visibility diminishes or if the heat drives the RIC down to the floor, the RIC fire fighters should continue moving the downed fire fighter out with the low drag.

The two-rescuer drag/push technique is the most common two-rescuer technique for removing a downed fighter taught in rapid intervention. It is simple and, if coordinated, allows the RIC to move the downed fire fighter to safety

FIGURE 8-6 Two RIC fire fighters dragging the downed fire fighter using his collar as a grab point.

FIGURE 8-7 Two RIC fire fighters using the high drag technique.

© Joe Nedder/Jones & Bartlett Learning

rapidly. To perform the two-rescuer drag/push, follow the steps as demonstrated in **SKILL DRILL 8-3** :

1 The first fire fighter takes the lead position at the downed fire fighter's head and positions himself as if he is doing the one-rescuer drag.

2 The second fire fighter, while on the same side as the first fire fighter:

1. Grabs the cuff of the downed fire fighter's leg and raises the leg up. **(STEP 1)**

2. He then positions himself with his arm wrapped tightly around the downed fire fighter's thigh, positioning himself as low as he can go, and allows the downed fire fighter's lower leg to rest on his back in line with the rescuer's SCBA cylinder. **(STEP 2)**

3. Finally, he positions his legs in the same manner as in the one-rescuer drag—with his front knee on the floor as a pivot point and his back leg in a position such that his foot is planted firmly on the floor, ready and cocked like a piston to exert energy by pushing off, and signals his partner that he is ready by calling out "All Set."

3 The first fire fighter then initiates the "Ready, Ready, Go!" commands.

4 The first fire fighter drags the downed fire fighter, as in the one-rescuer drag technique, and the second fire fighter pushes against the downed fire fighter's thigh, using his rear leg as a piston. **(STEP 3)**

5 The two rescuers should move the downed fire fighter a few feet, stop, and reinitiate the "Ready, Ready, Go!" command, continuing until they reach safety or cannot physically continue.

Crew Notes

Get as low on the downed fire fighter's thigh as possible. You must be below the downed fire fighter's knee joint, giving you more leverage and a better pushing point. If you do not get low enough (i.e., if you are above the knee joint), you will accomplish nothing but to bend the downed fire fighter's leg over, possibly causing injury, and provide no physical assistance to the first rescuer trying to drag the downed fire fighter.

Three-Rescuer Drag/Push

This technique combines the skills of the two-rescuer drag/push and adds a third rescuer at the head to help drag. This technique is excellent for moving a larger fire fighter, or if the original two rescuers are getting fatigued. The key to using this technique is for the two rescuers pulling not

---

SKILL DRILL 8-3   Two-Rescuer Drag/Push
NFPA 1407, 7.12(1)

1 The second fire fighter, while on the same side as the first fire fighter grabs the cuff of the downed fire fighter's leg and raises the leg up.

2 The rescuer then positions himself with his arm wrapped tightly around the downed fire fighter's thigh, with the downed fire fighter's lower leg on his back.

(Continued)

SKILL DRILL 8-3   Two-Rescuer Drag/Push (*Continued*)

3   The first fire fighter then initiates the "Ready, Ready, Go!" commands. The first fire fighter drags the downed fire fighter, as in the one-rescuer drag technique, and the second fire fighter pushes against the downed fire fighter's thigh, using his rear leg as a piston. The two rescuers should move the downed fire fighter a few feet, stop, and reinitiate the "Ready, Ready, Go!" command, continuing until they reach safety or cannot physically continue.

© Joe Nedder/Jones & Bartlett Learning

to advance more than the one rescuer who is pushing can reasonably accomplish. The "Ready, Ready, Go!" commands are between the lead rescuer and the rescuer in the pushing position. The additional rescuer pulling should inform the lead rescuer with an "All Set," at which point the lead rescuer will commence with the commands **FIGURE 8-8**.

■ Utilizing the Drag Rescue Device

In 2007, the National Fire Protection Agency (NFPA) included the <u>drag rescue device (DRD)</u> as a mandatory part of a fire fighters turnout ensemble under NFPA 1971, *Standard on Protective Ensembles for Structural Firefighting and Proximity Firefighting*. The DRD is built into the coat drag harness that engages the fire fighter under his arms into loops, so that the strap can be used to drag him along. The concept of the DRD is a good one; however, it has limitations. Fire fighters need to understand these limitations and know what works and what does not. The DRD was designed to assist with rescuing an unresponsive fire fighter. It allows one fire fighter the opportunity to rescue another fire fighter by dragging him to safety. The DRD is designed for use in a straight-line horizontal drag

FIGURE 8-8 Three RIC fire fighters in the proper position to initiate a three-rescuer drag/push.

© Joe Nedder/Jones & Bartlett Learning

(along the floor) only and without any lifting. When the DRD is used for a vertical lift, the system will begin to pull upward, causing the coat to begin to pull inside out while still on the downed fire fighter; this means that the DRD *cannot* be used for a stair rescue.

The advantage of the DRD is that it gives the rescuing fire fighters a grab point to assist in removal. Using the DRD takes practice, and you must be able to deploy it for use under extreme conditions with gloves on. To deploy the DRD:

1. Locate the DRD tab in between the shoulders of the coat.
2. Open the flap and remove the strap **FIGURE 8-9**.

What makes this a challenge is that the DRD tab is usually found under or partially obstructed by the SCBA **FIGURE 8-10**. This means you will have to loosen the shoulder straps and reposition the cylinder or squeeze your hand under the SCBA and locate the tab. It is advised that if the downed fire fighter is still wearing his SCBA, the DRD not be used. Instead, utilize the shoulder straps or the drag handle built into the SCBA frame if so equipped. Again, know your equipment. If the fire fighter has removed his SCBA for whatever reason, the DRD can be used for horizontal drags. Fire fighters should practice deploying the DRD while wearing their PPE, with gloves on, and with their vision blacked out. This is the reality of the RIC—to practice and train under other conditions will not develop and build the skills you need.

FIGURE 8-10 Many times the SCBA can obstruct access to the DRD tab.

A.

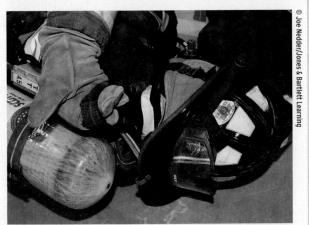

B.

FIGURE 8-9 The DRD. **A.** The DRD tab being opened; and **B.** the DRD being pulled out for use.

Safety

Fire fighters must take great care of the DRD. It must be properly reinstalled after your PPE has been washed. You must also ensure that the closure tab is properly and securely closed. DRDs not properly secured have been known to get caught on items and deploy without the fire fighter's knowledge, creating an entanglement hazard.

Removal Technique Safety Concerns

There has been a recent trend to create new RIC drag techniques. Any time improvements can be made to techniques designed to save one of our own, it should be given serious consideration; however, many of these "improvements" have the following errors or endangerments:

- One technique shows the lead fire fighter "bear hugging" the downed fire fighter around the chest while in a standing position, and then the rescuer proceeds to walk backwards. It is imperative to *always face* in the direction you are going.
- Another technique shows the DRD being used in a vertical manner; however, the DRD will not function properly in anything but a *horizontal* pull.

It is easy, when training on the apparatus floor in a situation without heat, smoke, anxiety, or obstacles such as furniture, stairs, and hoselines, to believe these techniques are easier and will work. Rapid intervention is usually carried out under very adverse and difficult conditions. When developing new techniques or adapting existing ones, common sense should be applied and safety should be the top priority. Improvements should always be researched, not just for the sake of change, but rather to improve both the fire service and the safety and survival of our own.

■ Harness Conversion: Converting the SCBA Waist Strap to an RIC Rescue Drag Harness

In an RIC operation, while attempting to move the downed fire fighter quickly, the SCBA shoulder straps will occasionally loosen to the point that they are coming off the downed fire fighter.

This is a hindrance to the rescue operation, and the crew needs to stop to adjust and tighten it. Many of the RIC maneuvers and techniques rely on the SCBA harness staying in place during the rescue. The questions then, are: when should the SCBA be converted to a harness? Are there guidelines? And, is it necessary to always convert?

The SCBA harness conversion will assist in the following types of RIC rescues:

- One-rescuer drag
- Two-rescuer drag/push
- Rescues up stairs
- Rescues down stairs

It can never be said that the SCBA should *always* converted to a harness. Determining if a harness is needed or if a harness conversion can be completed depends on many factors:

- The condition of the harness
- The proximity of the exit point
- The size of the downed fire fighter
- Whether or not the rescuers have the skills

If the SCBA harness conversion is required, it needs to be accomplished quickly.

The SCBA waist strap–RIC harness conversion is not an NFPA-rated harness. Its sole purpose is to help keep the SCBA in place on the downed fire fighter while dragging and carrying them to safety. *This type of harness should never be used as a harness to assist in lowering or lifting a fire fighter.* The safety concerns are great, and the waist straps may not hold when lifting an unresponsive person. If it does not hold or fails because either the buckle was not securely connected or for some other reason, the downed fire fighter could fall to his death, depending on how high off the ground he is when the failure occurs.

To convert the SCBA waist strap to an RIC rescue drag harness, perform the following steps as demonstrated in **SKILL DRILL 8-4**:

1. Working from the high side of the fire fighter, loosen (never undo until ready to convert to a harness) the high side waist strap. (**STEP 1**)
2. Grab the loosened strap and, using it to pull the fire fighter towards you, rotate him on the SCBA cylinder just enough so that you can place your hand under the low side and loosen the waist strap. (**STEP 2**) Use you other hand to help pull the fire fighter towards you. **Note:** If you cannot move the downed fire fighter yourself, have another member of the RIC help to execute this maneuver.
3. After the waist strap is loosened on both sides, return the downed fire fighter to his side.
4. With the waist straps completely loosened, reposition yourself between the downed fire fighter's legs, grabbing one leg by the cuff, and raising it onto your shoulder. (**STEP 3**)
5. With the leg resting on your shoulder, push the leg back into as close to a vertical position as possible without causing injury. At this point, place both your hands on the waist strap by the buckle. (**STEP 4**) Having the leg on your shoulder frees up both hands for the maneuver.
6. When ready, disconnect the waist strap buckle and, with each hand holding one side of the waist strap,

pull one strap around the leg that is in the vertical position, then connect the buckle together. (**STEP 5**)
7. Lower the leg and tighten the high side waist strap. (**STEP 6**)

If the fire fighter is large, you might find it is not possible to make the harness conversion. An option would be for you and your partner to:

- Loosen both shoulder and waist straps and roll the fire fighter face down.
- Working together, shift the SCBA lower on his back—as far down as you can position it.
- Roll the fire fighter onto his back and retry to convert to the harness using the method taught above.

If you are still unable to make the conversion, resecure the waist strap and retighten the shoulder straps as much as possible before moving the fire fighter.

■ Life Safety Harnesses Built Into Your PPE

There has been a dramatic increase in Class II harnesses installed within or on PPE. The major advantage of this is that, coming as a complete ensemble from the manufacturer, it is a NFPA-compliant ensemble. There are different styles of harnesses, each having their own advantages.

A Class II harness that is installed over the bunker pants by the manufacturer gives you a compliant harness that can, if needed, be added to your bunker pants by the PPE manufacturer at a later date **FIGURE 8-11**. A Class II harness installed by the manufacturer inside the bunker pants gives you a harness that is NFPA-compliant and, because it is contained within the gear, has little chance of the leg

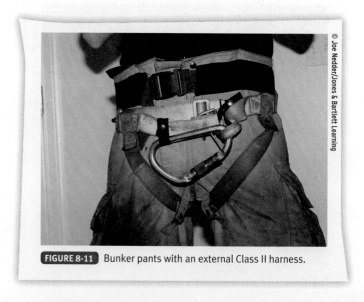

FIGURE 8-11 Bunker pants with an external Class II harness.

SKILL DRILL 8-4 Converting the SCBA Waist Strap to an RIC Rescue Drag Harness
NFPA 1407, 7.12(1)

1 Working from the high side of the fire fighter, loosen (never undo until ready to convert to a harness) the high side waist strap.

2 Grab the loosened strap and, using it to pull the fire fighter towards you, rotate him on the SCBA cylinder just enough so that you can place your hand under the low side and loosen the waist strap. After the waist strap is loosened on both sides, return the downed fire fighter to his side.

3 With the waist straps completely loosened, reposition yourself between the downed fire fighter's legs, grabbing one leg by the cuff, and raising it onto your shoulder.

4 With the leg resting on your shoulder, push the leg back into as close to a vertical position as possible and place both your hands on the waist strap by the buckle.

(Continued)

SKILL DRILL 8-4   Converting the SCBA Waist Strap to an RIC Rescue Drag Harness (*Continued*)

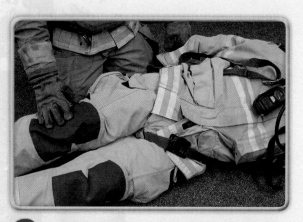

6 Lower the leg and tighten the high side waist strap.

5 When ready, disconnect the waist strap buckle and, with each hand holding one side of the waist strap, pull one strap around the leg that is in the vertical position, then connect the buckle together.

© Joe Nedder/Jones & Bartlett Learning

straps becoming caught or damaged at an emergency scene **FIGURE 8-12**. The primary purpose of these harnesses is to be used by fire fighters during an emergency escape, technical rescue, and also as a ladder belt. They can be used in rapid intervention to lower or raise a downed fire fighter and also, in some cases, to assist in dragging or hauling a fire fighter with a rope to safety.

■ Tethering Harnesses

A rescue fire fighter must use great caution when tethering or connecting himself to the downed fire fighter via his Class II harness or his SCBA. Imagine you are in a hostile, dangerous environment with little or no visibility. By tethering yourself you are:

- Creating a classic "you go–we go" scenario. If you experience a floor or ceiling collapse while tied or tethered to an unresponsive fire fighter, your ability to self-rescue becomes greatly diminished.
- Overexerting yourself. Trying to position yourself by straddling the downed fire fighter or crawling, or dragging him, while tethered to him in bad conditions, is not an easy feat.

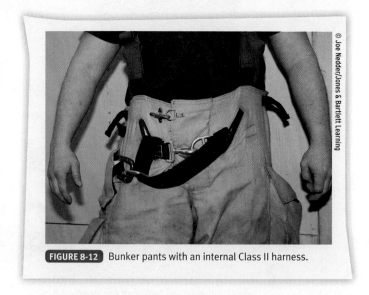

FIGURE 8-12 Bunker pants with an internal Class II harness.

© Joe Nedder/Jones & Bartlett Learning

For an RIC to be effective, they need to remember the word *rapid* and that an activated RIC has two primary goals: (1) to get the RIC company back out alive and (2) to locate and begin the rescue of the downed fire fighter, in that order.

Additional Rescue/Removal Skills

If conditions are not poor, other removal skills can also be used, including the basic Fire Fighter I/II simple victim carries such as the blanket drag, extremities carry, seat carry, and chair carry. All these skills, with the exception of the blanket drag, require the rescue fire fighters to be in a standing position, and rescue fire fighters should avoid standing in poor conditions at all costs. The blanket drag requires a blanket, which is usually not readily available to the RIC. These rescue removal skills are not designed for RIC rescues under high heat and heavy smoke conditions, but using the RIC rescue skills presented here as your first alternatives will assist you to rapidly remove the downed fire fighter to safety. To review the basic Fire Fighter I/II rescue skills, refer to your *Fundamentals of Fire Fighter Skills* textbook.

Changing the Downed Fire Fighter's Body Orientation for Rescue

The skills presented so far are for moving a downed fire fighter forward, with the head oriented towards the avenue of escape. There are instances when there is a need to rotate (or spin) the fire fighter to change the direction of his body orientation.

1. You have dragged the fire fighter to a window for extraction but he needs to be oriented feet first for the window/ladder rescue.
2. While dragging the fire fighter, you need to move around a sharp or tight corner, furniture, or other some other obstacle.
3. When found, the downed fire fighter is not oriented for a drag rescue and needs to be reoriented or turned around.

Whatever the reason for spinning the fire fighter, we need to have coordinated skill so it can be done rapidly and efficiently.

■ Downed Fire Fighter Change of Direction Techniques

There are two spin techniques that can be utilized: the one-rescuer and the two-rescuer techniques. If at all possible, it is always easier for two fire fighters working together to complete this maneuver.

To perform the one-rescuer spin, follow the steps as demonstrated in **SKILL DRILL 8-5** :

1. Determine the direction you need to spin the downed fire fighter.
2. Grab both of the downed fire fighter's legs by the cuffs and begin to raise his legs together. (**STEP 1**)

SKILL DRILL 8-5   One-Rescuer Spin

NFPA 1407, 7.12(1)

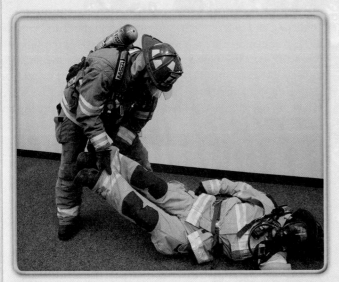

1 Determine the direction you need to spin the downed fire fighter. Grab both of the downed fire fighter's legs by the cuffs and begin to raise his legs together.

2 Raise both of the downed fire fighter's legs at least 90 degrees and place them on your shoulder, wrapping both arms around his legs.

(Continued)

SKILL DRILL 8-5   One-Rescuer Spin *(Continued)*

3 Staying as low as possible, lift up on the downed fire fighter's legs and drag the fire fighter in the direction of the spin, using his shoulder area or cylinder as a pivot point.

4 Once the downed fire fighter is properly positioned, place his legs on the ground.

© Joe Nedder/Jones & Bartlett Learning

3 Raise both of the downed fire fighter's legs at least 90 degrees and place them on your shoulder, wrapping both arms around his legs. (**STEP 2**)

4 Staying as low as possible, lift up on the downed fire fighter's legs and drag the fire fighter in the direction of the spin, using his shoulder area or cylinder as a pivot point. (**STEP 3**)

5 Once the downed fire fighter is properly positioned, place his legs on the ground. (**STEP 4**)

Crew Notes

Raising the downed fire fighter's legs at least 90 degrees makes the spin easier by shifting the downed fire fighter's weight towards the shoulder, decreasing his body surface area, and reducing the amount of equipment that is against the floor.

In order for two RIC fire fighters, working together, to understand which direction to move the downed fire fighter, terms such as clockwise and counter-clockwise can be confusing and will not work. A simple command and control voice command is for the lead rescuer (possibly the officer), when the two fire fighters are ready to execute the skill, to indicate the spin direction by shouting out "Head to Me" or "Head to You." Either of these indicators is clear, and there can be no question which way to move the downed fire fighter **FIGURE 8-13**. To perform the two-rescuer spin, follow the steps as demonstrated in **SKILL DRILL 8-6** :

1 Determine the direction you need to spin the downed fire fighter.

2 The two rescuers position themselves on opposite sides of the downed fire fighter, each grabbing the leg cuff closest, with the other hand placed inside the downed fighter's collar (not the SCBA straps). (**STEP 1**) **Note:** you might have to *slightly* open the coat to gain good access to the collar.

3 Raise the downed fire fighter's legs to at least 90 degrees. (**STEP 2**) The lead rescuer then gives the command, "Head to Me" (or "Head to You"), (**STEP 3**) and the two rescuers spin the downed fire fighter around to the position needed. (**STEP 4**)

SKILL DRILL 8-6 Two-Rescuer Spin
NFPA 1407, 7.12(1)

1 Determine the direction you need to spin the downed fire fighter. The two rescuers position themselves on opposite sides of the downed fire fighter, each grabbing the leg cuff closest, with the other hand placed inside the downed fighter's collar (not the SCBA straps).

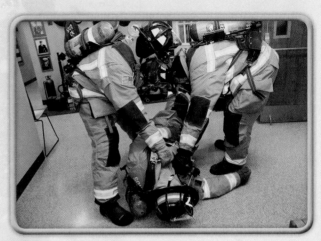

2 Raise the downed fire fighter's legs to at least 90 degrees.

3 The lead rescuer then gives the command, "Head to Me" (or "Head to You").

4 The two rescuers spin the downed fire fighter around to the position needed.

© Joe Nedder/Jones & Bartlett Learning

■ Use of Webbing When Moving a Downed Fire Fighter

When operating in an RIC mode, every advantage that can help make the rescue rapid and doable needs to be found and used. One such item, which is an important and irreplaceable tool for any RIC, is 1-inch (25-mm) tubular webbing. There are numerous applications and ways that webbing can be used, including:

- Dragging a fire fighter who has removed his SCBA
- When a single rescuer needs to use a door frame to gain mechanical advantage or leverage to pull a downed fire fighter

© Joe Nedder/Jones & Bartlett Learning

FIGURE 8-13 Two RIC fire fighters in position and ready to spin the downed fire fighter after the command "Head to Me."

- When lifting a large fire fighter up a flight of stairs
- When lifting a large fire fighter out of a window

The 1-inch (25 mm) tubular webbing should be 20 to 25 feet (6 to 8 m) in length and rated at 4,000 pounds (1,814 kg)tensile strength. This is a significant and strong tool, and the length will give you great versatility as to what you can do with it. The webbing should be tied into one loop using a water knot. The water knot is designed to be used with webbing, making it strong and easy to untie, and the loop needs to be continuous. There are currently rescue slings on the market that utilize numerous small loops. These tools are fine, but a single piece of webbing, tied into one loop with a water knot, will do everything you need to do and also give you more flexibility for other rescue techniques. How to tie webbing with a water knot, how to carry it ready for deployment, and how to easily deploy it was discussed in the chapter *Planning for a Prepared Rapid Intervention Crew*.

The advantage of carrying the webbing rolled and with a water knot is that you can deploy your webbing for use and have total control in a matter of two or three seconds. In technical rescue, it is taught to carry webbing "daisy-chained," but this is not recommended for RIC. In limited visibility, under adverse conditions, trying to deploy webbing that is stored daisy-chained could very well result in the webbing being pulled the wrong way, causing the webbing to pull together into a difficult knot to untie. In an RIC rescue, your webbing should be carried as demonstrated in the chapter *Planning for a*

Crew Notes

The use of webbing is not only for RIC applications. It can also be used to harness and rescue unconscious civilians found during a search operation.

Prepared Rapid Intervention Crew, for fast and effortless deployment when you have only seconds to react when trying to save one of our own.

■ Webbing Harnesses

The use of webbing as a harness is an important skill and tool. It has been documented that lost fire fighters have removed their SCBA, and the primary reason for this is carbon monoxide poisoning. When a downed fire fighter runs out of air, if he is conscious, he will remove the regulator from the mask. To not do so would quickly lead to suffocation. The fire fighter is then exposed to the smoke and gases present. Carbon monoxide is one of the first gases that will affect him, causing disorientation and loss of the ability to make rational decisions and choices. Many times the fire fighter has removed his SCBA thinking it would help. If and when an RIC locates this downed fire fighter, a quick means to rapidly remove him is required. A webbing harness is just such a tool. There are two webbing harnesses that are both easy, rapid, and effective: the webbing shoulder harness and the webbing cinch harness. Each harness will give you the ability to rapidly put the downed fire fighter into the harness and remove them quickly.

To use the webbing shoulder harness, follow the steps as demonstrated in **SKILL DRILL 8-7**:

1. The rescuer pulls or pushes the downed fire fighter to a seated position and takes a position behind him, using his knee to hold the fire fighter upright. (**STEP ❶**)
2. Remove the webbing, release it, place the excess on the ground directly in front of you, and grab the webbing with both hands, palms down, positioned behind the downed fire fighter's neck. (**STEP ❷**)
3. Drape the webbing around the fire fighter's neck (like wrapping a scarf) (**STEP ❸**) and use your right hand to grab both pieces at the chest. (**STEP ❹**)
4. With your left hand, grab the fire fighter's left arm and "pull" it through the webbing. (**STEP ❺**)
5. Transfer hands holding the webbing at the chest. **Caution!** *Never* let go of the webbing—*always* make sure one hand has contact and control.
6. Pull the fire fighter's right arm through the webbing. (**STEP ❻**)
7. Place your right hand above your left hand, which is holding the webbing, and slide it up into the "V" being formed. (**STEP ❼**) This will assist you in zero-visibility conditions.
8. With your right hand in the "V," grab the webbing and slide your hand to behind the fire fighter's neck, keeping the webbing pulled back. (**STEP ❽**)
9. With your left hand, grab the webbing on the ground in front of you and push it up through the loop your right hand is creating. (**STEP ❾**)
10. Pull the webbing snug to complete the harness. (**STEP ❿**)

SKILL DRILL 8-7 The Webbing Shoulder Harness
NFPA 1407, 7.12(1)

1 The rescuer pulls or pushes the downed fire fighter to a seated position and takes a position behind him, using his knee to hold the fire fighter upright.

2 Remove the webbing, release it, place the excess on the ground directly in front of you, and grab the webbing with both hands, palms down, positioned behind the downed fire fighter's neck.

3 Drape the webbing around the fire fighter's neck (like wrapping a scarf).

4 Use your right hand to grab both pieces at the chest.

(Continued)

SKILL DRILL 8-7   The Webbing Shoulder Harness (*Continued*)

5 With your left hand, grab the fire fighter's left arm and "pull" it through the webbing. Transfer hands holding the webbing at the chest.

6 Pull the fire fighter's right arm through the webbing.

7 Place your right hand above your left hand, which is holding the webbing, and slide it up into the "V" being formed.

8 With your right hand in the "V," grab the webbing and slide your hand to behind the fire fighter's neck, keeping the webbing pulled back.

9 With your left hand, grab the webbing on the ground in front of you and push it up through the loop your right hand is creating.

10 Pull the webbing snug to complete the harness.

To use the webbing cinch harness, follow the steps as demonstrated in **SKILL DRILL 8-8** :

1 The rescuer pulls or pushes the downed fire fighter to a seated position and takes a position behind the him, using his knee to hold the fire fighter upright. (**STEP 1**)

2 Remove your webbing, release it, place the excess on the ground directly in front of you, and grabbing a bight of webbing, pass it around the fire fighter's chest. (**STEP 2**)

3 Continue the wrap, and then take the opposite end of the webbing and pull it through. (**STEP 3**)

4 Pull the webbing snug to complete the harness. (**STEP 4**)

Safety

Caution! The cinch harness can cause compression on the chest, making it difficult for the downed fire fighter to breathe. Great caution must be taken if this type of harness is used.

SKILL DRILL 8-8

The Webbing Cinch Harness
NFPA 1407, 7.12(1)

1 The rescuer pulls or pushes the downed fire fighter to a seated position and takes a position behind the him, using his knee to hold the fire fighter upright.

2 Remove your webbing, release it, place the excess on the ground directly in front of you, and grabbing a bight of webbing, pass it around the fire fighter's chest.

3 Continue the wrap, and then take the opposite end of the webbing and pull it through the bight.

4 Pull the webbing snug to complete the harness.

■ Other RIC Webbing Uses

A single rescuer, to gain leverage when trying to move a downed fire fighter, can also use the webbing harnesses. In most cases, average room sizes are not very large. In a room that is 12 × 12 feet (4 × 4 m), you can pull the downed fire fighter to safety using a cinch harness around his or her feet. To do so, make the harness and deploy the webbing, repositioning yourself at the doorway in a seated position with your foot against the door jamb. This position will provide additional leverage to assist in pulling the downed fire fighter out of the room **FIGURE 8-14** . An inexpensive piece of webbing, without any gimmicks or complications, is a tool that is almost unlimited in use.

FIGURE 8-14   The fire fighter holding the webbing in two hands.

Crew Notes

Always remember that rescuing and moving a downed fire fighter needs to done rapidly. Having skill sets that are common, practiced, and proven will allow the RIC to rapidly rescue the fire fighter.

Wrap-Up

Chief Concepts

- Understand the importance of having coordinated and common skill techniques to remove a downed fire fighter.
- Understand the importance of drilling and practicing skills frequently.
- There are three ways you will encounter a downed fire fighter: (1) witnessed as part of the team or located after becoming missing; (2) after he has called a MAYDAY; or (3) it is discovered that he is missing/unaccounted for.
- RIC must be *prepared and ready* to take immediate action to *rapidly* remove the fire fighter to safety and medical care.
- There are four ways to move someone: drag (pull), lift, push, or carry.
- A fire fighter in full PPE has some basic grab points that we utilize in an RIC rescue: the SCBA shoulder strap, the bunker coat collar, and the cuff of the bunker pants.

- If you see your partner go down, move your partner to safety first; once you are clear of the fire room, then declare a MAYDAY. If fire conditions do not warrant an emergency move, declare the MAYDAY first and then move.
- When two or more rescuers are involved in any rescue, coordination and teamwork between the rescuers is mandatory.
- The DRD is designed for use in a straight-line horizontal drag only.
- When removing a downed fire fighter, always face in the direction you are going.

Hot Terms

<u>Basic grab points</u>   Various points on a fire fighter's SCBA and turnout gear that are easily grabbed under adverse conditions in order to move the downed fire fighter.

<u>Drag rescue device (DRD)</u>   A device integrated into the turnout coat that can be utilized to drag a fire fighter.

<u>One-rescuer drag</u>   The technique of one fire fighter moving another fire fighter who is downed.

SCBA harness conversion A technique for turning the straps of a fire fighter's SCBA into a harness for use in dragging the downed fire fighter.

Two-rescuer drag The technique of two fire fighters moving another fire fighter who is downed.

Two-rescuer drag/push The technique of two fire fighters moving a downed fire fighter—one is dragging and while the other is pushing.

Two-rescuer side-by-side drag A technique used to move a downed fire fighter by two fire fighters dragging side-by-side.

Webbing cinch harness A technique used to rapidly make a harness around a downed fire fighter utilizing personal webbing (caution should be used as a cinch harness will squeeze the chest when tensioned is applied).

Webbing shoulder harness A technique used to make a harness from personal webbing around a downed fire fighter that does not compress the chest.

References

International Association of Fire Chiefs, National Fire Protection Association. *Fundamentals of Fire Fighter Skill, Third edition.* Burlington, MA: Jones & Bartlett Publishing; 2014.

National Fire Protection Agency (NFPA) 1407, *Standard for Fire Service Rapid Intervention Crews.* 2015. http://www.nfpa.org/codes-and-standards/document-information-pages?mode=code&code=1407. Accessed February 11, 2014.

National Fire Protection Agency (NFPA) 1971, *Standard on Protective Ensembles for Structural Firefighting and Proximity Firefighting.* 2013. http://www.nfpa.org/codes-and-standards/document-information-pages?mode=code&code=1971. Accessed October 30, 2013.

RAPID INTERVENTION CREW MEMBER
in action

Your company is short a fire fighter and, due to budget cuts, you must operate understaffed. While en route to a structure fire, you are assigned to the second floor to assist with a search. You are a two-man company, as the chauffer of your truck is outside. Upon entering the second floor, you find the downed fire fighter in a room moments from flashover; the space is cramped, and you need to rapidly remove the fire fighter out of the immediate area *now*.

1. How can one fire fighter effectively move a downed fire fighter in a tight area?
   A. Using the one-rescuer drag.
   B. One fire fighter cannot move another fire fighter by him or herself.
   C. With a pre-rigged haul system.
   D. By attaching a rope to the fire fighter and pulling from outside of the structure.

2. Knowing that conditions are deteriorating rapidly and flashover is imminent, you should call for the MAYDAY first, then move the fire fighter out of harm's way.
   A. True
   B. False

3. What points on the downed member's gear could you grab easily to move him?
   A. Collar of the jacket.
   B. Shoulder pads.
   C. Arm cuffs.
   D. Bellows pockets of the pants.

4. When should you not take the time to convert the fire fighter's air pack to a harness?
   A. When moving the fire fighter any significant distance.
   B. If conditions allow the time to.
   C. If you need lift the fire fighter.
   D. If the conditions are too severe to take the time.

5. NFPA requires all turnout gear to be fitted with a DRD in the turn out coat. What is a major drawback of a DRD?
   A. A DRD can only be used for a horizontal pull.
   B. A DRD is typically placed in a spot where the SCBA would make it difficult to deploy.
   C. Using a DRD is difficult with a gloved hand.
   D. All of the above.

RIC Stair Rescue Techniques: Variables, Challenges, and Skills

© Photos.com

© Joe Nedder/Jones & Bartlett Learning

Knowledge Objectives

After studying this chapter you will be able to:

- Understand the variables of stairs and how they can make a rescue more complex. (p 153–155)
- Understand the basics of stair construction. (p 153–155)
- Understand why a rescue over stairs is difficult and demanding. (NFPA 1407, 7.12(2) , p 153–155)
- Understand the advantages and disadvantages of a rescue over stairs. (NFPA 1407, 7.12(2) , p 153–155)
- Understand why a stair rescue must be done rapidly. (NFPA 1407, 7.4(3), 7.4(7), 7.12(2) , p 155)

Skills Objectives

After studying this chapter and with hands-on training, you will be able to perform the following skills:

- Positioning the downed fire fighter for the upward stair rescue. (NFPA 1407, 7.12(2), 7.12(8) , p 155–157)
- Two-fire fighter upward stair rescue. (NFPA 1407, 7.12(2), 7.12(8) , p 157–159)
- Three-fire fighter upward stair rescue. (NFPA 1407, 7.12(2), 7.12(8) , p 159–161)
- Four-fire fighter upward stair rescue. (NFPA 1407, 7.12(2), 7.12(8) , p 162–163)
- Using webbing to assist in the upward stair rescue. (NFPA 1407, 7.12(2) , p 163)
- One-fire fighter downward stair rescue. (NFPA 1407, 7.12(2) , p 164–165)
- Two-fire fighter downward stair rescue using the self-contained breathing apparatus (SCBA) straps. (NFPA 1407, 7.12(2) ,p 165–166)
- Two-fire fighter downward stair rescue using webbing. (NFPA 1407, 7.12(2) , p 166)

As a member of the rapid intervention crew (RIC), your company has answered the MAYDAY, located the unresponsive fire fighter in the basement, and has dragged him to the base of the stairs. The fire in the basement is gaining in size and ferocity. Smoke conditions have made visibility all but zero, and the heat is beginning to become unbearable. You have one chance to rescue the unresponsive fire fighter and get the RIC out to safety.

1. How will you decide which technique will do the job quickly?
2. Why is it important for all members of the RIC to have the set of skills to make a stair rescue?

Introduction

Bringing a downed fire fighter to safety up or down stairs is one of the most challenging predicaments an RIC can face. At first glance, stairs do not seem to present any obstacles. However, there are many variables that the stairs themselves present that can affect the rescue attempt. Variables of stair rescues include:

- The stair <u>tread plate</u> nose
- Open or closed <u>riser</u>
- Open or closed stairway
- <u>Balusters</u>
- Width of the stairs
- Straight run or multiple <u>landings</u> (wraparound staircase)
- Number of hoselines currently in place on the stairs
- Number of companies trying to move up or down the staircase
- Quality of construction of the stairs
- The chance of collapse if the stairs have been impinged or engaged with fire

The size of the downed fire fighter and the physical capabilities of the rescuers are additional variables that can also affect the success of the rescue. Here, we will look at these variables and present skill-based solutions that will, with training and practice, enable your RIC to make the rescue. The techniques and skills include rescues up stairs, namely the two-, three-, four-fire fighter rescues and using webbing for a stair rescue; and rescues down stairs, namely the one- and two-fire fighter rescues.

Rescuing a Downed Fire Fighter Up a Staircase

Rescuing and removing a downed fire fighter up or down a set of stairs is one of the most challenging scenarios an RIC can face. You are working against time trying to rapidly remove the fire fighter to safety and have no choice but to maneuver either up or down stairs. Before you can learn the needed skills to make this rescue coordinated and efficient, you need to understand what can affect your efforts and the rescue.

■ Advantages and Disadvantages of Stair Rescues

Under normal circumstances, we would always use stairs to gain access or egress; however, in a burning building, things change. First, remember that your mission is to rapidly remove the downed fire fighter to safety and medical care if needed. To achieve this, the RIC Officer must evaluate the situation and determine what will be the fastest way out. The answer might be through a first floor window or over a ladder from the second floor or higher, or it might be using the interior staircase. The advantages of the rescue over stairs is:

- It is the most obvious way out.
- Two to four fire fighters might be able to assist in the rescue.
- The downed fire fighter does not have to be carried over a ladder.

The disadvantages are:

- The stairs function as a "chimney," so as you ascend or descend, if the fire is below you, the rescue will be done in severe heat and smoke conditions, with the possibility of fire drawing upon you.
- The stairs might have numerous hoselines running up or down, making it difficult to maneuver.
- Other fire fighters are using the stairs to enter and exit.
- The stairs might have become unstable due to fire.
- The quality of the stairs' construction, especially basement stairs, to support the amount of weight and stress (live load) being placed upon it by an RIC.

To better understand stair rescues, you must understand some basics of stair construction and its components, including the tread, riser, and <u>stringer</u> **FIGURE 9-1**, and how each of the stair components can affect the rescue. The rise or riser is how high each step is and how high you must lift the downed fire fighter. The tread typically has a nose that overhangs the riser. It is this nosing that the self-contained breathing apparatus (SCBA) cylinder valve can get caught on while attempting to lift and cause difficulties for the fire fighters making the rescue. Some stairs have metal nose guards that can tear out and entangle the cylinder stem **FIGURE 9-2**. The handrails or banisters in a residential setting are typically secured with small screws, which can make them prone to pulling out under the stress of an RIC using them as leverage to pull. In order to

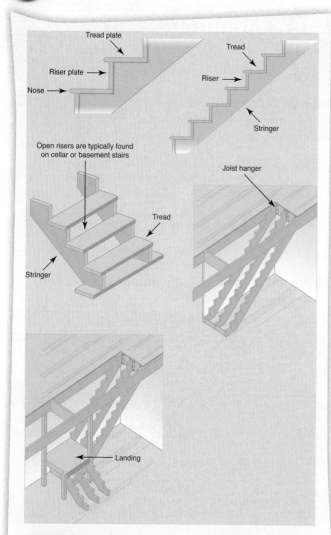

FIGURE 9-1 The various components of a set of stairs: the tread, the riser, and the stringer.

execute a rapid and successful rescue up or down the stairs, the RIC needs to have coordinated skills that overcome the typical obstacles encountered while staying proactive and thinking one or two steps ahead.

Stair rescues can be broken into two groups—up the stairs and down the stairs—and there are numerous techniques and skills that can be used for a rescue over stairs. Which technique you use will be determined by the following:

- How wide are the stairs? Is there enough room for two rescuers to stand side-by-side or is the staircase only 24 inches (61 cm) wide?
- Is it a straight run of stairs or is it a wraparound staircase with multiple landings? Each landing will require a change in direction, and this can and will slow you down **FIGURE 9-3**.
- Are the stairs currently clogged with numerous hoselines? Hoselines present tripping and safety hazards, especially in an environment with poor visibility **FIGURE 9-4**.
- Is this a large operation with numerous companies going up and down the staircase? This can present a clear egress problem. How many times have you or

FIGURE 9-3 Stairs with multiple landings will require a change in direction during a rescue.

FIGURE 9-2 Stairs with metal hose guards that could tear out and entangle the cylinder stem.

© Joe Nedder/Jones & Bartlett Learning

FIGURE 9-4 Hoselines present tripping and safety hazards on stairs.

which uses ropes, carabiners, and/or a pulley to create a 2-to-1 or better mechanical advantage. To utilize a rope hauling system, you will need to either have the equipment with you or call for it to be brought in. In addition, the hauling system will have to be set up, anchored, rigged, and deployed, all of which takes time and manpower. Finally, when a hauling system is used, the downed fire fighter is "dragged" up the stairs. If he is on his back, the SCBA cylinder will continuously catch on the stair nose and tread, slowing the rescue down. If he is dragged on his chest his SCBA regulator, flashlight, radio, and anything else attached to his body will also catch, hindering the rescue. Remember, RIC needs to be *rapid*. Time is against you and the downed fire fighter. Keep things simple and move the downed fire fighter out as quickly as possible to safety and medical care.

Rescues Up the Stairs

There are two scenarios in which you would rescue a fire fighter by carrying him up the stairs. In the first you are below grade, in a basement, and going up is the only way out. The second reason is not as obvious. Suppose you are on the second floor and the staircase is unstable, burned through, or blocked by numerous lines or companies—or if the floor you are on has no window openings. Your solution might be to go up a flight of stairs, if available, and call for a ladder carry rescue (ladder carry techniques and rescues are discussed in detail in the chapter *Rescuing Fire Fighters Through Windows, Over Ladders, and From Restrictive, Limited Space*) from that floor. Regardless of the reason or situation, carrying a fire fighter up a set of stairs is difficult and challenging. In order to be able to perform this rescue, practice of dedicated skills is a necessity.

> **Safety**
>
> Always use the safest and fastest way out. Sometimes the obvious way out is not the fastest. A great RIC is always anticipating and staying proactive!

Every rescue over stairs is broken into three parts:
1. Getting the downed fire fighter to the stairs
2. Positioning the downed fire fighter for the stair rescue
3. Moving the downed fire fighter up or down the stairs

The techniques for moving a downed fire fighter have been discussed elsewhere, in the chapter *Rapidly Moving a Downed Fire Fighter*, so the skill sets discussed here will begin with the positioning of the downed fire fighter.

■ Initial Positioning of the Downed Fire Fighter

Once the RIC has moved the downed fire fighter to the stairs, we must position him or her onto the first few stair tread plates to execute the rescue maneuver. As you drag the fire fighter towards the stairs, get as close as possible. Once the downed fire fighter is at the staircase, position him in a seated position and have two rescuers move him to the base of the stairs, then lift him upward two or three treads. This will place the downed fire fighter in the optimum position for going up the stairs rapidly. It

your company tried to navigate a stairway only to be blocked by other companies? No matter how much you yell "Make a hole," the message and action needed seems to take forever.

- Are the stairs still structurally strong or have they been compromised by fire? If they are compromised, you will need to find another way out.
- Are you exiting from a lower level, below grade, with limited access, under extreme smoke and fire conditions?

■ Keeping the Rescue Rapid

Since the 1990s, rapid intervention has received the widespread attention it deserves, and many ideas have been introduced. Some ideas have worked, but others only complicate or prolong the rescue. The first of these was trying to use a Stokes basket for a stair rescue. The use of a Stokes is not suggested or recommended. A great use of a Stokes is to carry the teams tools on the fireground, but it has no place within the structure for use as a dragging device or a tool transport. The basket is large, bulky, and lacks any mobile features needed to work in tight and close quarters. If anything, it will complicate and drastically slow down the removal and rescue of the downed fire fighter.

Fire fighters have designed simple mechanical advantage rope systems to use to haul the downed fire fighter up the stairs. Typically, it is a 2-to-1 mechanical advantage system, is practically impossible for a single rescuer to attempt to place or carry a downed fire fighter up a set of stairs alone.

To position the downed fire fighter for the upward stair rescue, follow the steps as demonstrated in **SKILL DRILL 9-1**:

1. Using the grab techniques taught, move the downed fire fighter as close to the base of the stairs as possible. (**STEP ❶**)

2. The rescue fire fighter at the downed fire fighter's feet will grab the SCBA shoulder straps of the downed fire fighter, pull him into a seated position, and grab him under the knees while the rescue fire fighter at the head assists with the shoulder strap. (**STEP ❷**)

3. The rescue fire fighter at the head of the downed fire fighter will now place his hands under the downed fire fighter's SCBA straps at the shoulder as close to the SCBA frame as possible. (**STEP ❸**)

4. With the rescue fire fighter at the head position calling out the preparatory command "To the stairs," and then using the "Ready, Ready, Go" sequence, both rescue fire fighters will lift the downed fire fighter and move him to the very base of the stairs. (**STEP ❹**)

5. Maintaining their grab positions and with the preparatory command of "On the stairs," followed by the "Ready, Ready, Go" sequence, the two rescue fire fighters lift the downed fire fighter up and place him on the first or second tread of the stairs. If the downed fire fighter is large or if the RIC is getting fatigued, they could lift him one step at a time. (**STEP ❺**) **Note:** If the rescue fire fighter at the feet is tall, it is advised to then place the downed fire fighter onto the third tread using the same system.

Crew Notes

Most SCBA shoulder straps use the parachute-type buckle **FIGURE 9-5**. These types of buckles have the tendency to loosen when you pull on the shoulder straps during an RIC rescue. It has been suggested that if you tighten the straps and then tie an overhand knot at the fastener, you will avoid this. However, keeping the word *rapid* in mind, how long will it take you and your partner in poor conditions to find and tie both straps into an overhand knot? What if you have to untie it? By grabbing the shoulder straps near the SCBA back frame you will limit, but not totally prevent, the unavoidable extension of the shoulder straps without having to tie the straps.

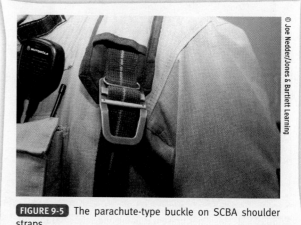

© Joe Nedder/Jones & Bartlett Learning

FIGURE 9-5 The parachute-type buckle on SCBA shoulder straps.

SKILL DRILL 9-1

Positioning the Downed Fire Fighter for the Upward Stair Rescue
NFPA 1407, 7.12(2), 7.12(8)

1. Using the grab techniques taught, move the downed fire fighter as close to the base of the stairs as possible.

2. The rescue fire fighter at the downed fire fighter's feet will grab the SCBA shoulder straps of the downed fire fighter, pull him into a seated position, and grab him under the knees while the rescue fire fighter at the head assists with the shoulder strap.

(Continued)

SKILL DRILL 9-1   Positioning the Downed Fire Fighter for the Upward Stair Rescue (*Continued*)

3 The rescue fire fighter at the head of the downed fire fighter will now place his hands under the downed fire fighter's SCBA straps at the shoulder as close to the SCBA frame as possible.

4 With the rescue fire fighter at the head position calling out the preparatory command "To the stairs," and then using the "Ready, Ready, Go!" sequence, both rescue fire fighters will lift the downed fire fighter and move him to the very base of the stairs.

5 Maintaining their grab positions, the two rescue fire fighters lift the downed fire fighter up and place him on the first or second tread of the stairs. If the downed fire fighter is large or if the RIC members are getting fatigued, they could lift him one step at a time.

© Joe Nedder/Jones & Bartlett Learning

■ Two-Rescuer Upward Stair Rescue

The two-rescuer upward stair rescue technique works best if you are faced with a narrow staircase where two fire fighters have difficulty standing side-by-side.

With the downed fire fighter positioned on the steps, perform the two-rescuer upward stair rescue as demonstrated in **SKILL DRILL 9-2**:

1 The rescue fire fighter at the downed fire fighter's head positions himself directly behind the downed fire fighter and, getting into a lifting position with knees bent, grabs the downed fire fighter's SCBA shoulder straps as close to the frame as possible. (**STEP 1**) Remember that as you lift by the straps, they have a tendency to loosen.

2 The rescue fire fighter at the downed fire fighter's feet squats down and positions himself in-between the legs of the downed fire fighter, placing the legs over his shoulders with the knee joints resting on his shoulders. (**STEP 2**)

3 With both rescuers in position, the fire fighter at the head will shout the preparatory command "Up the stairs," and then using the "Ready, Ready, Go" sequence, the rescue fire fighter at the head will lift directly straight up, and the rescue fire fighter at the feet will, from a weight lifter's squat, lift straight up. *Then* both fire fighters will start to climb the stairs. (**STEP ③**)

④ Depending on the size of the downed fire fighter and the physical capabilities of the RIC fire fighters, you might only be able to lift and move one step at a time, in which case you should repeat step 3 until the downed fire fighter has reached safety. Alternately, you might be able to move right up the stairs. If you can keep going, the head fire fighter should keep calling out "GO! GO!"

Crew Notes

It is very important that the downed fire fighter's legs are positioned with the knee joint resting on the rescue fire fighter's shoulders, as this position provides the needed lifting point for the rescue. If the downed fire fighter's lower leg or calf is resting on the rescue fire fighter's shoulders, the rescue fire fighter will have no leverage for the lift, allowing the weight of the downed fire fighter to fall forward, causing the other rescue fire fighter to lift almost the entire weight of the downed fire fighter on his own.

Crew Notes

It is imperative that the first actions of both rescuers be to lift straight up at the same time. Your tendency will be to lift and step forward at the same time. If the rescuer at the downed fire fighter's feet does this, he will force the rescuer at the downed fire fighter's head backwards, with a high probability that the downed fire fighter's cylinder stem will get hung up on the nose of the tread plate (**FIGURE 9-6**). Remember, lift straight up and *then* step forewards.

© Joe Nedder/Jones & Bartlett Learning

FIGURE 9-6 The downed fire fighter's cylinder stem getting caught on the nose of the tread plate.

SKILL DRILL 9-2   Two-Rescuer Upward Stair Rescue
NFPA 1407, 7.12(2), 7.12(8)

① The rescue fire fighter at the downed fire fighter's head positions himself directly behind the downed fire fighter and, getting into a lifting position with knees bent, grabs the downed fire fighter's SCBA shoulder straps as close to the frame as possible.

② The rescue fire fighter at the downed fire fighter's feet squats down and positions himself in-between the legs of the downed fire fighter, placing the legs over his shoulders with the knee joints resting on his shoulders.

(Continued)

SKILL DRILL 9-2   Two-Rescuer Upward Stair Rescue (*Continued*)

3 With both rescuers in position, the fire fighter at the head will shout the preparatory command "Up the stairs," and then using the "Ready, Ready, Go!" sequence, the rescue fire fighter at the head will lift directly straight up, and the rescue fire fighter at the feet will, from a weight lifter's squat, lift straight up. *Then* both fire fighters will start to climb the stairs.

© Joe Nedder/Jones & Bartlett Learning

As this is a physically difficult and demanding rescue, if the RIC is unable to move the fire fighter rapidly to safety, request additional help or a fresh crew. Do not wait to call for help! It has been proven that once RIC fire fighters fatigue and start having difficulties with the rescue, no matter how hard they try, they will be unable to complete the rescue. Calling for help even before you need it could make the difference in the speed and success of the rescue.

The rescuer lifting the downed fire fighter by his shoulder straps can use an alternative technique for lifting. The alternative shoulder straps lifting technique involves loosening the downed fire fighter's shoulder straps completely, passing them under his arms, and then using the loops created to lift. This technique typically will not work when trying to rescue a larger fire fighter, because the shoulder straps are not long enough. The advantage of this technique, however, is that when the straps are completely loosened, they will not expand further when lifting. To perform the alternative shoulder straps lifting technique follow the steps in **SKILL DRILL 9-3**:

1 With the downed fire fighter in a seated position, the rescuer loosens both shoulder straps completely. The second rescuer can assist if needed. (**STEP 1**)

2 Once the straps are loosened, the rescuer will pass both the shoulder straps under the arms of the downed fire fighter. (**STEP 2**)

3 The rescuer now grabs onto the shoulder straps creating a handle for lifting. (**STEP 3**)

■ Three-Rescuer Upward Stair Rescue

The three-rescuer upward stair rescue utilizes two fire fighters at the head position and one at the legs. The advantage of this rescue is that, with two fire fighters lifting from the top, it will make the lift easier and should also make the rescue faster. If the stairs are narrow, the two fire fighters at the head will have to face each other, but if the stairs are wide enough, the two fire fighters will get more leverage standing side-by-side. If the downed fire fighter is large or if the RIC is becoming physically exhausted, the three fire fighter upward rescue should be attempted. Remember, you have one chance: if you cannot execute the rescue, get more help right away.

With the downed fire fighter positioned on the steps, perform the three-rescuer upward stair rescue as demonstrated in **SKILL DRILL 9-4**:

1 The two rescue fire fighters at the downed fire fighter's head will position themselves directly behind the downed fire fighter and, getting into a lifting position with knees bent, grab the downed fire fighter's SCBA shoulder straps as close to the SCBA frame as possible. (**STEP 1**)

SKILL DRILL 9-3   Alternative Shoulder Straps Lifting Technique
NFPA 1407, 7.12(2)

1 With the downed fire fighter in a seated position, the rescuer loosens both shoulder straps completely. The second rescuer can assist if needed.

2 Once the straps are loosened, the rescuer will pass both the shoulder straps under the arms of the downed fire fighter.

3 The rescuer now grabs onto the shoulder straps creating a handle for lifting.

© Joe Nedder/Jones & Bartlett Learning

2 The rescue fire fighter at the downed fire fighter's feet squats down and positions himself in-between the legs of the downed fire fighter, placing the downed fire fighter's legs over his shoulders with the knee joints resting on his shoulders (as in the two-rescuer upward stair rescue). (**STEP 2**)

3 With all the rescuers in position, the lead fire fighter or officer at the head will shout the preparatory command "Up the stairs," and then using the "Ready, Ready, Go" sequence, the rescue fire fighters at the head position will lift directly straight up, and the lower fire fighter will, from a weight lifter's squat, lift straight up. Then, all three rescue fire fighters will start to climb the stairs. (**STEP 3**)

4 Depending on the size of the downed fire fighter and the physical capabilities of the RIC, you might only be able to lift and move the downed fire fighter one or two steps at a time, in which case you should continue to repeat step 3 until the downed fire fighter has reached safety. Alternately, you might be able to move right up the stairs. If you can keep going, the head fire fighter should keep calling out "GO! GO!"

SKILL DRILL 9-4 Three-Rescuer Upward Stair Rescue
NFPA 1407, 7.12(2), 7.12(8)

1 The two rescue fire fighters at the downed fire fighter's head will position themselves directly behind the downed fire fighter and, getting into a lifting position with knees bent, grab the downed fire fighter's SCBA shoulder straps as close to the SCBA frame as possible.

2 The rescue fire fighter at the downed fire fighter's feet squats down and positions himself in-between the legs of the downed fire fighter, placing the downed fire fighter's legs over his shoulders with the knee joints resting on his shoulders (as in the two-rescuer upward stair rescue).

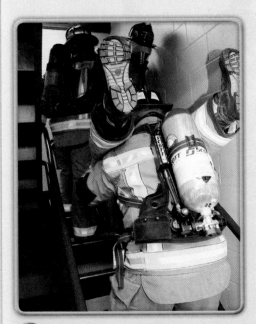

3 With all the rescuers in position, the lead fire fighter or officer at the head will shout the preparatory command "Up the stairs," and then using the "Ready, Ready, Go" sequence, the rescue fire fighters at the head position will lift directly straight up, and the lower fire fighter will, from a weight lifter's squat, lift straight up. Then, all three rescue fire fighters will start to climb the stairs.

Utilizing Webbing for an Upward Stair Rescue

As with the rescue drag skills, the 1-inch (25-mm) tubular webbing can be utilized to assist in an upward stair rescue. Reasons for using the webbing include:

- The downed fire fighter was found without a SCBA on.
- The SCBA was removed due to entanglement or another issue that would affect the rapid rescue.
- The downed fire fighter is large, or the crew does not have the physical capabilities to lift him with standard RIC techniques.
- Three or four RIC fire fighters are performing the rescue.

Safety

Caution! The drag rescue device (DRD) was not designed for or intended to be a lifting device. Moving a fire fighter up the stairs requires lifting. Do not waste time—use your webbing.

■ Four-Rescuer Upward Stair Rescue With Webbing

The four-rescuer upward stair rescue is used when the staircase is wide and strong enough to permit the technique, the manpower

is available to rapidly execute the maneuver, or the downed fire fighter is very heavy and no other option is available.

The four-rescuer upward stair rescue with webbing technique utilizes two fire fighters at the head position and two fire fighters at the leg position. With the downed fire fighter positioned on the steps, perform the following steps as demonstrated in **SKILL DRILL 9-5**:

1. The two fire fighters at the downed fire fighter's head position themselves directly behind the downed fire fighter in a lifting position with knees bent and grab the SCBA shoulder straps as close to the SCBA frame as possible. (**STEP 1**)

2. The two fire fighters at the legs will use their webbing to wrap both legs together and create a "handle" on each side of the legs. They will then take a position opposite each other on the outside of the downed fire fighter's legs. Utilizing these handles, the two rescue fire fighters at the legs can each grab one end of the webbing and wrap it around the palm of their hands like a "rodeo grip" and use it to lift the downed fire fighter when the command is given, or they can grab the handles created and lift. (**STEP 2**)

3. With all rescuers in position, the lead fire fighter or officer at the head will shout the preparatory command "Up the stairs" and then use the "Ready, Ready, Go" commands. The fire fighters at the head will lift directly

SKILL DRILL 9-5   Four-Rescuer Upward Stair Rescue with Webbing
NFPA 1407, 7.12(2), 7.12(8)

1 The two fire fighters at the downed fire fighter's head position themselves directly behind the downed fire fighter in a lifting position with knees bent and grab the SCBA shoulder straps as close to the SCBA frame as possible.

2 The two fire fighters at the legs will use their webbing to wrap both legs together and create a "handle" on each side of the legs. Utilizing these handles, the two rescue fire fighters at the legs can each grab one end of the webbing and wrap it around the palm of their hands like a "rodeo grip" and use it to lift the downed fire fighter when the command is given, or they can grab the handles created and lift.

(Continued)

SKILL DRILL 9-5 Four-Rescuer Upward Stair Rescue with Webbing (*Continued*)

3 With all rescuers in position, the lead fire fighter or officer at the head will shout the preparatory command "Up the stairs" and then use the "Ready, Ready, Go" commands. The fire fighters at the head will lift directly straight up, the two fire fighters at the legs will both lift using the webbing handle straight up, and the team will carry the downed fire fighter up the stairs.

© Joe Nedder/Jones & Bartlett Learning

straight up, the two fire fighters at the legs will both lift using the webbing handle straight up, and the team will carry the downed fire fighter up the stairs. (**STEP 3**)

4 Depending on the size of the downed fire fighter and the physical capabilities of the RIC fire fighters you might only be able to lift and move the downed fire fighter one or two steps at a time, in which case you should continue to repeat step 3 until the downed fire fighter has reached safety. Alternately, you might be able to move right up the stairs. If you can keep going, the head fire fighter should keep calling out "GO! GO!"

■ Using Webbing to Assist the RIC Fire Fighter at the Head

Webbing can be used to create a webbing harness around the downed fire fighter's chest. There might be a need for this if the downed fire fighter removed his SCBA prior to

rescue or if the RIC removed the downed fire fighter's SCBA for extreme reasons. With this type of harness, the fire fighter at the head can place the running end of the webbing over his shoulder and, from a squatting position, "lift" the downed fire fighter up and forward. The webbing can also be wrapped around the downed fire fighter's chest two or three times—creating a harness with two handles, one on either side of the downed fire fighter for the two rescuers at the head to use as a lift handle.

Safety

Caution: As with any harness that wraps around a fire fighter's chest, you must be cautious that you are not compressing the fire fighter's chest to the point it is causing respiratory distress. A better choice might be the webbing shoulder harness.

■ Using a Tool to Assist in the Upward Stair Rescue

It has been suggested that if there are two rescue fire fighters at the head position and they are having difficulty with the SCBA shoulder straps, they can use a tool placed under the straps to create a handle for lifting. Typically, if you are having difficulty with the shoulder straps, the use of the tool might not solve the problem and will only complicate matters. In fact, you will experience the same loosening of the straps if you are using a tool, and when using a tool placed under the shoulder straps to "assist" with the lift, the two rescuers will have a tendency to lift up to different heights. This will cause the tool to be on a angle, and the shoulder straps will slip and slide to the lower height, putting more weight and burden on that rescuer. Instead of a tool, try grabbing the shoulder straps closer to the frame first, and if that does not work, a better solution than a tool would be to utilize webbing.

Rescuing a Fire Fighter Down a Staircase

There are three rescue techniques to bring a fire fighter down a set of stairs to safety that will be discussed here. They are all relatively easy and fast, and the variables for this type of rescue depend on whether the downed fire fighter still has his SCBA harness on and if the stairs are wide enough for two fire fighters to walk downward side-by-side. The three descending stair rescue techniques are:

- One-fire fighter rescue using the SCBA straps
- Two-fire fighter rescue using the SCBA straps
- Two-fire fighter rescue using webbing

FIGURE 9-7 A single rescuer positioning the downed fire fighter for a downward stair rescue.

© Joe Nedder/Jones & Bartlett Learning

When rescuing a fire fighter down a set of stairs, the rescuers first need to get the downed fire fighter into position. This can be accomplished easily by dragging him or her to the top of the stairs (landing) headfirst. Once there, position the downed fire fighter with his or her shoulders at the edge of the stairs. This position will allow a fast and safe rescue **FIGURE 9-7**.

With the downed fire fighter positioned properly, a single fire fighter can execute the one-rescuer downward stair rescue by performing the following steps as demonstrated in **SKILL DRILL 9-6**:

1. Place one arm under the downed fire fighter's neck to support it, and grab the SCBA shoulder strap with your hand palm up. (**STEP 1**)

SKILL DRILL 9-6 One-Rescuer Downward Stair Rescue
NFPA 1407, 7.12(2)

1. Place one arm under the downed fire fighter's neck to support it, and grab the SCBA shoulder strap with your hand palm up.

2. With your other arm, reach across the downed fire fighter's body and grab the SCBA waist strap.

(Continued)

SKILL DRILL 9-6 One-Rescuer Downward Stair Rescue (*Continued*)

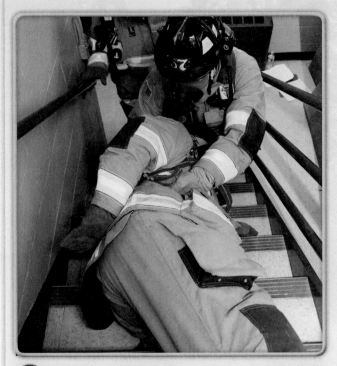

3 Pull the downed fire fighter towards you and begin to descend the stairway *one step at a time*. As you descend, be cautious and check to make sure the next tread plate is there and safe!

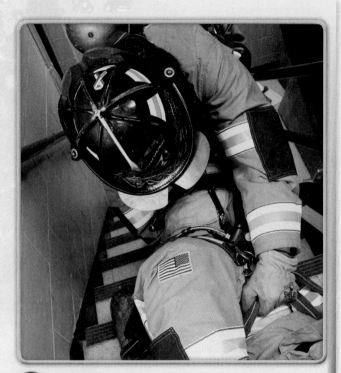

4 Drag the downed fire fighter downward using your body—you can lean into the fire fighter as a "brake" to control the descent. Continue down the stairs.

© Joe Nedder/Jones & Bartlett Learning

2 With your other arm, reach across the downed fire fighter's body and grab the SCBA waist strap. (**STEP 2**)

3 Pull the downed fire fighter towards you and begin to descend the stairway *one step at a time*. As you descend, be cautious and check to make sure the next tread plate is there and safe! (**STEP 3**)

4 Drag the downed fire fighter downward using your body—you can lean into the fire fighter as a "brake" to control the descent. (**STEP 4**)

5 Continue down the stairs.

An advantage of this technique is that the rescuer is also in a position to provide protection to the downed fire fighter from falling debris or spalling, simply by leaning over the body and providing a shield **FIGURE 9-8**.

The two-fire fighter downward stair rescue using the SCBA straps can be accomplished if the stairs are wide enough to allow the two rescue fire fighters to stand side-by-side. To perform this rescue, follow the steps as demonstrated in **SKILL DRILL 9-7**:

1 Position the downed fire fighter at the top of the stairs as described above.

© Joe Nedder/Jones & Bartlett Learning

FIGURE 9-8 A single rescuer can use his body as a shield to protect the downed fire fighter.

2 The two rescue fire fighters should position themselves to face forward as they descend the stairs, each grabbing one of the downed fire fighter's SCBA

SKILL DRILL 9-7

Two-Fire Fighter Downward Stair Rescue Using the SCBA Straps
NFPA 1407, 7.12(2)

1 Position the downed fire fighter at the top of the stairs as described in below. The two rescue fire fighters should position themselves to face forward as they descend the stairs, each grabbing one of the downed fire fighter's SCBA shoulder straps with their inward hand. Lift the downed fire fighter by his SCBA shoulder straps into a seated or semi-seated position.

2 Using the "Ready, Ready, Go!" commands, the two rescuers will walk down the stairs, dragging the downed fire fighter with them.

© Joe Nedder/Jones & Bartlett Learning

shoulder straps with their inward hand. (**Note:** make sure the SCBA shoulder straps are tight before beginning downward.) Lift the downed fire fighter by his SCBA shoulder straps into a seated or semi-seated position. (**STEP 1**)

3 Using the "Ready, Ready, Go!" commands, the two rescuers will walk down the stairs, dragging the downed fire fighter with them. (**STEP 2**)

A variation of the previous skill would be to use webbing if the downed fire fighter was without his SCBA. The skill steps for the two-fire fighter downward stair rescue using webbing are identical as the two-fire fighter downward stair rescue using the SCBA straps, except a harness is created for a downward stair rescue. To perform the two-fire fighter downward stair rescue using webbing, follow the steps as explained in **SKILL DRILL 9-8** [NFPA 1407, 7.12(2)]:

1 Raise the downed fire fighter to a seated position.

2 Wrap your webbing two or three times around the downed fire fighter's chest, under the arms, making sure to create a handle on one side. The additional webbing can be wrapped around the

second fire fighter's hand (rodeo wrap) to deal with the excess and create a second handle.

3 Follow the skill steps as used in the two-fire fighter downward stair rescue utilizing the webbing in lieu of the SCBA straps.

Crew Notes

The DRD is not intended for this technique, because this is not a horizontal drag.

The rescue of a downed fire fighter is a difficult and strenuous job. The skills outlined here are coordinated, and they work; however, in dealing with whatever situation presents itself, the RIC must be proactive and have the ability to adapt some other skill or try something different in order to get the fire fighter out quickly.

Wrap-Up

Chief Concepts

- Rescuing and removing a downed fire fighter up or down a set of stairs is one of the most challenging scenarios an RIC can face.
- There are many variables that must be considered when deciding to use stairs to rescue a fire fighter.
- The variables of the RIC must also be considered when deciding to use stairs.
- A stairway is the most obvious way out of a below-grade rescue.
- A stairway can act as a chimney and can quickly become untenable for rescuers.
- A wooden stairway can burn.

Hot Terms

<u>Baluster</u> An upright support piece used to hold the railing of an open stairway.

<u>Landing</u> An area where one stairway ends and another begins.

<u>Riser</u> The near-vertical element in a set of stairs, forming the space between steps.

<u>Stringer</u> The part of the stairway used to support the steps and risers.

<u>Tread plate</u> The part of a stairway that is walked on.

Reference

National Fire Protection Agency (NFPA) 1407, *Standard for Fire Service Rapid Intervention Crews*. 2015. http://www.nfpa .org/codes-and-standards/document-information-pages? mode=code&code=1407. Accessed February 11, 2014.

RAPID INTERVENTION CREW MEMBER
in action

Your company has been activated to rescue a downed fire fighter in the basement of a residential home. There is fire in the basement and it is not fully controlled by the remaining attack crew. It appears that the stairs are the only feasible way to remove the downed member.

1. What are some advantages of rescuing over stairs?
   - **A.** It is the most obvious way out.
   - **B.** Two to four fire fighters might be able to assist in the rescue.
   - **C.** The downed fire fighter does not have to be carried over a ladder.
   - **D.** All of these are advantages.

2. What are some disadvantages of rescuing over stairs?
   - **A.** The stairs function as a chimney. As you descend, if the fire is below you, the rescue will be done in severe heat and smoke conditions.
   - **B.** The stairs might have many hoselines running up or down.
   - **C.** The stairs may have become unstable due to fire.
   - **D.** All of these are disadvantages.

3. A stairway is usually the most obvious way out of a below grade or basement area.

   - **A.** True
   - **B.** False

4. What are the limitations of using a mechanical advantage system to assist in bringing the fire fighter up a set of stairs?
   - **A.** A mechanical advantage system is difficult to work with in low visibility conditions.
   - **B.** It takes extra manpower to rig a mechanical advantage system.
   - **C.** It takes more time to set up.
   - **D.** All of these are limitations.

5. Stairs should *never* be used when rescuing a downed fire fighter if there is fire in a basement or below grade.
   - **A.** True
   - **B.** False

CHAPTER

10

Rescuing Fire Fighters Through Windows, Over Ladders, and From Restrictive, Limited Space

Knowledge Objectives

After studying this chapter you will be able to:

- Understand the importance of having the skills to make a ladder rescue. (NFPA 1407, 7.12(4), p 169)
- Explain what three basic steps are involved in a window rescue. (NFPA 1407, 7.12(5), p 169)
- Understand the interior and exterior rapid intervention crew (RIC) variables of a window rescue. (NFPA 1407, 7.12(5), p 169)
- Understand why good communication is critical in a window rescue. (NFPA 1407, 7.12(5), p 169–170)
- Realize that window rescues still need to be rapid and know what skills will help you to keep them rapid. (NFPA 1407, 7.4(3), 7.12(5), p 169–170)
- Know the variables to assess when a downed fire fighter is wearing a self-contained breathing apparatus (SCBA). (p 170)
- Know the variables to assess when the downed fire fighter is not wearing a SCBA. (p 176)
- Understand that the fire fighter playing the role of the downed fire fighter from an elevated emergency egress should be secured by a belay line that complies with NFPA 1983. (NFPA 1407, 7.10.1, 7.10.2(1), p 185)

Skills Objectives

After studying this chapter and with hands-on training, you will be able to perform the following skills:

- The two-rescuer feet-first window lift for a downed fire fighter wearing a SCBA. (NFPA 1407, 7.12(4), 7.12(5), p 170–172)
- The three-rescuer feet-first window lift for a downed fire fighter wearing a SCBA. (NFPA 1407, 7.12(4), 7.12(5), p 172–174)
- The two-rescuer head-first window lift for a downed fire fighter wearing a SCBA. (NFPA 1407, 7.12(4), 7.12(5), p 174–176)
- The removal of a SCBA from an unresponsive fire fighter. (p 177–178)
- The two-rescuer Chicago Lift window lift for a fire fighter without a SCBA. (NFPA 1407, 7.12(5), p 178–180)
- The two-rescuer Chicago Lift window lift for a fire fighter without a SCBA utilizing webbing. (NFPA 1407, 7.12(5), p 180–183)
- The three-rescuer window lift for a fire fighter without a SCBA. (NFPA 1407, 7.12(5), p 183–185)
- The one interior fire fighter window rescue using a long board and webbing. (NFPA 1407, 7.12(3), p 185–187)
- The proper positioning and stabilizing of an RIC rescue ladder. (NFPA 1407, 7.12(4), p 188–189)
- The one-rescuer cradle carry when the downed fire fighter is presented feet first. (NFPA 1407, 7.12(5), p 190–192)
- The one-rescuer cradle carry when the downed fire fighter is presented head first. (NFPA 1407, 7.12(5), p 192–193)
- The one-rescuer Chicago carry for a downed fire fighter who is not wearing a SCBA. (NFPA 1407, 7.12(5), p 194–195)
- The two-ladders–two-rescuers cradle carry rescue. (NFPA 1407, 7.12(5), p 194, 196–197)
- The Denver drill technique. (NFPA 1407, 7.12(9), p 198–203)

Additional NFPA Standards

- NFPA 1983, *Standard on Life Safety Rope and Equipment for Emergency Services*

Your RIC is moving to safety as fast as possible with the downed fire fighter you have located. You encounter a partial floor collapse that prevents you from getting to the interior stairs; your only option is to exit via a window.

1. What technique you will use and how will you make the decision based upon the facts before you?
2. What information needs to be transmitted to the RIC Operations Chief or to the Incident Commander?
3. What additional resources will be needed to assist your company?

Introduction

Removing and rescuing fire fighters through windows and from a confined, restrictive space are key skills needed in rapid intervention. The line-of-duty death (LODD) of Mark Langvart of the Denver Colorado Fire Department was such an incident that is widely referred to and will be discussed in more detail in this chapter. The RIC's job is to get the downed fire fighter to safety and medical care as quickly as possible, and often the fastest, and in some cases the easiest, way out is through a window. Various window lifts and ladder carries will provide the options needed for the fastest and most doable rescue. Here, basic window rescues and ladder carries will be discussed, followed by rescues from restrictive and limited space, also known as the Denver drill.

Rescuing a Downed Fire Fighter Through a Window

Besides the interior stairs, the use of a window to rescue a downed fire fighter is the most common. Unfortunately, if fire fighters were polled regarding their abilities or training when it comes to window rescues and ladder carries, their lack of skill and practice devoted to this type of rescue is disturbing, especially since the window rescue is also a skill needed to save civilians. How two or three rescuers coordinate and lift a downed fire fighter out a window makes a difference and is dependent on whether or not the downed fire fighter is wearing a self-contained breathing apparatus (SCBA).

A window rescue involves three basic steps:
- Lifting the downed fire fighter to the window sill.
- Transferring the downed fire fighter to a ladder rescuer.
- Carrying the downed fire fighter down a ladder.

These three steps relate to numerous skills that are needed to execute the rescue safely and rapidly, which are divided into two categories: (1) the interior RIC lifting the downed fire fighter to the window, and (2) the exterior or second RIC responsible for the ladder rescue. Window rescues are dangerous and require physical ability and practiced skills to be successful.

Why Window Rescues Are Important

When thinking of rescuing a downed fire fighter, most will think of exiting through the front door; however, when it comes to fire fighter rescue and survival, you cannot limit yourselves to your skills' comfort zone. In many instances, being prepared to deal with the given situation will mean using a window as the fastest and safest way out for the rescue.

Variables of Window Rescues

Rescuing a downed fire fighter through a window has numerous variables that will affect the rescue, the lifting technique used, and the ladder carry. These variables are important and must be reviewed first. The window rescue variables are broken in to two categories: (1) the interior (lift) variables and (2) the exterior (carry) variables. Window rescue interior RIC variables include:
- Fire fighter with a SCBA
- Fire fighter without a SCBA
- Fire fighter connected to an RIC air supply
- Size of the fire fighter
- Physical abilities of the crew
- Size of the window
- Height to the sill
- Fire and smoke conditions

Window rescue exterior (ladder rescue crew) variables include:
- Fire fighter with a SCBA
- Fire fighter without a SCBA
- Fire fighter connected to an RIC air supply
- Size of the fire fighter
- Positioning of the ladder
- Ability of the ladder rescuer to perform a ladder carry.
- Physical abilities of the rescuer

Window Lifts of a Downed Fire Fighter

Good Communications in Window Rescue

As the RIC, you have exerted a lot of energy finding and dragging the downed fire fighter to a window. The rescue is just moments away if you can lift the fire fighter to the window rapidly and place him out the window to the rescuer on the ladder. However, for a window rescue to be successful it takes *coordinated efforts* and *good communications* among all the rescuers involved. RIC Operations first needs to be informed that the rescue will be out a window. The RIC Officer needs to identify the window to be used or at least a geographic location so a ladder can be positioned. This gives a target to start a second RIC with a ladder. When you enter the room and select the window to use, break it out, clean it out, and let the outside ladder crew know. This communication can be through the Incident Commander (IC), the RIC Operations Chief, by calling the second RIC directly on the radio, or by signaling the crew outside by using your flashlight or by waving your arms when visible contact has been established. Most times the broken glass and cleaned out window will be the indication as to which window you have chosen.

Lifting a Fire Fighter With a SCBA Through a Window

The first requirement of any window rescue is to get the downed fire fighter to the window using the drags presented in the chapter *RIC Skills: Rapidly Moving a Downed Fire Fighter*. Once at the window, the RIC needs to decide if the downed fire fighter is going out the window head first or feet first. If it is feet first, reposition the fire fighter with his feet towards the window using the skills taught in the chapter *RIC Skills: Rapidly Moving a Downed Fire Fighter*. The most common rescue involves a downed fire fighter wearing a SCBA, which, although most common, requires well-coordinated skills and physical abilities. The ladder rescuer will most often receive the downed fire fighter in the cradle carry position. The different techniques for ladder carries that support the window rescues presented here will also be discussed. To begin a window rescue with the fire fighter "on air," the following variables need to be assessed:

- Is the downed fire fighter "on air" but tethered to a RIC air pack? If so, you need to consider how you will manage this second piece of equipment going out the window and down the ladder with the rescuer. How much extra time is it going to take to get the fire fighter and the RIC air pack out the window, keeping in mind that you will also need to assist in getting the air pack down to the ground without interfering with the ladder rescue? Remember that the unit weighs between 30 and 50 pounds (14 and 23 kg) and is tethered or tied to the downed fire fighter **FIGURE 10-1**.
- Is the tether with the low-pressure line and <u>mask-mounted regulator (MMR)</u> or is the universal air connection (UAC) line connected? If it is the UAC, disconnect it! Remember, the UAC is not meant to attach and leave on—use it and disconnect it. If an MMR is connected to an RIC air supply, you will either have to assist with lowering the RIC air supply down or disconnect the MMR that is tethered when the downed fire fighter is placed out the window onto the ladder.

FIGURE 10-1 The downed fire fighter tethered to an RIC rescue air unit by a mask-mounted regulator (MMR) swap.

- What are the fire, smoke, and heat conditions in the room being used for the rescue? If it is high heat and you disconnect the MMR inside the room, the fire fighter, if still breathing, will have his airway and lungs exposed to searing heat and toxic smoke.
- What are the fire, smoke, and heat conditions in the room being used for the rescue and at what danger level is the RIC?
- Can two rescuers lift the downed fire fighter or do we need a third to assist?
- Are you on a floor or at a window that a ground ladder can reach?

The two-rescuer <u>feet-first window lift</u> rescue technique is designed to place a fire fighter, on air, out a window either feet-first or head-first to a ladder rescuer who will use the cradle carry **FIGURE 10-2**. The lift is designed as a two-step lift, which will help to minimize the overall weight being lifted and assist in getting the rescue done rapidly. Before the fire fighter can be placed out of the window, remember to communicate with the RIC Operations Chief so that a rescue ladder and a second RIC can be placed to help support your efforts; then perform the two-rescuer feet-first window lift for a fire fighter wearing a SCBA as demonstrated in **SKILL DRILL 10-1**:

1. Using the drag skills previously taught, bring the fire fighter to the window that is to be used. Once at the window reposition the fire fighter with his feet towards the window. (**STEP 1**)
2. Check out the window to ensure that the ladder is in place and the rescuer is positioned on the ladder. Inform the ladder rescuer as to whether the fire fighter will be coming out feet- or head-first. (**STEP 2**)
3. Roll the fire fighter onto his stomach. (**STEP 3**)
4. Each rescuer, one on either side of the downed fire fighter, will grab a shoulder strap and the cuff of the

pants and, with the preparatory command "To the window" and using the "Ready, Ready, Go!" commands, slide and lift the downed fire fighter along the floor to the window. Bend his knees so that his feet are straight up and against the wall. (**STEP ④**)

⑤ Each rescuer will grab the SCBA waist strap and the cuff of the bunker pants and, with the preparatory command "Out the window" and using the "Ready, Ready, Go!" commands, lift the downed fire fighter and begin to exit him through the window, placing his waist onto the sill. (**STEP ⑤**)

⑥ The two rescuers will reposition their hands, grabbing the downed fire fighter's shoulder and waist straps. Again, using the proper commands, continue to lift and exit the fire fighter out into the arms of the rescuer. (**STEP ⑥**)

⑦ If the downed fire fighter is tethered to an RIC rescue air unit using an MMR, or by the emergency buddy breathing system (EBBS), disconnect it. If this is not possible, you must assist in lowering the RIC air supply to the ground. This is difficult and if mismanaged can cause the tethered lines to go taut and possibly interfere with the ladder carry. One method that might work is attaching a rope to the RIC rescue air unit and lowering it to the ground slowly as the ladder rescue is executed. Remember, however, that the conditions in the room must be tenable for you to have the time to lower the air unit.

⑧ Once the ladder rescuer is ready, he will descend down the ladder.

FIGURE 10-2 The rescue fire fighter on the ladder receiving the downed fire fighter for the cradle carry.

Safety

The ladder rescuer cannot manage both the downed fire fighter and the RIC rescue air unit alone.

SKILL DRILL 10-1   Two-Rescuer Feet-First Window Lift for a Fire Fighter Wearing a SCBA
NFPA 1407, 7.12(4), 7.12(5)

① Bring the fire fighter to the window that is to be used and reposition the fire fighter with his feet towards the window.

② Check out the window to ensure that the ladder is in place and the rescuer is positioned on the ladder. Inform the ladder rescuer as to whether the fire fighter will be coming out feet- or head-first.

(Continued)

SKILL DRILL 10-1 Two-Rescuer Feet-First Window Lift for a Fire Fighter Wearing a SCBA (*Continued*)

3 Roll the fire fighter onto his stomach.

4 Each rescuer, one on either side of the downed fire fighter, will grab a shoulder strap and the cuff of the pants and slide and lift the downed fire fighter along the floor to the window. Bend his knees so that his feet are straight up and against the wall.

5 Each rescuer will grab the SCBA waist strap and the cuff of the bunker pants and lift the downed fire fighter and begin to exit him through the window, placing his waist onto the sill.

6 The two rescuers will reposition their hands, grabbing the downed fire fighter's shoulder and waist straps. Again, using the proper commands, continue to lift and exit the fire fighter out into the arms of the rescuer. Once the ladder rescuer is ready, he will descend down the ladder.

© Joe Nedder/Jones & Bartlett Learning

The three-rescuer feet-first window lift begins the same as steps 1 through 4 of the two-rescuer feet-first window lift for a fire fighter wearing a SCBA. Perform the three-rescuer feet-first window lift for a fire fighter wearing a SCBA as demonstrated in **SKILL DRILL 10-2**:

1 Using the drag skills previously taught, bring the fire fighter to the window that is to be used. Once at the window, reposition the fire fighter with his feet towards the window. (**STEP 1**)

2 Check out the window to ensure that the ladder is in place and the rescuer is positioned on the ladder. Inform the ladder rescuer as to whether the fire fighter will be coming out feet- or head-first. (**STEP 2**)

3 Roll the fire fighter onto his stomach. (**STEP 3**)

4 Two rescuers, one on each side of the downed fire fighter, will grab a shoulder strap and the cuff of the downed fire fighter's pants and move him to the window with his knees bent, so that his feet are straight up and against the wall. If needed, the third rescuer can assist. (**STEP 4**)

5 The third rescuer will now position himself at the head of the downed fire fighter and grab his shoulder straps. The other two rescuers will each grab the SCBA waist strap and the cuff of the downed fire fighter's

bunker pants, and with the preparatory command "To the window" and using the "Ready, Ready, Go!" commands, all three rescuers will lift the downed fire fighter and begin to exit him through the window, placing his waist onto the sill. (**STEP 5**)

6 The two rescuers will reposition their hand closest to the window and grab the downed fire fighter's waist strap and place their other hand under the downed fire fighter's chest for support. (**STEP 6**)

SKILL DRILL 10-2 Three-Rescuer Feet-First Window Lift for a Fire Fighter Wearing a SCBA
NFPA 1407, 7.12(4), 7.12(5)

1 Bring the fire fighter to the window that is to be used and reposition the fire fighter with his feet towards the window.

2 Check out the window to ensure that the ladder is in place and the rescuer is positioned on the ladder. Inform the ladder rescuer as to whether the fire fighter will be coming out feet- or head-first.

3 Roll the fire fighter onto his stomach.

4 Two rescuers, one on each side of the downed fire fighter, will grab a shoulder strap and the cuff of the downed fire fighter's pants and move him to the window with his feet are straight up and against the wall. If needed, the third rescuer can assist.

(Continued)

SKILL DRILL 10-2   Three-Rescuer Feet-First Window Lift for a Fire Fighter Wearing a SCBA (*Continued*)

5 The third rescuer will now position himself at the head of the downed fire fighter and grab his shoulder straps. The other two rescuers will each grab the SCBA waist strap and the cuff of the downed fire fighter's bunker pants, and all three rescuers will lift the downed fire fighter and begin to exit him through the window, placing his waist onto the sill.

6 The two rescuers will reposition their hand closest to the window and grab the downed fire fighter's waist strap and place their other hand under the downed fire fighter's chest for support. The rescuer at the head will continue to grab the downed fire fighter's shoulder straps. The three rescuers will lift and, coordinating with the ladder rescuer, continue to exit the fire fighter out to the rescuer. If the downed fire fighter is tethered to an RIC air supply, reach out and disconnect it. Once the ladder rescuer is ready, he will descend down the ladder.

© Joe Nedder/Jones & Bartlett Learning

7 The rescuer at the head will continue to grab the downed fire fighter's shoulder straps. Again, using the "Ready, Ready, Go!" commands, the three rescuers will lift and, coordinating with the ladder rescuer, continue to exit the fire fighter out to the rescuer.

8 If the downed fire fighter is tethered to an RIC air supply, reach out and disconnect it.

9 Once the ladder rescuer is ready, he will descend down the ladder.

Putting a fire fighter out of a window feet-first is usually an easier position for the ladder rescuer to gain control. However, in always being proactive, it is important to have the skills to execute the lift head first as well. The two-rescuer head-first window lift is more physically demanding than the feet-first method, but with practice can also be done rapidly. To perform the two-rescuer head-first window lift for a fire fighter wearing a SCBA, follow the steps as demonstrated in **SKILL DRILL 10-3** :

1 Using the drag skills previously taught, bring the fire fighter to the window that is to be used. Position him as close to the window as possible and roll him onto his stomach. (**STEP 1**)

2 Check out the window to ensure that the ladder is in place and the rescuer is positioned on the ladder. If possible, inform the ladder rescuer as to whether the fire fighter will be coming out feet or head first. (**STEP 2**)

3 The two rescuers position themselves on either side of the downed fire fighter and each grab his shoulder and waist straps. (**STEP 3**)

4 With the preparatory command "Out the window" and using the "Ready, Ready, Go!" sequence, the two rescuers lift the fire fighter up and place him on the sill chest down. If the downed fire fighter is tethered to an RIC air supply, the MMR is now disconnected. (**STEP 4**)

5 The two rescuers reposition their hands and grab the downed fire fighter's waist strap and cuff of his pants. Using "Ready, Ready, Go!" commands, they then begin to move the fire fighter further out the window into the arms of the ladder rescuer. The ladder rescuer communicates with the inside rescuers to get the downed fire fighter's body positioned safely into his arms for a cradle carry. (**STEP 5**)

6 Once the ladder rescuer is ready he will descend down the ladder.

Crew Notes

At the completion of step 4, the downed fire fighter's arms will probably still be at his side or hanging down. Do not waste time trying to get them out the window at this point.

SKILL DRILL 10-3

Two-Rescuer Head-First Window Lift for a Fire Fighter Wearing a SCBA
NFPA 1407, 7.12(4), 7.12(5)

1 Bring the fire fighter to the window that is to be used, position him as close to the window as possible, and roll him onto his stomach.

2 Check out the window to ensure that the ladder is in place and the rescuer is positioned on the ladder. If possible, inform the ladder rescuer as to whether the fire fighter will be coming out feet- or head-first.

3 The two rescuers position themselves on either side of the downed fire fighter and each grab his shoulder and waist straps.

4 The two rescuers lift the fire fighter up and place him on the sill chest down. If the downed fire fighter is tethered to an RIC air supply, the MMR is now disconnected.

(Continued)

5 The two rescuers reposition their hands and grab the downed fire fighter's waist strap and cuff of his pants, and then begin to move the fire fighter further out the window into the arms of the ladder rescuer. The ladder rescuer communicates with the inside rescuers to get the downed fire fighter's body positioned safely into his arms for a cradle carry. Once the ladder rescuer is ready he will descend down the ladder.

© Joe Nedder/Jones & Bartlett Learning

■ Lifting a Fire Fighter, Without SCBA, Through a Window: The Chicago Lift and Carry

There will be times when the downed fire fighter will be lifted out a window without a SCBA on, including if the fire fighter was found without his SCBA and *not breathing*. In this instance, no time would be taken to secure him to an air supply. If the fire fighter was found heavily entangled, out of air and unresponsive, or possibly in full cardiac arrest, the SCBA would be removed by the RIC to expedite the rescue and get him to medical care rapidly. Another example might be if the fire fighter was very heavy. In this case, removing his SCBA at the window just before the lift would expedite the rescue, including the ladder rescue, significantly.

There is a school of thought that you should never remove a fire fighter from his air supply, and generally this is a true statement. However, the word *never* excludes flexibility, which is a key component in rapid intervention. A quick decision based upon assessment of the following is often required:

- What is the downed fire fighter's current situation? Is he:
  - On air?
  - Breathing?
  - Responsive or unresponsive?
  - In cardiac arrest?
  - Experiencing significant trauma?
- What level of medical care does he need immediately?
- If you remove the air supply just before going out the window, will it expedite the speed of the rescue and

support the above statements without doing significant harm to the downed fire fighter?

The key to the Chicago lift and carry is the removal of the SCBA right before putting the downed fire fighter out the window. This skill is called the unresponsive fire fighter SCBA removal and, with practice, will be the fastest method to get a fire fighter out the window. It is also a great technique when trying to lift someone who is very large. The ladder carry portion, which will be presented later in this chapter, is by far the easiest way for a rescue fire fighter to carry a very large fire fighter down a ladder.

Crew Notes

This technique is referred to here as the Chicago lift and carry. It is so named because this lift and carry was a part of the rapid intervention skills developed, enhanced, and taught by Bob Hoff and Rick Kolomay, along with their Chicago Fire Department training team, to every member of the Chicago Fire Department in 2001. The Chicago lift and carry, when properly used, is one of the fastest methods to get a downed fire fighter into a safe environment. With *practice* and *coordination*, an RIC should be able to position, remove the SCBA from, and place the downed fire fighter out the window in about 60 seconds, with the exposure to the environment without bottled air at only approximately 15 to 30 seconds maximum. The technique can use either two or three fire fighters inside for the lift, and the ladder carry is specifically designed to bring the downed fire fighter without a SCBA down a ladder quickly and safely.

■ Removing an Unconscious Fire Fighter's SCBA

This SCBA removal technique will work with almost any brand SCBA, is designed to be fast—without having to cut straps or sit the downed fire fighter up, and will allow you to remove the SCBA from the body of the downed fire fighter while still maintaining the low-pressure regulator mounted on the mask. If the brand SCBA you have has an MMR that cannot be removed, you will have to consider cutting the low-pressure hose to separate the pack from the mask, but if the low-pressure line on this brand SCBA is cut, it is *critical* that you switch the mask-mounted switch from bottled air to ambient air. Failure to do this may cause the downed fire fighter to suffocate in his mask. Currently there is one brand of SCBA on the U.S. market (ISI) with this type of arrangement. If you are a user of this SCBA, consult your manufacturer for further information.

The removal of the SCBA needs to be done quickly and can usually be accomplished by one fire fighter. If the downed fire fighter is very heavy, the assistance of a second rescuer is advisable. To perform the unconscious fire fighter SCBA removal technique, follow the steps demonstrated in **SKILL DRILL 10-4** :

1 Move the downed fire fighter close to the window and position him feet first towards the window.

2 With the downed fire fighter on his side, a rescuer positions himself behind the downed fire fighter, releases the waist strap buckle, and places the high waist strap towards the cylinder stem. (**STEP 1**)

3 The rescuer loosens the high shoulder strap by pulling it up and activating the fastener so that the strap loosens. Holding the downed fire fighter's strap straight up, take the downed fire fighter's arm and place his hand through the strap opening and place it over his heart. Pull the strap towards the back and drop. (**STEP 2**)

4 Place the downed fire fighter's lower arm straight over his head in an EMS log roll position. Roll the downed fire fighter onto his stomach, and once face down, pull the SCBA away from his back and up and over the extended arm, freeing the SCBA. Remember that the downed fire fighter is still on air, so place the SCBA by the downed fire fighter's side. (**STEP 3**) This will allow you to move the fire fighter closer to the window with minimal dragging of the SCBA.

5 Roll the downed fire fighter onto his back, preparing him for the rescue lift. (**STEP 4**)

6 The MMR is not removed until the downed fire fighter is in position by the window and ready to be lifted out. At that point he is seconds away from being placed onto the ladder and into clean air.

SKILL DRILL 10-4   Unconscious Fire Fighter SCBA Removal Technique

1 With the downed fire fighter on his side, a rescuer positions himself behind the downed fire fighter, releases the waist strap buckle, and places the high waist strap towards the cylinder stem.

2 The rescuer loosens the high shoulder strap by pulling it up and activating the fastener so that the strap loosens. Holding the downed fire fighter's strap straight up, take the downed fire fighter's arm and place his hand through the strap opening and place it over his heart. Pull the strap towards the back and drop.

(Continued)

SKILL DRILL 10-4   Unconscious Fire Fighter SCBA Removal Technique (*Continued*)

3 Place the downed fire fighter's lower arm straight over his head in an EMS log roll position. Roll the downed fire fighter onto his stomach, and once face down, pull the SCBA away from his back and up and over the extended arm, freeing the SCBA, and placing the SCBA by the downed fire fighter's side.

4 Roll the downed fire fighter onto his back, preparing him for the rescue lift. The MMR is not removed until the downed fire fighter is in position by the window and ready to be lifted out.

© Joe Nedder/Jones & Bartlett Learning

Crew Notes

Some brands of SCBAs, such as MSA, have a chest harness that must be unbuckled as part of step 1 of the SCBA removal technique on an unconscious fire fighter **FIGURE 10-3**. Review and know your equipment—a life depends on it!

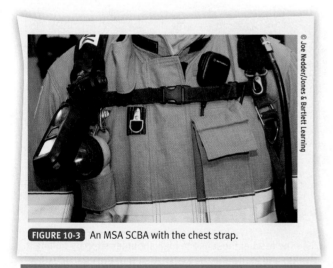

© Joe Nedder/Jones & Bartlett Learning

FIGURE 10-3   An MSA SCBA with the chest strap.

The two-rescuer window Chicago lift for a fire fighter without a SCBA can be performed by following steps as demonstrated in **SKILL DRILL 10-5**:

1 Using the drag skills previously taught, bring the downed fire fighter to the window that is to be used.

Once at the window, reposition him (fire fighter spin) with his feet towards the window.

2 Check outside the window to ensure that the ladder is in place and the rescuer is positioned on the ladder. Inform the ladder rescuer that the Chicago lift and carry will be used.

3 Remove the SCBA from the fire fighter, making sure to keep the regulator (MMR) mounted in the mask and the SCBA positioned near the downed fire fighter's head. (**STEP 1**)

4 With a rescuer on each side, grab inside the fire fighter's coat collar with one hand and, with the second hand, grab the cuff of the bunker pants. Using the "Ready, Ready, Go!" command and with the prompt "To the window," the two rescuers will slide the fire fighter to the wall below the window and position him with his buttocks against the wall and his legs sticking straight up. (**STEP 2**)

5 Just before lifting the downed fire fighter out the window, remove the regulator from the mask. (**STEP 3**)

6 Each rescuer will grab the inside collar of the bunker coat (grab point) and with the other hand grab under the fire fighter's buttocks. With the preparatory order "Out the window" and using the "Ready, Ready, Go!" command, the downed fire fighter is lifted up and placed in a seated position on the window sill. As you lift the fire fighter out the window, the two rescuers should each rotate one knee so that it is positioned under the fire fighter being lifted. (**STEP ④**) This simple maneuver will assist you by resting the fire fighter on your knees, instead of putting him back on the floor, if the entire lift cannot be completed. This reduces the follow-up lift, if necessary, by about 24 inches (61 cm).

⑦ Working with the ladder rescuer, place the downed fire fighter onto the ladder "seated" on the rescuer's shoulders. (**STEP ⑤**)

⑧ Once the ladder rescuer is ready, he will descend down the ladder.

SKILL DRILL 10-5   Two-Rescuer Window Chicago Lift for a Fire Fighter Without a SCBA
NFPA 1407, 7.12(5)

1 Remove the SCBA from the fire fighter, making sure to keep the regulator (MMR) mounted in the mask and the SCBA positioned near the downed fire fighter's head.

2 With a rescuer on each side, grab inside the fire fighter's coat collar with one hand and, with the second hand, grab the cuff of the bunker pants. The two rescuers then slide the fire fighter to the wall below the window and position him with his buttocks against the wall and his legs sticking straight up.

3 Just before lifting the downed fire fighter out the window, remove the regulator from the mask.

4 Each rescuer will grab the inside collar of the bunker coat (grab point) and with the other hand grab under the downed fire fighter's buttocks. The downed fire fighter is lifted up and placed in a seated position on the window sill. As you lift the fire fighter out the window, the two rescuers should each rotate one knee so that it is positioned under the fire fighter being lifted.

(Continued)

SKILL DRILL 10-5   Two-Rescuer Window Chicago Lift for a Fire Fighter Without a SCBA (*Continued*)

5 Working with the ladder rescuer, place the downed fire fighter onto the ladder "seated" on the rescuer's shoulders. Once the ladder rescuer is ready, he will descend down the ladder.

© Joe Nedder/Jones & Bartlett Learning

Safety

When you position the fire fighter onto the window sill, make sure that both rescuers maintain a firm hold on him so as to prevent him falling or slipping out **FIGURE 10-4** .

© Joe Nedder/Jones & Bartlett Learning

FIGURE 10-4 The downed fire fighter on the window sill being held securely by the two rescuers.

One option to assist in this rescue technique would be to use your webbing. Webbing could be used if you have a fire fighter who is heavy, or if the two rescuers are near exhaustion. The webbing will give you leverage and a better grip at the waist area.

The two-rescuer window Chicago lift for a fire fighter without a SCBA utilizing webbing can be performed by following steps as demonstrated in **SKILL DRILL 10-6** :

1 Using the drag skills previously taught, bring the downed fire fighter to the window that is to be used. Once at the window, reposition the downed fire fighter (fire fighter spin) with his feet towards the window.

2 Check out the window to ensure that the ladder is in place and the rescuer is positioned on the ladder. Inform the ladder rescuer that the Chicago lift and carry will be used.

3 Remove the SCBA from the fire fighter, making sure to keep the regulator mounted in the mask and the SCBA positioned near the side of the downed fire fighter. (**STEP ❶**)

4 With a rescuer on each side, grab inside the fire fighter's collar with one hand and with the second hand grab the cuff of the bunker pants. Using the "Ready, Ready, Go!" command and with the officer ordering "To the window," the two rescuers will slide the fire fighter to the wall below the window and position him with his buttocks against the wall and his legs sticking straight up. (**STEP ❷**)

5 A rescuer should deploy his webbing and, holding a bight, pass the webbing in-between the downed fire fighter's legs and the wall to the second rescuer. (**STEP ❸**)

6 The two rescuers then take the webbing and pull it to the floor, using a sawing, back-and-forth motion. (**STEP** **4**)

7 Position the webbing under the fire fighter's hips. The webbing should be wrapped "rodeo style" around the rescuers hand as close to the fire fighter's waist as possible. (**STEP** **5**) This will create a solid handle and leverage for the lift. With the command "Out the window" and using the "Ready, Ready, Go!" commands, the downed fire fighter is lifted up and out the window in a seated position onto the sill. (**STEP** **6**)

8 As you lift the downed fire fighter out the window, the two rescuers should each place one knee under the fire fighter being lifted. (**STEP** **7**)

SKILL DRILL 10-6 Two-Rescuer Window Chicago Lift for a Fire Fighter Without a SCBA Utilizing Webbing
NFPA 1407, 7.12(5)

1 Remove the SCBA from the fire fighter, making sure to keep the regulator mounted in the mask and the SCBA positioned near the side of the downed fire fighter.

2 With a rescuer on each side, grab inside the fire fighter's collar with one hand and with the second hand grab the cuff of the bunker pants. The two rescuers will then slide the fire fighter to the wall below the window and position him with his buttocks against the wall and his legs sticking straight up.

3 A rescuer should deploy his webbing and, holding a bight, pass the webbing in-between the downed fire fighter's legs and the wall to the second rescuer.

4 The two rescuers then take the webbing and pull it to the floor, using a sawing, back-and-forth motion.

(Continued)

SKILL DRILL 10-6 Two-Rescuer Window Chicago Lift for a Fire Fighter Without a SCBA Utilizing Webbing (*Continued*)

5 Position the webbing under the fire fighter's hips. The webbing should be wrapped "rodeo style" around the rescuers hand as close to the fire fighter's waist as possible.

6 The downed fire fighter can the be lifted up and out the window in a seated position onto the sill.

7 As you lift the downed fire fighter out the window, the two rescuers should each place one knee under the fire fighter being lifted.

8 Once the fire fighter is placed on the window sill, safety and caution should be utilized to prevent him from falling or slipping out.

9 Working with the ladder rescuer, place the downed fire fighter onto the ladder "seated" on the rescuer's shoulders. Once the ladder rescuer is ready, he will descend down the ladder.

9 This simple maneuver will assist you by resting the fighter on your knees instead of putting him back on the floor, if the entire lift cannot be completed. This reduces the follow-up lift, if necessary, by about 24 inches (61 cm). Once the fire fighter is placed on the window sill, safety and caution should be utilized to prevent him from falling or slipping out. (**STEP 8**)

10 Working with the ladder rescuer, place the downed fire fighter onto the ladder "seated" on the rescuer's shoulders. Once the ladder rescuer is ready, he will descend down the ladder. (**STEP 9**)

Safety

Caution! Although the webbing will greatly assist in the lifting of the downed fire fighter, you must be careful when lifting him so as not to propel him in a dangerous way out the window.

The three-rescuer window lift for a fire fighter without a SCBA can be used if the downed fire fighter is very large, if the RIC is physically exhausted or incapable, or if an extra RIC member is readily available. To perform the three-rescuer window lift for a fire fighter without a SCBA, follow the steps as demonstrated in **SKILL DRILL 10-7** :

1 Using the drag skills previously taught, bring the fire fighter to the window that is to be used. Once at the window, reposition the downed fire fighter (fire fighter spin) with his feet towards the window.

2 Check out the window to ensure that the ladder is in place and the rescuer is positioned on the ladder. Inform the ladder rescuer that the Chicago lift and carry will be used.

3 Remove the SCBA from the fire fighter, making sure to keep the regulator mounted in the mask and the SCBA positioned near the downed fire fighter's side. (**STEP 1**)

4 With a rescuer on each side, grab inside the fire fighter's collar with one hand and, with the second hand, grab some material under the bunker pants. With the order "To the window" and using the "Ready, Ready, Go!" commands, the two rescuers will slide the fire fighter to the wall below the window and position him with his buttocks

---

SKILL DRILL 10-7

Three-Rescuer Window Lift for a Fire Fighter Without a SCBA
NFPA 1407, 7.12(5)

1 Remove the SCBA from the fire fighter, making sure to keep the regulator mounted in the mask and the SCBA positioned near the downed fire fighter's side.

2 With a rescuer on each side, grab inside the fire fighter's collar with one hand and, with the second hand, grab some material under the bunker pants. The two rescuers will then slide the fire fighter to the wall below the window and position him with his buttocks against the wall and his legs sticking straight up. The third fire fighter can assist if needed by positioning at the head and pushing towards the window.

(Continued)

SKILL DRILL 10-7   Three-Rescuer Window Lift for a Fire Fighter Without a SCBA (*Continued*)

3 With a rescuer on each side, grab inside the downed fire fighter's collar with one hand and, with the second hand, grab some material under the bunker pants. The third rescuer positions himself at the downed fire fighter's head and, from weight lifter's squat, grabs under the downed fire fighter's arms.

4 The three rescuers lift the downed fire fighter up and out and sit him on the sill. Once the downed fire fighter is placed on the sill, safety and caution should be utilized to prevent him from falling or slipping out.

5 Working with the ladder rescuer, place the downed fire fighter onto the ladder, "seated" on the rescuer's shoulders. Once the ladder rescuer is ready, he will descend down the ladder.

© Joe Nedder/Jones & Bartlett Learning

against the wall and his legs sticking straight up. The third fire fighter can assist if needed by positioning at the head and pushing towards the window. (**STEP ②**)

5 With a rescuer on each side, grab inside the downed fire fighter's collar with one hand and, with the second hand, grab some material under the bunker pants. The third rescuer positions himself at the downed fire fighter's head and, from weight lifter's squat, grabs under the downed fire fighter's arms. (**STEP ③**)

⑥ The fire fighter at the head gives the order "Out the window" and with the "Ready, Ready, Go!" command, the three rescuers lift the downed fire fighter up and out and sit him on the sill. As you lift the downed fire fighter out the window, the two rescuers should rotate their knees so that they are positioned under the fire fighter being lifted. Once the downed fire fighter is placed on the sill, safety and caution should be utilized to prevent him from falling or slipping out. (**STEP ④**) This lift can also use the webbing as previously taught.

⑦ Working with the ladder rescuer, place the downed fire fighter onto the ladder, "seated" on the rescuer's shoulders. (**STEP ⑤**)

⑧ Once the ladder rescuer is ready, he will descend down the ladder.

The technique for removing a downed fire fighter utilizing a long board and webbing can be used from a first floor or ground window. It can also be used from an aerial platform, but is strongly not recommended if a ladder (either ground or aerial) is involved. This rescue is designed if there is only one rescuer inside the window and he is unable to lift the downed fire fighter up and out by himself. It can be performed whether or not the downed fire fighter is wearing a SCBA, and a long board and webbing are needed. To perform this technique, follow the steps in **SKILL DRILL 10-8**:

① Advise Command as to what window you going to use and request help with a long board.

② Drag the fire fighter to within 3 or 4 feet (91 cm or 1 m) of the window. (**STEP ①**)

③ Check outside the window to make sure a rescuer is in place.

④ Deploy your webbing and secure it to the downed fire fighter's SCBA shoulder straps (**STEP ②**) or create a harness if there is no SCBA. Pass the running end outside the window. (**STEP ③**)

⑤ The inside rescuer pulls the fire fighter into a seated position and calls outside that he is ready for the long board. (**STEP ④**)

⑥ The outside rescuers slide the board over the window sill and place it as close to under the downed fire fighter as possible. (**STEP ⑤**)

⑦ The outside rescuer then grabs the webbing, and the inside rescuer lowers the fire fighter onto the board while still grabbing the shoulder straps. (**STEP ⑥**)

⑧ With the outside rescuer calling the "Ready, Ready, Go!" commands, he will pull on the webbing while the inside rescuer assists by using the shoulder straps and placing the downed fire fighter onto the long board. (**STEP ⑦**)

⑨ Once on the board the inside rescuer can lift up the board and, working with the outside rescuer, slide the board over the sill. (**STEP ⑧**)

⑩ Additional rescuers will assist in grabbing the board and lowering the fire fighter to the ground.

Safety

Putting a downed fire fighter on a board and "sliding" him out is very dangerous, as the downed fire fighter can slide off and incur severe injury. Using a long board is a skill to meet the NFPA 1407 requirements, but should be used with significant *caution* and care. If the downed fire fighter is to be placed onto an aerial platform, he should be properly secured to the board *before* exiting the window and being placed onto the platform. Once on the platform, the board should be properly secured to avoid any danger of the downed fire fighter falling off.

Safety

Great caution must be used when moving a downed fire fighter using a long board and webbing from a first floor window if the fire fighter is not secured to the board.

Ladder Carries to Rescue a Downed Fire Fighter

Ladder rescues are an important part of any RIC training program. As such, they must be frequently practiced to maintain the skills so that if needed they can be executed rapidly and safely.

Ground ladders, in general, have been and still are of concern on many firegrounds. It is still not uncommon to see a lack of egress ladders being thrown on firegrounds, with the general excuse being "lack of manpower." However, when we think of how many fire fighters it really takes to throw a 24-, 28-, or even 35-foot (7-, 9-, or even 11-m) ladder, the excuse wears thin. Throwing emergency egress ground ladders is being proactive with fire fighter life safety and can be done by the RIC.

Safety

When you drill on any ladder carries, always exercise extreme safety and caution—especially for the fire fighter playing the role of the victim. This means that in order to meet the requirements of NFPA 1407, the fire fighter playing the role of the downed fire fighter must be in a *minimum* of an escape belt and be secured to a belay line that complies with National Fire Protection Association (NFPA) 1983 *Standard on Life Safety Rope and Equipment for Emergency Services*. We strongly suggest that a life safety harness such as a Class III, with an anchor point in the back and above the shoulders, always be used as it provides the most protection and will prevent the "victim" from flipping parallel or, worse yet, upside down with his feet up in the air if he falls! As always, the belay line needs to be properly rigged and operated by a fire fighter who is trained and competent with belay lines.

SKILL DRILL 10-8

Using a Long Board and Webbing from a First Floor Window
NFPA 1407, 7.12(3)

1 Drag the fire fighter to within 3 or 4 feet (91 cm or 1 m) of the window. Check outside the window to make sure a rescuer is in place.

2 Deploy your webbing and secure it to the downed fire fighter's SCBA shoulder straps.

3 Create a harness if there is no SCBA, and pass the running end outside the window.

4 The inside rescuer pulls the fire fighter into a seated position and calls outside that he is ready for the long board.

5 The outside rescuers slide the board over the window sill and place it as close to under the downed fire fighter as possible.

6 The outside rescuer then grabs the webbing, and the inside rescuer lowers the fire fighter onto the board while still grabbing the shoulder straps.

(Continued)

SKILL DRILL 10-8 Using a Long Board and Webbing from a First Floor Window (*Continued*)

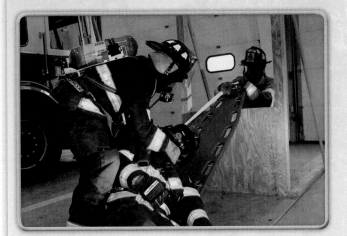

7 With the outside rescuer will pull on the webbing while the inside rescuer assists by using the shoulder straps and placing the downed fire fighter onto the long board.

8 Once on the board, the inside rescuer can lift up the board and, working with the outside rescuer, slide the board over the sill. Additional rescuers will assist in grabbing the board and lowering the fire fighter to the ground.

© Joe Nedder/Jones & Bartlett Learning

Rapid intervention has created a new opportunity to once again be proactive and aggressive with ground ladder placement. Every member of every RIC needs to be proficient and capable in rapidly moving, raising, and positioning ground ladders for a ladder rescue.

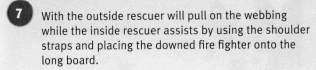

Safety

In this text we are focusing on carrying the downed fire fighter to safety over ladders. There are other techniques used in RIC, such as the high-rung lowering technique, that are not covered here. The high-rung technique relies on the downed fire fighter's SCBA waist strap being converted into a harness, but since the majority of fire fighters do not have Class II life safety harnesses built into their personal protective equipment (PPE), converting a SCBA waist strap into a harness is dangerous and not an intended use by the SCBA manufacturers. Not only is the SCBA not designed for this type of use, it is not a rated or tested harness. By using the SCBA in this way, you are relying on the buckle staying together while you lower the fire fighter out the window to the ground, and if the harness fails, the fire fighter's fall could cause severe injury or death. The use of a ladder to carry a fire fighter down to safety is usually a much safer and faster method of rescue.

RIC Ladder Rescues

When a ladder rescue is needed, the RIC must notify RIC Operations of the need and request a ladder to a specific location. This was discussed earlier in this chapter. Once that information is transmitted, the RIC Operations Chief needs to assign another RIC or other competent crew to the ladder rescue. For fire departments that are usually minimally staffed, this might translate to only two fire fighters. The most important considerations and variables with the ladder rescue are:

- Assigning the tactic to another RIC or, if not available, to a company that is RIC capable right away. This is very important, because the interior RIC will be communicating with the ladder crew, and they must understand and know what is going on and what is required of them.
- Knowing where the ladder is going—what side and what floor.
- Knowing what size ladder is needed. This should be known because of where the RIC requested the ladder (what floor number and building side) and the information that was obtained from the 360-RECON.
- Knowing if there are any obstructions (branches, wires, propane tanks, AC compressors) by the selected window that the interior RIC cannot see or know about that will inhibit or prevent the rescue. If so, is there another window that can be used? If this is the case, that information must be relayed to the interior RIC.

When the tactic is assigned, the Company Officer of the RIC that was given the ladder rescue assignment must select what size ladder is needed and assist the crew to rapidly get it in place. There can be no mistakes or errors with this assignment.

Positioning the Ladder for the RIC Rescue

When a ladder is deployed and put in place for a ladder rescue, it needs to be positioned properly. This means that the tip of the ladder must be at or slightly below the sill **FIGURE 10-5**. A ladder tip that extends above the sill creates two significant issues:

1. As a downed fire fighter is passed out over the ladder, if the tip is protruding above the sill, it creates an entanglement hazard where gear can catch. This will make it harder for the rescuers to place the downed fire fighter on the ladder.
2. A tip protruding over the sill will increase the overall lifting height that the rescuers must overcome while passing the downed fire fighter out.

A properly positioned ladder will help to avoid these types of issues and any subsequent delay to the RIC and ladder rescue.

Stabilizing the Ladder for the RIC Rescue

All fire service training textbooks teach that a ladder needs to be positioned at a 75-degree angle in order to be safe and provide the maximum ladder capacity. However, in the real world, this ladder angle cannot always be achieved. Many times the ladder can only be positioned at a shallower angle due to ground terrain, obstructions near the building such as trees, shrubbery, outside stairs, or HVAC equipment, or an unusual building detail **FIGURE 10-6**.

The 75-degree angle is designed for climbing—in RIC it has been found that a shallower angle of approximately 60 degrees will make the ladder position more advantageous for carrying an unresponsive fire fighter down **FIGURE 10-7**. A shallower angle allows the rescuer to feel more balanced and safe on the ladder, and when the downed fire fighter is passed out to him, the weight of the fire fighter will not be as "pressing downward" as it would with the traditional 75-degree angle. The shallower angle also allows the rescuer to step up and into the unresponsive fire fighter being passed to him, so as to better control the descent, which translates to a safer and more rapid rescue.

Every ladder that a fire fighter climbs needs to be properly heeled or footed for safety, and for rapid intervention, the ladder needs to be aggressively heeled. *Aggressive heeling* refers to two fire fighters working together by facing the ladder and stabilizing it to prevent it from kicking out **FIGURE 10-8**. In fire fighter basic training, you are taught to foot the ladder by standing under or in front of it **FIGURE 10-9**; however, this technique should *never* be used in rapid intervention. Using the aggressive heeling technique:

- Ensures that the ladder cannot kick out because two fire fighters are holding it in place.

FIGURE 10-6 An outside stairway can interfere with ladder placement against a building.

FIGURE 10-7 A ladder at the proper angle for RIC operations.

FIGURE 10-5 A ladder properly positioned with the tips at or slightly below the window sill.

FIGURE 10-8 Two fire fighters aggressively heeling a rescue ladder.

Apply Safety When Climbing a Ladder

When a rescuer climbs a ladder, the ladder should be climbed using a hand-to-rung, not hand-to-beam, technique, and when descending the ladder with the downed fire fighter, the rescuer should always use the hand-to-beam method as if carrying a tool. Foot placement is also critical for safety. Too often fire fighters are seen climbing ladders on the balls of their feet, which is not safe **FIGURE 10-10**. Placing the ball of your foot on the rung allows your foot to slip if the rung is wet or iced over. The preferred method is to always "lock" your heel to the rung **FIGURE 10-11**. This gives your foot a very secure position on the rung and will prevent you from slipping. Proper climbing skills comes from practice, and for an RIC operation to be successful, these basic skills need to be automatic.

RIC Ladder Carries

The ladder carry skills presented here are in direct support of the window lifts presented earlier. The three ladder carries discussed are the one-rescuer cradle carry, the one-rescuer

FIGURE 10-10 A fire fighter whose feet are improperly placed on the ladder rungs.

FIGURE 10-11 A fire fighter whose feet are properly placed on the ladder rungs, with his heels "locked in."

FIGURE 10-9 A fire fighter performing a typical ladder heeling, which should never be used in rapid intervention.

- Places two additional fire fighters who can assist with getting the downed fire fighter off the ladder when at ground level.
- Positions the two fire fighters, who are facing the ladder and also the building, where they can both observe the rescue and be on the watch for something that might go wrong.

Chicago carry, and the two-ladder–two-rescuers cradle carry. Each of these techniques is designed to support the interior RIC rescue. The keys to the success of these types of rescues are: (1) that all members, both inside and outside, are trained in the same set of skills and can properly and rapidly execute them; and (2) that there is excellent communications between the RIC inside and the ladder rescue RIC or assisting company.

■ Cradle Carry Rescue

The cradle carry rescue is used to support the window lift with a fire fighter wearing a SCBA. For the cradle carry, the rescue fire fighter supports the downed fire fighter's chest with one arm and the other arm is placed between the downed fire fighter's legs. Both of the rescuer's hands grab the ladder beams. The friction of the downed fire fighters body running over the ladder beams, plus the rescue fire fighter pushing against the body, will help to control the decent. This carry is difficult and must be practiced to be proficient. The variables of this technique include:

- If the downed fire fighter is presented feet first or head first.
- The size of the downed fire fighter—if the ladder rescuer is not capable of carrying the downed fire fighter, the Chicago carry must be considered.

After the ladder is properly placed, positioned with aggressive heeling, and the fly is tied off, the one-rescuer cradle carry when the downed fire fighter is presented feet first can be performed by following steps as demonstrated in **SKILL DRILL 10-9** :

1 The rescuer climbs the ladder to the window and calls out "Rescuer is at the window!" or "Ladder man at the window!" (**STEP 1**) This is an important action because you cannot rely on the inside RIC seeing you—they need to know you are in place before they begin to lift the downed fire fighter to safety.

2 The interior RIC will come to the window and indicate how the downed fire fighter is coming out (feet first or head first). (**STEP 2**) If it is going to be feet first and a cradle carry is to be used, the ladder rescuer should indicate which way he wants the legs to go (to his left or right), if possible (this is an opportunity for the ladder rescuer to favor his strong side).

3 As the downed fire fighter's legs start coming out the window, the ladder rescuer will place one arm in-between the downed fire fighter's legs. The interior RIC will then continue to pass the downed fire fighter out. With his hand between the downed fire fighter's legs, the ladder rescuer will grab hold of the ladder beam. He will then place his other hand on the opposite beam and have the interior RIC place the fire fighter's chest onto his arm. (**STEP 3**)

4 The ladder rescuer steps up and leans into the fire fighter to securely seat him on his arms. (**STEP 4**)

5 The ladder rescuer descends the ladder one rung at a time. As he descends, he needs to make sure to keep leaning into the fire fighter. This action will help to control the speed of the descent by creating more friction. (**STEP 5**)

6 At the bottom of the ladder, the two fire fighters heeling the ladder will assist and lift the downed fire fighter off the rescuer's arms and bring him to ground. (**STEP 6**)

SKILL DRILL 10-9   One-Rescuer Cradle Carry When the Downed Fire Fighter Is Presented Feet First
NFPA 1407, 7.12(5)

1 The rescuer climbs the ladder to the window and calls out "Rescuer is at the window!" or "Ladder man at the window!"

2 The interior RIC will come to the window and indicate how the downed fire fighter is coming out (feet first or head first).

(Continued)

SKILL DRILL 10-9 One-Rescuer Cradle Carry When the Downed Fire Fighter Is Presented Feet First (*Continued*)

3 As the downed fire fighter's legs start coming out the window, the ladder rescuer will place one arm in-between the downed fire fighter's legs. The interior RIC will then continue to pass the downed fire fighter out. With his hand between the downed fire fighter's legs, the ladder rescuer will grab hold of the ladder beam. He will then place his other hand on the opposite beam and have the interior RIC place the fire fighter's chest onto his arm.

4 The ladder rescuer steps up and leans into the fire fighter to securely seat him on his arms.

5 The ladder rescuer descends the ladder one rung at a time. As he descends, he needs to make sure to keep leaning into the fire fighter. This action will help to control the speed of the descent by creating more friction.

6 At the bottom of the ladder, the two fire fighters heeling the ladder will assist and lift the downed fire fighter off the rescuer's arms and bring him to ground.

© Joe Nedder/Jones & Bartlett Learning

After the ladder is properly placed, positioned with aggressive heeling, and the fly is tied off, perform the one-rescuer cradle carry when the downed fire fighter is presented head first by following the steps as demonstrated in **SKILL DRILL 10-10**:

① The rescuer climbs the ladder to the window and calls out "Rescuer is at the window!" or "Ladder man at the window!" This is an important action because you cannot rely on the inside RIC seeing you—they need to know you are in place before they begin to lift the downed fire fighter to safety.

② The interior RIC will come to the window and indicate how the downed fire fighter is coming out (feet first or head first). (**STEP ①**) If it is going to be head first and a cradle carry is to be used, the ladder rescuer should indicate, if possible, which way he wants the head to go as it comes out the window (to his left or right).

③ As the downed fire fighter's head starts to come out the window, the ladder rescuer will grab hold of the ladder beams on both sides. The interior RIC will continue to feed the downed fire fighter out, placing the fire fighter's chest onto the ladder rescuer's designated arm. (**STEP ②**)

④ After the downed fire fighter's chest is on the ladder rescuer, the interior RIC will the place the downed fire fighter's legs out the window. The ladder rescuer should place his hand in-between the fire fighter's legs and grab hold of the ladder beam. (**STEP ③**)

⑤ The ladder rescuer then steps up and leans into the fire fighter to securely seat him on his arms.

⑥ The ladder rescuer now descends the ladder one rung at a time, being sure to continue to lean into the fire fighter. This action will help to control the speed of the descent. (**STEP ④**)

⑦ At the bottom of the ladder, the two fire fighters heeling the ladder will assist and lift the downed fire fighter off the rescuer's arms and bring him to ground. (**STEP ⑤**)

■ Chicago Ladder Carry Rescue

The Chicago ladder carry is specifically designed to carry a fire fighter, *without* a SCBA, down a ladder and is also excellent for carrying a large fire fighter or a civilian. This technique has the interior RIC placing the downed fire fighter's legs onto the

---

SKILL DRILL 10-10 | One-Rescuer Cradle Carry When the Downed Fire Fighter Is Presented Head First
NFPA 1407, 7.12(5)

1 The interior RIC will come to the window and indicate how the downed fire fighter is coming out (feet first or head first).

2 As the downed fire fighter's head starts to come out the window, the ladder rescuer will grab hold of the ladder beams on both sides. The interior RIC will continue to feed the downed fire fighter out, placing the fire fighter's chest onto the ladder rescuer's designated arm.

(Continued)

SKILL DRILL 10-10   One-Rescuer Cradle Carry When the Downed Fire Fighter Is Presented Head First (*Continued*)

3 After the downed fire fighter's chest is on the ladder rescuer, the interior RIC will then place the downed fire fighter's legs out the window. The ladder rescuer should place his hand in-between the fire fighter's legs and grab hold of the ladder beam. The ladder rescuer then steps up and leans into the fire fighter to securely seat him on his arms.

4 The ladder rescuer now descends the ladder one rung at a time, being sure to continue to lean into the fire fighter. This action will help to control the speed of the descent.

5 At the bottom of the ladder, the two fire fighters heeling the ladder will assist and lift the downed fire fighter off the rescuer's arms and bring him to ground.

© Joe Nedder/Jones & Bartlett Learning

shoulders of the ladder rescuer. This a good carry because the ladder rescuer is using the muscles in his legs to support the fire fighter's weight and the friction of the fire fighter's body against the ladder for control. Almost any rescuer will be able to carry a fire fighter of any size down a ladder using this technique. The only variable with this rescue is that the SCBA must have been removed before the fire fighter is placed out onto the ladder.

Crew Notes

Sometimes the back brim of the traditional-style helmets will catch in-between ladder rungs during a Chicago carry ladder rescue. To avoid this, have the interior RIC lean the downed fire fighter's head forward when they place him on the ladder or remove the helmet if possible.

After the ladder is properly placed, positioned with aggressive heeling, and the fly is tied off, the one-rescuer Chicago carry for a downed fire fighter not wearing a SCBA can be performed by following the steps as demonstrated in **SKILL DRILL 10-11**:

1 The rescuer climbs the ladder to the window and calls out "Rescuer is at the window!" or "Ladder man at the window!" This is an important action because you cannot rely on the inside RIC seeing you—they need to know you are in place before they begin the lift to safety.

2 The interior RIC will come to the window and indicate it is the Chicago carry that is to be used. **(STEP 1)**

3 The ladder rescuer positions himself below the sill, otherwise he will probably get hit with the downed fire fighter's feet on their way out. **(STEP 2)**

4 The interior RIC will lift and place the downed fire fighter onto the sill in a seated position. The ladder rescuer should step up and place one of the fire fighter's thighs onto one of his shoulders and then the other thigh on his other shoulder. Both hands should now be grabbing the ladder beams. Once this is done, the ladder rescuer should instruct the interior RIC to continue to pass the downed fire fighter completely out onto his shoulders. **(STEP 3)** If needed, the ladder rescuer can take a step down to assist in getting the downed fire fighter onto the ladder in a seated position and onto his shoulders.

5 The ladder rescuer should keep his body close to the ladder for better control of the downed fire fighter and begin to descend one rung at a time. **(STEP 4)**

6 At the bottom of the ladder, the two fire fighters heeling the ladder will assist and lift the downed fire fighter off the rescuer's arms and bring him to ground. **(STEP 5)**

The two-ladders–two-rescuers cradle carry rescue involves the use of two ladders placed side-by-side and a ladder rescuer on each. This technique is used primarily when the downed fire fighter is very large **FIGURE 10-12**. The downed fire fighter is passed out the window and laid across the arms of both rescuer fire fighters. This technique is used for second-floor and higher rescues where ground ladders can reach. The two-ladders–two-rescuers technique is designed to support the rescue of a very large or tall fire fighter when one rescuer is unable to make the carry. It is also an excellent choice to support the Denver drill rescue, in which the fire fighter is passed out the window quickly and the interior rescuers are heavily relying on the ladder rescuers to place him as they are unable to do so because of their position.

Upon notification that the downed fire fighter will be extracted and that two ladder rescuers are needed, the RIC Operations Chief needs to assign the ladder rescue to one or two crews. In reality, you need a rescuer and two fire fighters to aggressively heel each ladder, for a total of six fire fighters. Again, as with any other RIC rescue, those being sent for

---

SKILL DRILL 10-11   One-Rescuer Chicago Carry for a Downed Fire Fighter Not Wearing a SCBA
NFPA 1407, 7.12(5)

1 The interior RIC will come to the window and indicate it is the Chicago carry that is to be used.

2 The ladder rescuer positions himself below the sill, otherwise he will probably get hit with the downed fire fighter's feet on their way out.

(Continued)

SKILL DRILL 10-11   One-Rescuer Chicago Carry for a Downed Fire Fighter Not Wearing a SCBA
(*Continued*)

3 The interior RIC will lift and place the downed fire fighter onto the sill in a seated position. The ladder rescuer should step up and place one of the fire fighter's thighs onto one of his shoulders and then the other thigh on his other shoulder. Both hands should now be grabbing the ladder beams. The ladder rescuer should then instruct the interior RIC to continue to pass the downed fire fighter completely out onto his shoulders.

4 The ladder rescuer should keep his body close to the ladder for better control of the downed fire fighter and begin to descend one rung at a time.

5 At the bottom of the ladder, the two fire fighters heeling the ladder will assist and lift the downed fire fighter off the rescuer's arms and bring him to ground.

© Joe Nedder/Jones & Bartlett Learning

FIGURE 10-12 The two-ladders–two-rescuers team preparing for the rescue.

the ladder rescue must be trained in the RIC skills they are supporting and assisting, otherwise the rapid rescue will be in danger of failing. The ladders should be placed directly next to each other with enough space between the two inner beams for the rescue fire fighters to be able to grasp them.

The ladders should also be placed at the same angle so the two rescuers can work in a coordinated manner. The skills of this rescue are almost identical to the two cradle carries previously presented, the difference being that the two rescuers will each handle about one-half of the downed fire fighter and it does not matter if he comes out feet first or head first.

The skill can also be used if the downed fire fighter is without a SCBA; however, this technique is typically slower, and the toxic smoke and high heat could affect the downed fire fighter's respiratory system if he is breathing. In this instance, the Chicago carry should be considered instead, using additional interior rescuers to assist in lifting the downed fire fighter out the window. After the two ladders are properly placed, positioned with aggressive heeling, and the fly is tied off, perform the two-ladders–two-rescuers cradle carry rescue by following steps as demonstrated in **SKILL DRILL 10-12**:

1. The two rescuers climb the ladder and tell the interior RIC that they have two rescuers in place.

2. The interior crew will come to the window and inform the ladder rescuers what they have. (**STEP 1**)

3. The interior RIC begins to pass the downed fire fighter out the window. (**STEP 2**) The two rescuers, working together, place the downed fire fighter's chest across the arms of one rescuer and the second rescuer places his outside arm through the downed fire fighter's legs. Both rescuers must grasp the beams with both hands. (**STEP 3**)

4. Working together in a coordinated manner, the rescuers lower the fire fighter down the ladder one rung at time using the coordinated command of "Ready, step." (**STEP 4**)

5. At the bottom of the ladder, the fire fighters who have been aggressively heeling the ladders step up and assist the downed fire fighter off the rescuers' arms and bring him to the ground. (**STEP 5**)

SKILL DRILL 10-12 Two-Ladders–Two-Rescuers Cradle Carry Rescue
NFPA 1407, 7.12(5)

1. The two rescuers climb the ladder and tell the interior RIC that they have two rescuers in place. The interior crew will come to the window and inform the ladder rescuers what they have.

2. The interior RIC begins to pass the downed fire fighter out the window.

(Continued)

SKILL DRILL 10-12   Two-Ladders–Two-Rescuers Cradle Carry Rescue (*Continued*)

3 The two rescuers, working together, place the downed fire fighter's chest across the arms of one rescuer and the second rescuer places his outside arm through the downed fire fighter's legs. Both rescuers must grasp the beams with both hands.

4 Working together in a coordinated manner, the rescuers lower the fire fighter down the ladder one rung at time using the coordinated command of "Ready, step."

5 At the bottom of the ladder, the fire fighters who have been aggressively heeling the ladders step up and assist the downed fire fighter off the rescuers' arms and bring him to the ground.

© Joe Nedder/Jones & Bartlett Learning

Rescuing a Downed Fire Fighter With Limited, Restrictive Space and an Elevated Window Opening

When discussing restrictive space rescues or elevated window heights, the discussion will always include the <u>Denver drill</u>, which is based on the LODD of a Denver, Colorado fire fighter summarized here. The NFPA 1407, *Standard for Fire Service Rapid Intervention Crews*, section 7.12 (9) specifically calls for training in "moving a downed fire fighter through an elevated (restricted size) window in a room with limited space for crew movement." Even though the LODD occurred in 1992, fire fighters must always expect that a similar situation could happen again and must be prepared. The Denver drill is one of the most physical and demanding RIC skills described here, but it works!

As you review the synopsis of this tragedy, think of the areas you protect—regardless if you are a major city urban department or a small, rural department. Every fire department has a potential situation like this. It might not be a file storage room, but the situation could be similar with bathroom windows, windows in mobile homes, or hallway windows on landings in apartment buildings. To say that this was a "one in a million situation" or, worse yet, "It will never happen here" is complacency, and in the fire service, complacency kills.

■ The LODD of Engineer Mark Langvardt

On September 28, 1992, Denver, Colorado Fire Engineer Mark Langvart died in the line of duty. This synopsis is based on an article written by David McGrail and Jack Rogers that appeared in *Fire Engineering Magazine* (April 1993) and also from *Firefighter Rescue and Survival* by Richard Kolomay and Robert Hoff.

> On September 28, 1992, at 0230 hours, the Denver Fire Department responded to a fire in a two-story commercial occupancy. The building was of ordinary construction and approximately 50 feet (15 m) wide by 60 feet (18 m) deep. Upon arrival, heavy smoke was showing, but no visible fire. Truck work was done to force entry at two doors on the lower floor, where heavy fire was found. Engine 16 knocked down the fire, but as additional horizontal ventilation was being performed, another fire was located, indicating the definite possibility of arson. Fire was observed in the southwest corner of the upper floor. Engine 21, assisted by Truck 16 (Langvardt's truck company) was assigned to the fire attack on the upper floor. The companies found a large body of fire along with significantly intensified heat and smoke. The fire was stubborn and intense, and progress was slow. About this time, Langvardt became separated from his partner and the other crews operating. At about 0237, Chief 3 observed what appeared to be a flashlight beam of light from a second floor window and yelled out "Do you need help?" There was no answer and the flashlight disappeared from view. Exterior fire fighters recognized this as a distress signal, and Members of Rescue 1 grabbed two ladders and hand and cutting tools to make entry through the metal grate-covered window where the beam of light was last seen. An interior rescue was also attempted, but was not successful due to a partial collapse of the upper floor. Members of Rescue 1 removed the metal grate from the window and broke out the glass. Dense smoke poured from the window and was so thick that the fire fighters on the two ladders placed side-by-side could not see each other even though the distance apart was only about 20 inches (51 cm).
>
> The two fire fighters from the ladders dove headfirst into the window that was 20 inch (51 cm) wide (due to a file cabinet up against the window restricting the space) and 42 inches (1 m) off the floor, landing onto Langvardt. They found the victim facedown, wedged in a fetal position with his head (helmet in place) pressed against the interior portion of the front wall. The two rescuers found themselves in a confined space—the room measured 6 feet (2 m) wide by 11 feet (3 m) deep, but was filled with filing cabinets and business equipment, leaving an aisle space of only 28 inches (71 cm)! This confined space was so tight the two rescuers could barely fit together or reach into their pockets for any hand tools. There was enough room for one fire fighter to bend over the victim and try and get leverage to lift him up and out. The standard size door of the room was blocked with equipment. A bifold door that was the only way in or out, but this path of egress was blocked by the floor collapse.

> Smoke in the room was dense, and the rescuers could not evaluate Langvardt's condition, although it was apparent that he was unresponsive. Attempts were made to lift him up to the window, but they could not do it. There was no room to work with traditional skills, and there was near-zero visibility.
>
> At this point, the advancing fire was threatening the rescuers' position. Several hand lines were deployed to push the fire away from the room and the front of the building. Numerous rescue attempts were made by rotating rescue teams to remove the trapped fire fighter out of the window, but none were successful. Some of the fire fighters working the rescue attempts thought that Langvardt was stuck or pinned, but he was not. The rescuers could not lift him up and out of the window because of the narrow 28-inch (71-cm) wide aisle and the 42 inch (1 m) distance to the sill. The most they could lift him was about 12 inches (30 cm). One fire fighter said that it was as if Langvardt was "tied to a thousand pounds (454 kg) of concrete." While this operation was going on, a second rescue attempt was being made from the interior, but with the southwest roof in a state of imminent collapse, all fire fighters working that area were pulled out.
>
> Access to Langvardt was finally achieved by breaching through the interior foyer wall. This operation was difficult because of heavy heat and smoke that had penetrated the stairwell in which the crews were working. The crews in the foyer worked off ground ladders using saws and other cutting tools to breach the wall, but even then, storage shelves and equipment blocked access and had to be removed. Langvardt was removed from the room at approximately 0330, after a 55-minute rescue operation.

■ Denver Drill Rescue Technique

The Denver drill is one of the most physically challenging and difficult RIC rescue techniques. In order for you to gain skill and competency with this technique, you will need initial and ongoing training, as well as the use of a Denver or multi-purpose prop that accurately recreates the scenario **FIGURE 10-13**. The use of the correct prop is crucial in order to provide the confinement and the significant challenges that the Denver Fire Department was faced with that night. To assist you in your training and with building an accurate prop, drawings for the multi-purpose prop can be found in the appendix to this text.

Safety

When the Denver drill is practiced, always have a few extra fire fighters functioning as "spotters" to help ensure the safety of the fire fighter playing the victim when he is lifted up and out. The spotters can then assist the victim once "out" back to the ground.

Using a Denver or multi-purpose prop, place a fire fighter in full PPE inside the prop in the position the downed fire fighter **FIGURE 10-14**. The Denver drill rescue technique requires at least three fire fighters, but four is preferable: two as inside rescuers and the additional two as the outside rescuers. It is highly advisable that the two-ladder–two-rescuer cradle carry technique be used for a Denver extraction. While

A.

B.

42 inches
(107 cm)

28 inches
(71 cm)

C.

FIGURE 10-13 The multi-purpose prop **A.** with a width of 28 inches (71 cm) **B.** and a lift of 42 inches (107 cm) **C.**

FIGURE 10-14 A "downed fire fighter" in position for the Denver drill; note that the available space is only 28 inches (71 cm) wide.

the multi-purpose prop is very limited. Therefore, to effectively illustrate the steps of the Denver drill and to provide a better understanding of the positioning of the two rescuers, one wall from the multi-purpose prop was removed in the Skill Drill images below. It is extremely important, however, that when training for this rescue, the prop be used in its entirety, creating the confinement necessary to accurately recreate the scenario.

For the Denver drill rescue, as with all window rescues, the exterior rescuers must be in place before beginning the extraction drill, and then perform the following steps as demonstrated in **SKILL DRILL 10-13**:

1. The first rescuer (rescuer #1) approaches the window, reaches in to determine if the downed fire fighter is there, and slides over the sill, entering the space. (**STEP 1**) As he enters, he locks his feet onto the sill before dropping completely in and then crawls over the downed fire fighter just past his feet. (**STEP 2**)

2. Rescuer #1 turns and faces the window he entered, leans over the downed fire fighter, tightens the fire fighter's shoulder straps, and then, by grabbing the two SCBA shoulder straps, pulls the downed fire fighter into a seated position. (**STEP 3**)

3. If needed, slight adjustments can be made to the downed fire fighter's body position to gain better access to the SCBA straps. After pulling him into a seated position, rescuer #1 squats or kneels down and calls the second rescuer (rescuer #2) in. (**STEP 4**)

4. Rescuer #2 enters the window in the same manner as rescuer #1 and positions himself with his back to the window. (**STEP 5**)

5. Rescuer #2 then squats down, with his back to the window, placing his SCBA cylinder into the "corner." (**STEP 6**) This will allow him more leverage when lifting. (**STEP 7**)

6. Rescuer #1 raises the downed fire fighter's knees to an upright, bent position, then grabs the downed fire fighter's SCBA shoulder straps low on the chest, placing one of his feet in between the downed fire fighter's legs, with his knees slightly bent. Rescuer #2 grabs the downed fire fighter under the SCBA cylinder with both hands. (**STEP 8**)

in training with this skill, the fire fighters entering must not be allowed to place their hands on the top of the walls of the prop for leverage or balance, as this would defeat the need and skills of the drill. Because of the limited space, there is little use of tools, although webbing may be considered to assist from the window with the final lift. **Note:** The space within

7 Using the "Ready, Ready, Go!" commands from rescuer #2, rescuer #1 pulls the downed fire fighter up and towards him in a "rocking motion" while rescuer #2 also lifts from under the cylinder. (**STEP 9**) Note the position of rescuer #1's leg against the downed fire fighter's foot. (**STEP 10**)

8 Rescuer #1 pushes the downed fire fighter back onto the knees in a seated position of rescuer #2, while rescuer #2 pulls the downed fire fighter into him on his knees. (**STEP 11**)

9 Rescuer #1 then gets on the floor in a weight lifter's squat, places the knees of fire fighter onto his shoulders and wraps his arms around the fire fighter's thighs. Rescuer #2 maintains his grasp on the bottom of the cylinder. (**STEP 12**)

10 Using the "Ready, Ready, Go!" command from rescuer #2, rescuer #1 lifts (by standing) straight up. (**STEP 13**)

11 At the same time, rescuer #2 lifts up and begins to pass the downed fire fighter over his head and out the window. For this final step, assistance is given from one of the ladder rescuers. (**STEP 14**)

This skill is challenging and physically demanding; however, with practice it becomes a skill that will work in an RIC environment.

Crew Notes

If rescuer #1 is unable to pull up on the shoulder straps and place the fire fighter onto rescuer #2's knees, an option would be for rescuer #1 to straddle the legs of the downed fire fighter, pull his knees upward, and then grab under the knees. Once in position, rescuer #1 can lift the downed fire fighter onto the knees of rescuer #2.

Crew Notes

When using the Denver drill extraction, once the two rescuers are in place for the lift and exit, they must move quickly! Remember the area you are working in is confining and your body position will be very uncomfortable and strength draining. If the crew cannot complete the lift and rescue, they need to give way immediately so that another RIC can make entry and attempt the rescue.

■ Options for the Denver Drill Rescue

When a fire fighter is found in a situation that requires a Denver drill-type rescue, the RIC needs to be proactive and determine if there is another rapid option. One option might be to take the

---

SKILL DRILL 10-13   Denver Drill Rescue Technique
NFPA 1407, 7.12(9)

1 The first rescuer (rescuer #1) approaches the window, reaches in to determine if the downed fire fighter is there, and slides over the sill, entering the space.

2 As he enters, he locks his feet onto the sill before dropping completely in and then crawls over the downed fire fighter just past his feet.

(Continued)

SKILL DRILL 10-13 Denver Drill Rescue Technique (*Continued*)

4 If needed, slight adjustments can be made to the downed fire fighter's body position to gain better access to the SCBA straps. After pulling him into a seated position, rescuer #1 squats or kneels down and calls the second rescuer (rescuer #2) in.

3 Rescuer #1 turns and faces the window he entered, leans over the downed fire fighter, tightens the fire fighter's shoulder straps, and then, by grabbing the two SCBA shoulder straps, pulls the downed fire fighter into a seated position.

5 Rescuer #2 enters the window in the same manner as rescuer #1 and positions himself with his back to the window.

6 Rescuer #2 then squats down, with his back to the window, placing his SCBA cylinder linto the "corner."

(Continued)

SKILL DRILL 10-13   Denver Drill Rescue Technique (*Continued*)

7   This position allows rescuer #2 more leverage when lifting.

8   Rescuer #1 raises the downed fire fighter's knees to an upright, bent position, then grabs the downed fire fighter's SCBA shoulder straps low on the chest, placing one of his feet in between the downed fire fighter's legs, with his knees slightly bent. Rescuer #2 grabs the downed fire fighter under the SCBA cylinder with both hands.

9   Using the "Ready, Ready, Go!" commands from rescuer #2, rescuer #1 pulls the downed fire fighter up and towards him in a "rocking motion" while rescuer #2 also lifts from under the cylinder.

10   Note the position of rescuer #1's leg against the downed fire fighter's foot.

(Continued)

SKILL DRILL 10-13   Denver Drill Rescue Technique (*Continued*)

11 Rescuer #1 pushes the downed fire fighter back onto the knees in a seated position of rescuer #2, while rescuer #2 pulls the downed fire fighter into him on his knees.

12 Rescuer #1 then gets on the floor in a weight lifter's squat, places the knees of fire fighter onto his shoulders and wraps his arms around the fire fighter's thighs. Rescuer #2 maintains his grasp on the bottom of the cylinder.

13 Using the "Ready, Ready, Go!" command from rescuer #2, rescuer #1 lifts (by standing) straight up.

14 At the same time rescuer #2 lifts up and begins to pass the downed fire fighter over his head and out the window. For this final step, assistance is given from one of the ladder rescuers.

© Joe Nedder/Jones & Bartlett Learning

window and make it a "door." This is especially helpful if you are operating with a large fire fighter from a first floor window. The interior RIC would request that another RIC proceed to the window and use a saw to cut downward from the window to the floor sill plate on both sides. After the cut is completed, the wall can be pulled outward and clear. This will now allow the downed fire fighter to be dragged or slid out. This technique can also be used on an upper floor window operating from a platform device. The downed fire fighter would then be placed onto the platform and secured before descending.

It is not suggested, based upon speed and safety, that a Stokes basket in conjunction with an aerial ladder be used for this rescue. First, there is no room to bring in and use a Stokes basket in the severely confined space. If you were able to get the downed fire fighter into the Stokes basket, he would need to be properly strapped in and secured before going out the window onto the aerial ladder, and failing to do this would put him in peril of falling to his death. If a platform is used, it is safer to position the platform at the window and have the downed fire fighter passed out without a Stokes basket to two rescuers in the platform basket. It is because of these two main reasons that the use of an aerial ladder for a Denver drill extraction is not suggested—it would be too time-consuming, and RIC needs to be *rapid*!

Wrap-Up

Chief Concepts

- The job of the RIC is to rapidly get the downed fire fighter out to safety and medical care.
- Using a window may be the quickest way to accomplish this.
- For a window rescue to be successful, it takes coordinated efforts and good communications among all the rescuers involved.
- You should avoid lowering a downed fire fighter out a window while connected to an RIC air supply whenever possible.
- When you position the downed fire fighter onto the window sill, make sure that both rescuers maintain a firm holds on the downed fire fighter so as to prevent him falling or slipping out.
- The use of personal webbing can greatly increase the rescuers ability to lift a heavier fire fighter.
- When practicing window techniques, *always* have the "victim" on a belay line!
- Clear communication to the IC as to which window will be used to extract the downed fire fighter is paramount.
- Be sure to place the tip of the ladder used for rescue at or below sill height to avoid entailments.
- The ladder used for rescue *must* be aggressively heeled by a minimum of two fire fighters to ensure safety.
- Avoid standing under the ladder while the downed fire fighter is being removed.

Hot Terms

<u>Chicago ladder carry</u> A technique used to carry either a fire fighter without a SCBA or a large fire fighter down a ladder. It can also be used for civilian rescue.

<u>Chicago lift</u> A technique for lifting a fire fighter out of a window.

<u>Denver drill</u> Term used to describe the removal of a fire fighter from a restricted space.

<u>Feet-first window lift</u> A technique for removing a fire fighter out of a window feet first.

<u>Head-first window lift</u> A technique for removing a fire fighter out of a window head first.

<u>Mask-mounted regulator (MMR)</u> Low-pressure regulator that attaches to the mask section of an SCBA.

References

Kolomay, R. and B. Hoff. *Firefighter Rescue and Survival*. Tulsa, OK: Penwell; 2003: 204.

McGrail, D. and J. Rogers. Confined Space Claims Denver Firefighter in a Tragic Building Fire. *Fire Engineering Magazine*. April 1993: 59.

National Fire Protection Agency (NFPA) 1407, *Standard for Fire Service Rapid Intervention Crews*. 2015. http://www.nfpa.org/codes-and-standards/document-information-pages?mode=code&code=1407. Accessed February 11, 2014.

National Fire Protection Association (NFPA) 1983, *Standard on Life Safety Rope and Equipment for Emergency Services*. 2012. http://www.nfpa.org/codes-and-standards/document-information-pages?mode=code&code=1983. Accessed October 14, 2013.

As the RIC Officer, you have to make decisions based on ever-changing conditions in a fire building. Bringing a downed fire fighter to safety can pose many unique challenges. As a well-trained RIC member, there are many techniques you can use depending on the situation that presents you. Choosing a method of removing a downed fire fighter from a window can make or break the rescue.

1. Why is good communication between the RIC and the IC so important when a fire fighter is being rescued via a window?

   A. The IC needs to know that a window rescue will be made and needs to identify the general location of the window to deploy a ladder crew.

   B. The IC does not necessarily need to know, so long as the rescue is being made.

   C. The IC will need to know where to direct EMS.

   D. All fireground operations will need to stop and all help should be directed to the window to assist with the rescue.

2. What are some of the problems associated with removing a fire fighter's SCBA prior to placing the downed fire fighter out a window?

   A. Removing the SCBA takes extra time.

   B. High heat/smoke conditions could cause airway compromise.

   C. Removing a SCBA in zero visibility and or restricted space is difficult.

   D. All of the above.

3. The Denver drill technique was developed after the tragic death of a fire fighter in a unique situation that occurred in a restricted space. Why is this important skill to practice?

   A. Almost every bathroom in a residential setting is a restricted space.

   B. A rescue from a stairway landing with a narrow window could utilize the same techniques.

   C. The overall skills can be used in many different type of restrictive space rescues.

   D. All of the above.

4. It is always best to remove the RIC air supply prior to placing a downed fire fighter out a window.

   A. True

   B. False

5. The ladder(s) used for the rescue of a downed fire fighter should be:

   A. At or below the window sill.

   B. Slightly above the window sill.

   C. To the right of the window.

   D. Two rungs in the window.

Rescuing a Fire Fighter From Below Grade and Through a Hole in the Floor

© Photos.com

© Joe Nedder/Jones & Bartlett Learning

Knowledge Objectives

After studying this chapter you will be able to:

- Understand how modern building construction can increase the danger of a floor collapse. (p 207–208)
- Understand the fire condition variables. (p 208)
- Understand the condition of the fire fighter variables. (p 208)
- Understand the building construction variables. (p 207–208)
- Recognize that rescuing a fire fighter who has fallen through a floor can be a difficult, complex, and labor-intensive. (p 209–210)
- Understand that we need to know if there is another means of accessing the fire fighter. (p 209–210)
- Realize that we must protect the downed fire fighter in place. (p 210)
- Remember to always train safely. (p 210)
- Know the different types of rescue solutions for a responsive or unresponsive fire fighter extraction. (NFPA 1407, 7.4(3), 7.12(6), 7.12(8)), p 210–218)
- Understand the need and importance of protective hoselines for both the downed fire fighter and the rescuers. (NFPA 1407, 7.11(2)), p 210, 213–216)
- Understand the limitations of using a ladder for this type of rescue. (p 219)

Skills Objectives

After studying this chapter and with hands-on training, you will be able to perform the following skills:

- Nance drill hose extraction for a conscious fire fighter. (NFPA 1407, 7.12(6), 7.12(8)), p 210–213)
- Rescuer hose slide. (NFPA 1407, 7.12(6), 7.12(8)), p 214–215)
- Nance drill five-step rope extraction for an unresponsive fire fighter. (NFPA 1407, 7.12(6), 7.12(8)), p 216–218)

You Are the Rapid Intervention Crew Member

A MAYDAY call is in progress and it is for a fire fighter who has fallen through the floor into the lower level, which is involved in fire. Your 360-RECON and size-up have shown that there is no access to the lower level except through the interior stairs, which are now burning. As a rapid intervention crew (RIC) member, you are activated and begin to make entry.

1. What tools and equipment should the RIC have with them?
2. What important information does the RIC needs to know?
3. If the downed fire fighter is unresponsive, what actions must be taken to protect him in place?

Introduction

When fire fighters are working on the floor directly above a fire, they are exposing themselves to significant risk. Although fire fighting is a job full of risk, you should always consider the risk–benefit analysis at any operation. As a fire fighter and as a member of an RIC, you must be pretrained in the skills and techniques that will assist and aid in the rescue of a fellow fire fighter who has fallen through a floor.

Here we will discuss some of the causes of a fire fighter going through a floor, the variables of the rescue, and the different techniques that can be used to gain access and make the rescue. Rescuing a downed fire fighter through a hole in the floor is difficult and takes great coordination, plus command and control. The skills presented here, if practiced and retained, will provide your crew with the techniques that can make a difference in the survival of a fire fighter.

Understanding Floor Construction

The issue of fire fighters falling through the floor is not a new one, although today we have a greater awareness of this possible event and have been given the knowledge and abilities to be better prepared. Fire fighters are now taught to better understand building construction, including floor construction in the various types of buildings. Building construction types III, IV, and V are the most common; however, never assume anything—the building you are entering may have been modified with an addition.

Fire that is directly impinging on a floor from below will affect its structural integrity by burning through the support lumber and/or reducing the size of the lumber holding up the floor. Either way, a fire fighter or a crew of fire fighters on the weakened floor can cause the floor to give way, which has been proven time and time again through near-misses and tragedy. With today's new construction techniques, the use of <u>engineered lumber</u> is creating greater unsupported spans **FIGURE 11-1**.

The lumber being used is called a wood truss or I-beam, which is manufactured in such a way as to increase the strength of the lumber. This sounds great in theory, but this

A.

B.

FIGURE 11-1 Engineered lumber with wide floor spans **A.** and **B.** cross-section of a wood I-beam.

system, which is held together with glue and a few wire nails, is not designed to deal with the effects of the fire and heat. The glue, when exposed to heat, will probably melt and fail before the wood elements burn through. This means that as you enter spaces that are supported by these wooden I-beams, you need to always be very cautious, feel for sponginess as you move on the floor, and take note if the floor is off-gassing and/or if there are reports of a significant body of fire on the floor below. Having a good level of awareness of the floor's condition can make the difference as to your survival.

■ Variables of a Fire Fighter Through the Floor

When a fire fighter falls through a floor, the variables to be evaluated fall into three categories:

1. Fire conditions
2. Condition of the fire fighter
3. Construction of the floor and building

Each of these variables will play an important part of your RIC and situational size-up and how the rescue attempt will be conducted.

Fire condition variables include:

- What are the fire conditions that the fire fighter has fallen into, and how long has the fire been burning?
- What are the fire conditions of the floor where you are, and how long has the fire been burning?
- Is the smoke coming from below heavy, dark, and fuel-enriched?
- Are there any firefighting efforts currently in place below?
- Are there any firefighting efforts currently in place on the floor you are on?
- What is the smoke condition and visibility on the floor where you are?
- Is there any venting in place that is helping to pull the fire onto the downed fire fighter?
- Is there any venting in place that will help with visibility on the floor you are on?

Variables regarding the condition of the fire fighter include:

- Are you able to see or communicate with the fire fighter?
- What was the overall crew air status when the fire fighter fell through the hole?
- If you can communicate with the fire fighter, what is his condition and can he assist in his rescue?

Floor and building construction variables include:

- What type of building are you in—residential or commercial?
- Do you have any other quick access to the area below?
- If the fire fighter fell into the lower level or basement, do you have any other points of access?
- What building construction class is it?
- How high are the ceilings (therefore, how far below is the fire fighter)?
- Was it a floor collapse or did the fire fighter go through a hole?

- Is the remaining floor reasonably safe for an RIC to operate on?
- If you have to enlarge the hole, what tools will you need? (If you are thinking of using a gas saw, determine if there is enough oxygen for the saw to run.)

The Line-of-Duty Death (LODD) of John Nance

When discussions about rescues through a hole in a floor take place, it is almost a certainty that the story of John Nance will come up. It was his tragic death in the line of duty that brought a better awareness to the fire service about these types of incidents. The December issue of the *Columbus Monthly* ran an article by Michael Norman entitled "The Murder of John Nance." The following is a synopsis of the story.

On July 25, 1987, John Nance was an Acting Lieutenant on Columbus (Ohio) Engine 3. They responded to the Mithoff Building at 151 N. High Street. On arrival, fire fighters saw smoke coming out of the center of the four-story, 11,500 square feet (1068 square meter) building. Nance positioned the Engine 3 crew in the rear and reported heavy smoke coming from a first floor shoe store and a possible basement fire. As Engine 3's crew entered the first floor, the smoke conditions began to worsen and extended from floor to ceiling. The heavy smoke conditions made visibility poor. When the self-contained breathing apparatus (SCBA) started to run low, the crew backed out, changed cylinders, and began to reenter. Before entering Chief Lindsey ordered that the crew use not only the hand line for orientation but also a rope as a lifeline. Engine 3 reentered.

Shortly after that, Chief Lindsey ordered several other fire fighters to take power saws into the building to ventilate the basement through the floor. To this day, it is not known how Nance fell into the basement, but it is believed he was searching an area to cut the floor when he fell through a section of weakened floor. He was about 70 feet (21 m) from the back door where entry had been made.

Another fire fighter (Wilson, Engine 10) who had crawled in from the front door had also fallen into a hole and grabbed for anything to prevent himself from falling into the basement. As Wilson pulled himself out he heard Nance screaming for help. The heat coming from the hole was intense, and Wilson's legs were burned as he pulled himself to safety. He did notice that the hole seemed to be about 12 feet (4 m) deep and that there was an orange glow inside. Wilson later stated, "I answered John, then radioed that a man was down in the basement." Wilson's air had begun to run out and at that point another fire fighter, Tim Cave, had come to the hole. Cave stated, "I found the hole and my arm went down to search with a light. I asked John if he could see my light. He replied in a very calm manner that he could see it, like he was just standing there waiting for me to get him out of there." Cave then asked Nance, "OK, how far down are you? Can you reach my hand? Nance reached up and grabbed my hand. He must have been standing on some stock because it was a real deep basement. I was able to reach down and I could see his blue glove meet mine." Cave tried to pull Nance out of the hole with one hand, but as he pulled he felt himself slipping into the hole. Cave told Nance "I can't pull you out" and Nance calmly replied, "OK, give me my hand back."

In the meantime, things were getting confusing inside the building. Rescue attempts from other fire fighters who had reached the hole began. One of the first actions was to take a hoseline and put it in the hole to provide protection and cooling for Nance. But the hose pulled up short and could not reach Nance. The rope used as a lifeline was found and it was sent down the hole to pull Nance out. Three fire fighters attempted to pull him out, but according to Cave, "We had his whole weight on the rope, but when we got to within about 3 feet (1 m) of the hole, we just lost all his weight." A second attempt was tried with a bowline; once the rope was lowered, Nance even tied two more knots himself. But even with more fire fighters pulling, the attempt failed as Nance fell off the rope halfway up. Heat and anxiety were beginning to take their toll on Nance. A ladder was called for and as they waited for it axes were used to enlarge the hole. As the fire fighters worked, conditions throughout the building deteriorated as the fire had spread through the walls. Nance's air supply was running low, and he tried to use it sparingly. The ladder was brought in and was lowered into the hole at a 90-degree angle. Nance found the ladder and climbed up but was on the wrong side striking his head on the floor joist. A fire fighter reached down to assist him but the heat drove Nance back to the basement floor. Nance was then lowered a SCBA cylinder as he had little air left.

With the fire conditions becoming unbearable, a dramatic rescue was attempted when a fire fighter descended the ladder with a hoseline. He tried to drag Nance's body up the ladder while fighting the fire back. Running out of air and unable to bring Nance up, the attempt failed. A second attempt was made, but the volume of fire forced the fire fighter to retreat. One last attempt was made, against the wishes of the chief officers, but the fire fighter could not even make it past the hole as the flames drove him back. Other last ditch efforts were attempted, including breeching sidewalls of the building, but before anyone knew it, the building was completely engulfed in flames. All rescue attempts were stopped. The next afternoon, John Nance's body was recovered from the ruins. The unused second SCBA cylinder that had been lowered to him was found lying next to him.

The RIC rescue techniques that have been developed to rescue a fire fighter in a similar situation are called the Nance drills.

Rescuing a Fire Fighter Who Has Fallen Through a Hole in the Floor

When a fire fighter falls through a floor, one of the first things that needs to be established is if there is another means of accessing the fire fighter besides the hole in the floor. If the fire fighter has fallen from one above-grade floor to another (for example, from the third floor to the second floor), you will in all likelihood have additional access via stairs and windows. This access would usually aid different RICs or other companies with the ability to make entry at the level on which the fire fighter has fallen. If, however, the fall has been from the first floor into a subfloor or basement, access to the fire fighter

might be very limited. If the fire fighter is in a basement, is there access only from an interior staircase, or is the basement a walkout with "at grade" access? If it is a residential building, is there a bulkhead? While trying to determine access, you must also size-up the fire situation. Has the fire fighter fallen into an uncontrolled area of fire? Has the hole now become a "chimney" drawing the fire to where the fire fighter has fallen through, and is this creating a large volume of smoke and fire where you are? Also, consideration must be given to the structural integrity of the floor that rescuers will need to stand on— can it bear the weight of the rescuers?

The rescue techniques presented here are based upon a below-grade rescue, with no other immediate access, on a floor that is safe for the fire fighters doing the rescue. The Nance drill is a series of skills and techniques that are used to rescue a fire fighter who has fallen through a hole and has no other means of escape except through the hole that he fell through. The techniques are broken into two categories:

1. Rescuing a responsive fire fighter who can assist in his own rescue.
2. Rescuing a unresponsive fire fighter who cannot assist in his own rescue.

■ MAYDAY, MAYDAY, MAYDAY! Fire Fighter Through the Floor!

When a MAYDAY is issued for a fire fighter who has fallen through the floor, the RIC needs to make sure they have the proper equipment with them. Along with the standard equipment discussed in the chapter *Planning for a Prepared Rapid Intervention Crew*, the RIC must also make sure that they have the rescue rope bag. The Incident Commander (IC), in addition to activating the RIC, must determine where the fallen fire fighter's company was located. Have they also fallen through the hole, or are they able to locate the hole and assist in the rescue? Thinking rescue and protection, the IC, working in conjunction with the RIC Chief, needs to begin moving two hoselines into the area of the rescue. These hoselines will play a key role in the rescue and survival of the downed fire fighter. The IC should also consider assembling a second and third RIC *immediately* and stage them to make entry to assist when needed. This drill can require a lot of manpower depending on conditions.

When the RIC makes entry, they will, as always, deploy their rope-assisted search procedure (RASP) line, securing it to a nonmoving object before entering the building; if it is a large commercial building, the RASP line should be secured to a nonmoving object in the portion of the building involved in fire. As the RIC advances into the general area, they must exercise extreme caution and make sure that there is always a floor ahead of them and not the hole. Always remember that something happened structurally to make a portion of the floor collapse. If a Company is already at the fall-through point, they can assist in guiding the RIC to their location. When the hole is located, the first priority will be to try and establish communications, either vocally or via radio, with the downed fire fighter. If contact is made, determine if he is able to assist in his own rescue. The RIC will also have to consider trying to stabilize the floor to a certain degree. Taking doors from the interior and placing them on the floor around the hole in an attempt to help distribute weight can accomplish this.

Another consideration is the size of the hole. If the hole is small and will complicate trying to bring the fallen fire fighter up and through it, consider enlarging the hole, but *only* as much as needed. Using axes and halligans will probably be the best option for a tool, considering that gas saws require clean air to run, and the physical environment will probably not provide adequate clean air for the saw.

Protecting the Fire Fighter in Place

When a fire fighter falls through a hole into an area involved in fire, every effort must be made to place protective hoselines in place. If it is in a basement with limited or nonexistent access, a hoseline for protection can be passed down the hole to a responsive fire fighter while the rescue is being organized. If the fire fighter is unresponsive, then a second RIC fire fighter must be placed through the hole to man the protective line for the downed fire fighter and the first rescuer while the rescue takes place **FIGURE 11-2**. At the same time, consideration needs to be given to the safety of the rescuers above. Depending on the fire situation, an additional line might be needed to protect the rescuers above.

Rescuing a Responsive Fire Fighter Who Can Assist in His Own Rescue

This rescue is used if the fire fighter who has fallen through the hole is responsive and able to assist in his own rescue **FIGURE 11-3**. If he were not able to assist, the techniques for an unresponsive fire fighter rescue would be used.

The Nance drill extraction using a hoseline is a technique that is used to extract a responsive fire fighter who is able to assist in his own rescue. This technique utilizes a charged hoseline and at a *minimum* five additional personnel to

FIGURE 11-2 The second RIC fire fighter mans the protective line.

© Joe Nedder/Jones & Bartlett Learning

Safety

When practicing any of the Nance drill skills, any fire fighter who is inserted or extracted from the hole must be on a belay line that is rigged and operated by competent personal, and a Safety Officer should also be present below grade when practicing these rescues.

make the rescue. Once it is established that the trapped fire fighter can assist in his own rescue, perform the Nance drill extraction using a hoseline by performing the following steps as demonstrated in **SKILL DRILL 11-1**:

1 Evaluate the hole size—does it need to be enlarged? If so, rapidly enlarge it to a point that is adequate.

2 A charged hoseline (a 1¾-inch [4-cm] line works best) is brought into place, and a bight of the hose is sent down the hole to the trapped fire fighter. (**STEP 1**) When the trapped fire fighter has received the bight of hose, he calls up to the rescuers that he has it.

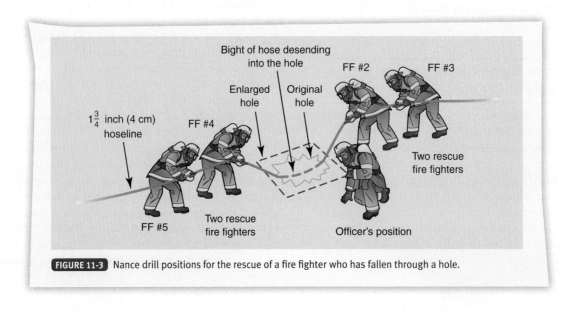

FIGURE 11-3 Nance drill positions for the rescue of a fire fighter who has fallen through a hole.

3 The five rescuers then position themselves with two fire fighters holding onto the hose at the two opposite corners, to create the widest opening for the rescue, and with the officer in-between. (**STEP ②**)

4 The trapped fire fighter stands on the bight of hose, "locking" his heels onto the hose, and then, crossing his arms, pulls the hose inwards towards him. He then calls up that he is ready. (**STEP ③**)

5 The two rescue fire fighters on the hoseline who are closest to the hole are the "front" and the two behind them are the "rear." (**STEP ④**) The front positions, with knees bent, grab the hose as low to the hole as possible. The rear positions, with knees bent, grab the hose close to the side of the front fire fighters.

SKILL DRILL 11-1   Nance Drill Extraction Using a Hoseline
NFPA 1407, 7.12(6), 7.12(8)

1 A charged hoseline (a 1³/₄-inch [4-cm] line works best) is brought into place, and a bight of the hose is sent down the hole to the trapped fire fighter.

2 The five rescuers then position themselves with two fire fighters holding onto the hose at the two opposite corners, to create the widest opening for the rescue, and with the officer in-between.

3 The trapped fire fighter stands on the bight of hose, "locking" his heels onto the hose, and then, crossing his arms, pulls the hose inwards towards him. He then calls up that he is ready.

4 The two rescue fire fighters on the hoseline who are closest to the hole are the "front" and the two behind them are the "rear." The officer prepares the team by calling out "Ready on my left?" touching the front man on his left, who then responds "Left ready!" and "Ready on my right?" touching the front man on his right, who responds "Right ready!".

(Continued)

6 The officer prepares the team by calling out "Ready on my left?" touching the front man on his left, who then responds "Left ready!" and "Ready on my right?" touching the front man on his right, who responds "Right ready!" These simple prompts will clearly establish the left and right crews for upcoming commands.

7 The officer squats low so that he can look down the hole. In this position the officer can: (1) command and control the crews; and (2) see the trapped fire fighter as he presents himself coming up from below. (**STEP 5**)

8 The officer then calls out "Pull" or "Haul." Both crews pull on the line and stop. (**STEP 6**)

9 The officer then gives the order "Front reset!" and only the two front fire fighters let go of the line and grab the hose once again, as close to the hole as possible. (**STEP 7**) Once done, the officer gives the order

---

SKILL DRILL 11-1 Nance Drill Extraction Using a Hoseline (*Continued*)

5 The officer squats low so that he can look down the hole. In this position the officer can: (1) command and control the crews; and (2) see the trapped fire fighter as he presents himself coming up from below.

6 The officer then calls out "Pull" or "Haul." Both crews pull on the line and stop.

7 The officer then gives the order "Front reset!" and only the two front fire fighters let go of the line and grab the hose once again, as close to the hole as possible.

8 Once done, the officer gives the order "Rear reset!" and the two rear fire fighters let go and grab the hose again, close to the front fire fighter. The officer then orders "Pull!" and both crews pull on the line and stop. This process is repeated as many times as needed to bring the fire fighter up and out.

(*Continued*)

9 When the officer can see and reach the trapped fire fighter, he should grab onto the trapped fire fighter's two shoulder straps and assist in the final pull to get the trapped fire fighter out of the hole.

© Joe Nedder/Jones & Bartlett Learning

"Rear reset!" and the two rear fire fighters let go and grab the hose again, close to the front fire fighter. (**STEP 8**)

10 The officer then orders "Pull!" and both crews pull on the line and stop.

11 This process is repeated as many times as needed to bring the fire fighter up and out.

12 When the officer can see and reach the trapped fire fighter, he should grab onto the trapped fire fighter's two shoulder straps and assist in the final pull to get the trapped fire fighter out of the hole. (**STEP 9**)

Crew Notes

Positioning the rescuers at opposite corners of the hole will increase the usable space within the opening for the rescue.

Safety

When the crew is hauling on either the hoseline or the rope for the unresponsive fire fighter rescue, it is extremely important that the rescuers reset only when instructed. This will avoid all the rescuers from letting go of the hose at the same time, which would result in the fire fighter being rescued falling back into the hole.

Crew Notes

When the hauling begins, it is important for the officer to start and keep a cadence. This will assist the rescuers who will not be waiting in anticipation for what is next.

Safety

As the trapped fire fighter is brought up, there is a great danger that he could bang his head on the floor joists or ceiling as he is being pulled through. The officer should be watching for this and ensure that when he has visibility of the fire fighter's head, it is positioned so this does not happen. This might mean repositioning the hoseline being used for the hauling within the hole with the fire fighter on it.

■ Rescuing an Unresponsive Fire Fighter

The rescue of an unresponsive fire fighter or a fire fighter who is responsive but unable to assist in his own rescue utilizes the same skills and techniques. Because the trapped fire fighter cannot assist, this rescue becomes a five-step process:

1. Put a rescue fire fighter down into the hole.
2. If the area is involved in fire, place a second fire fighter into the hole to man a hoseline and to protect the downed fire fighter in place.
3. Search for and locate the trapped fire fighter and, if needed, bring him back to the area below the hole.

4. Extract the trapped fire fighter.

5. Extract the rescue fire fighters.

Putting a Rescuer Down the Hole

The easiest and usually the quickest way to put a rescuer through the hole is to utilize a 1¾-inch (4-cm) hoseline and slide down it like a fire pole. This takes skill, coordination, and a mentality for keeping safety in mind. The hand line must be charged for this to work. Using the hand line has three purposes:

1. It provides an entry device.

2. It provides a hand line to use for protection.

3. It provides an exit device.

The rescuer hose slide is demonstrated in **SKILL DRILL 11-2**:

1 A charged hoseline is placed though the hole to the floor below, nozzle first.

2 At least two RIC fire fighters grab and secure the hose-line where the rescuer will make entry. (**STEP 1**)

3 The rescuer seats himself in the hole adjacent to the hoseline, grabs onto the hoseline above the floor with both hands, (**STEP 2**) and rolls his body onto the hose, with his legs wrapped around the hose. (**STEP 3**)

4 The rescuer then, one hand at a time, lets go and reaches under the floor, grabbing the hose below. (**STEP 4**)

5 As soon as both hands are passed under the floor and grabbing the hose, the rescuer slides down, in as controlled a manner as possible, to the floor below. (**STEP 5**)

6 Once on the floor, the rescuer maintains contact with the hose and reaches for the trapped fire fighter. (**STEP 6**)

7 If he cannot see or locate the downed fire fighter, the rescuer will tie off his personal rope to the hose above the nozzle and use it as a safety and search line. (**STEP 7**)

Tying the rope off to the hoseline is a key life safety point. If for whatever reason the rescuer becomes disoriented and cannot relocate the hose, the rope will always bring him back to the hoseline—it is his way out.

Placing a Second Rescuer Into the Hole

If the area is involved in fire, a second RIC fire fighter will need to go into the hole to use the line to protect the downed fire fighter in place. A key part of any RIC operation is to protect the downed fire fighter in place when possible, and this is especially important in a below-grade rescue with no other means of access. The second rescuer will slide down the hose in the same manner as the first rescuer, and when he reaches bottom, will communicate with the first rescuer and then call for more line to be sent down, if needed. Once the additional line has been received, the second rescuer will proceed to operate the nozzle as needed.

Crew Notes

When calling for more line, do not just yell, "More line!" Be specific and indicate exactly what is needed: "10 feet (3 m) more line!" Better communications make for a smarter operation.

Locating the Trapped Fire Fighter

Search for and locate the trapped fire fighter and, if needed, bring him back to the area below the hole. When a fire fighter falls through a floor, we cannot always expect to find him directly below the hole. He might have hit and rolled; he might have been semi-responsive and tried to find his way out; or he might have hit furniture, shelving, or another object when falling through the floor and been redirected. There are many different reasons why the downed fire fighter might not be directly below the hole, but regardless of the reason, the priority is still to locate him right away.

SKILL DRILL 11-2 | Rescuer Hose Slide
NFPA 1407, 7.12(6), 7.12(8)

1 A charged hoseline is placed though the hole to the floor below, nozzle first. At least two RIC fire fighters grab and secure the hoseline where the rescuer will make entry.

2 The rescuer seats himself in the hole adjacent to the hoseline and grabs onto the hoseline above the floor with both hands.

(Continued)

SKILL DRILL 11-2   Rescuer Hose Slide (*Continued*)

3 The rescuer rolls his body onto the hose with his legs wrapped around the hose. Note belay line used in training.

4 The rescuer then, one hand at a time, lets go and reaches under the floor, grabbing the hose below.

5 As soon as both hands are passed under the floor and grabbing the hose, the rescuer slides down, in as controlled a manner as possible, to the floor below.

6 Once on the floor, the rescuer maintains contact with the hose and reaches for the trapped fire fighter.

7 If he cannot see or locate the downed fire fighter, the rescuer will tie off his personal rope to the hose above the nozzle and use it as a safety and search line.

To conduct this search, the first fire fighter will utilize his personal rope that is tied off to the hoseline. Even if the hoseline is used by the second fire fighter for a protective line, the rope will still lead you to the way out. With the rope, the first fire fighter will search out 5 to 10 feet (2 to 3 m) and then complete a 360-degree sweep. Listen for the personal alert safety system (PASS) device sounding. If the fire fighter is entangled or buried under objects or debris, the PASS might be muted, but it will still be sounding. Pause, hold your breath, and listen. The use of a thermal imaging camera (TIC) will greatly increase the chances of successfully locating the downed fire fighter, provided that the rescuer is trained and competent in the use of the TIC and that there are not a lot of obstacles blocking the camera from "seeing." When the downed fire fighter is located, the rescuer will bring him back to the area directly under the hole and update the RIC above the hole with his situation, condition, and the type of extraction technique that will be needed.

Extracting the Trapped Fire Fighter

Extracting the trapped fire fighter through the hole will utilize skills similar to the hose extraction. Because the fire fighter is unresponsive, you cannot use the bight of hose method, because there is no way to secure him to it. In its place use the ½-inch (13-mm) life safety rope that is in the rescue rope bag.

This Nance drill extraction using a life safety rope skill takes a minimum of five rescuers, similar to the hose extraction technique. The fire fighters will perform in the same manner as in the hose drill, but using the life safety rope instead of the hose for the extraction. It is important for the rescuer who has entered, located the trapped fire fighter, and moved him to the general location of the hole to communicate with the RIC above that it is a rope extraction. The Nance drill extraction using a life safety rope is demonstrated in **SKILL DRILL 11-3**:

1. The RIC, using the rescue rope bag, takes the rope and pulls out an amount twice what is needed to reach the floor below. They then place a handcuff knot at the midpoint and lower the bight of rope with the handcuff knot to the rescuer below. **(STEP ①)** **Note:** The rope can also be sent down as just a bight, and the rescuer can tie the handcuff knot.

2. The rescuer below will take the handcuff knot and place it on the forearms of the downed fire fighter as close to, but not above, the elbows as possible. **(STEP ②)** This will allow the coat materials to gather and bunch, so as to help secure the rope, which will hold in place while the fire fighter is lifted up and out. After the handcuff knot is placed in position on both of the trapped fire fighter's arms, the rescuer will tighten the knot securely. **(STEP ③)**

3. While the rescuer below is preparing the victim, the five rescuers above will position themselves in a similar manner as in the extraction using a hoseline, with two fire fighters holding onto the rope at the two opposite corners to create the widest opening for the rescue, with the officer in-between. This is the same technique as is used in the Nance drill extraction using a hoseline.

4. When the rescuer below is ready, he will call up to the crew above indicating the downed fire fighter is ready to go and to take up slack in the rope. While slack is being taken up, the rescuer should position the fire fighter's hands over his head, in position for the rescue lift. **(STEP ④)**

5. The two fire fighters on the rope line who are closest to the hole are the front, and the two behind them are the rear. The front positions, with knees bent, grab the rope as low to the hole as possible, and the rear positions, with knees bent, grab the rope as close to the side of the front fire fighters.

6. The officer prepares the team by calling out "Ready on my left?", touching the front man on his left, who responds "Left ready!", and then "Ready on my right?", touching the front man on his right, who responds "Right ready!" These simple prompts will clearly establish the left and right crews for upcoming commands.

7. The officer then calls out "Pull" or "Haul." Both crews pull on the rope, stop, and hold in place.

8. During this extraction process, the rescuer below assists in guiding the unconscious fire fighter through hole above. **(STEP ⑤)** The officer squats low, looking down the hole. In this position, the officer can (1) command and control the crews and (2) see the trapped fire fighter as he presents himself coming up from below. **(STEP ⑥)**

9. The officer then gives the order "Front reset!" and the two front fire fighters let go of the rope and grab the rope once again, as close to the hole as possible. Once done, the officer gives the order "Rear reset!" and the two rear fire fighters let go and grab the rope again close to the front fire fighter.

10. The officer then orders "Pull!" and both crews pull on the rope, stop, and hold in place.

11. This process is repeated as many times as needed to bring the downed fire fighter up and out.

12. When the officer can see and reach the trapped fire fighter, he should grab onto the trapped fire fighter's two shoulder straps and assist in the final pull to get the trapped fire fighter out of the hole. **(STEP ⑦)**

Extracting the Rescue Fire Fighters

After the unresponsive fire fighter has been extracted to the floor above, you still have one or two rescuers who will need to be pulled to safety. This can be accomplished by utilizing the Nance drill extraction using a hoseline skill drill 11-1. The fire fighters above the hole can either send down a bight of hose or they can withdraw the protective line in the hole if it is not needed and send it back down with a bight. Either way, the same skill set is used for the extraction. After the first rescuer is removed, the hose is sent back in to extract the second rescuer.

SKILL DRILL 11-3

Nance Drill Extraction Using Life Safety Rope
NFPA 1407, 7.12(6), 7.12(8)

1 The RIC, using the rescue rope bag, takes the rope and pulls out an amount twice what is needed to reach the floor below. They then place a handcuff knot at the midpoint and lower the bight of rope with the handcuff knot to the rescuer below.

2 The rescuer below will take the handcuff knot and place it on the forearms of the downed fire fighter as close to, but not above, the elbows as possible.

3 After the handcuff knot is in position on both of the trapped fire fighter's arms, the rescuer will tighten the knot securely.

4 When the rescuer below is ready, he will call up to the crew above indicating the downed fire fighter is ready to go and to take up slack in the rope. While slack is being taken up, the rescuer should position the fire fighter's hands over his head, in position for the rescue lift.

(Continued)

SKILL DRILL 11-3   Nance Drill Extraction Using Life Safety Rope (*Continued*)

5   During this extraction process, the rescuer below assists in guiding the unconscious fire fighter through hole above.

6   The officer squats low, looking down the hole. In this position, the officer can (1) command and control the crews and (2) see the trapped fire fighter as he presents himself coming up from below.

7   When the officer can see and reach the trapped fire fighter, he should grab onto the trapped fire fighter's two shoulder straps and assist in the final pull to get the trapped fire fighter out of the hole.

© Joe Nedder/Jones & Bartlett Learning

Crew Notes

If the rescuers find the fire fighter is low on air, they can request that the RIC air supply unit be sent down. If they are not using the universal air connection (UAC), then a regulator swap might be warranted. If so, the spare SCBA will have to be secured to the unresponsive fire fighter as he is lifted out of the hole.

After the Extraction of the Downed Fire Fighter

This RIC skill is manpower-intensive and requires good communications and coordination. When an RIC rescue operation such as the Nance drill is in progress, the RIC Operations Chief needs to have additional teams ready to bring the downed fire fighter out to safety and medical care as soon as he is extracted from the hole. The team doing the extraction will continue to extract the rescuers down below, while the additional RIC will complete the rescue. This type of proactivity will bring the downed fire fighter out of the building faster, and in many cases, speed will be important. The RIC Operations Chief and the IC must plan for this additional manpower, even if that means pulling from existing companies on the scene. Larger departments will usually have a great many fire fighters on scene. Many smaller organizations will be manpower short, and with mutual aid response times often exceeding 20 minutes, the IC must find a way to utilize the manpower on scene to complete the rescue. The crew that will bring the downed fire fighter out of the building must take the required steps to prepare him for exit, including ensuring an air supply. Remember, if the UAC was used to supply air before the downed fire fighter was extracted, there might be a need to resupply his cylinder before the exit extraction begins.

Using Ladders for Through-the-Hole Rescues

The use of a ladder for a through-the-hole rescue at first glance might look to be the simplest solution, but it is most likely not the fastest solution. To say "never use a ladder" is a dangerous statement. You need to consider all of your options and use the ones that will make the rescue rapidly. In most cases the use of the Nance drills with either hose or rope will be the fastest.

When a MAYDAY is called for a fire fighter through a hole in the floor or collapse, actions need to happen rapidly. Imagine an RIC having to grab a ladder and bring it into the building to the proper location and then having to place it into the hole. How long will this take? And if there is a crew inside that witnessed the fall and they are waiting for you to bring in a ladder, how long will it take for you to get to them and what will there air status be when you arrive?

- The first question is: what size ladder is needed?
- Will a 12- or 14-foot (approximately 4-m) straight or roof ladder work?

- What about a pencil or attic ladder? Will it be tall enough?
- How are you going to get the ladder to the proper location? Can you put it through a window adjacent to the hole? Unless you are in a large, wide-open commercial structure, ladders will not go around corners.
- If you can get a ladder inside, will there be enough ceiling height to tip it up and drop it into the hole?
- Once in the hole, the ladder will need to be kept as vertical as possible. This will be easier if the hole size is as large as possible so that the fire fighters can fit **FIGURE 11-4**. How are you going to secure and keep the ladder vertical?

Each of these concerns will be a major obstacle to make the rescue successful, but utilizing the hose or rope, as presented in the Nance drills, will help keep the rescue rapid.

© Joe Nedder/Jones & Bartlett Learning

FIGURE 11-4 The ladder in the hole needs to remain as vertical as possible, so as to not restrict a fire fighter from getting out of the hole.

Wrap-Up

Chief Concepts

- When fire fighters are working on the floor directly above a fire, they are exposing themselves to significant risk.
- We must remember to always consider the risk–benefit analysis at any operation.
- You must be pretrained in skills and techniques for the rescue of a fellow fire fighter who has fallen through a floor.
- Fire that is directly impinging on a floor from below will affect its structural integrity by burning through the support lumber and/or reduce the size of the lumber holding up the floor.
- New construction techniques use engineered lumber, which is creating greater unsupported spans.
- Engineered lumber (truss or I-beam) is typically only held together by glue and a few small wire nails.
- Engineered lumber does not hold up well to heat; the glue will melt when exposed to heat.
- One of the very first things that need to be established when a fire fighter falls through a floor is: do we have another means of accessing the fire fighter besides the hole in the floor?
- When the hole where the fire fighter(s) have fallen through is located, the first priority will be to try and establish communications—vocal or via radio—with the downed fire fighter.
- The use of a ladder is not always the best option for retrieving a fire fighter who has fallen through a floor; however, we need to always consider all available options.

Hot Terms

<u>Engineered lumber</u> Larger spans of lumber used for floor or ceiling construction; typically glued together under pressure.

<u>Nance drill</u> A series of skills and techniques that are used to rescue a fire fighter who has fallen through a hole and who has no other means of escape except through the hole.

<u>Rescue rope bag</u> A rope bag containing 75–100 feet (23–30 m) of 9-mm (½-inch) rope for purposes of rescuing a fire fighter.

References

National Fire Protection Agency (NFPA) 1407, *Standard for Fire Service Rapid Intervention Crews*. 2015. http://www.nfpa.org/codes-and-standards/document-information-pages?mode=code&code=1407. Accessed February 11, 2014.

Norman, M. The Murder of Norman Nance. *Columbus Monthly*. 13:12; December 1987.

RAPID INTERVENTION CREW MEMBER
in action

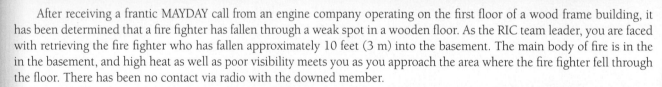

After receiving a frantic MAYDAY call from an engine company operating on the first floor of a wood frame building, it has been determined that a fire fighter has fallen through a weak spot in a wooden floor. As the RIC team leader, you are faced with retrieving the fire fighter who has fallen approximately 10 feet (3 m) into the basement. The main body of fire is in the in the basement, and high heat as well as poor visibility meets you as you approach the area where the fire fighter fell through the floor. There has been no contact via radio with the downed member.

1. Given that there has been no communication with the downed fire fighter, what will the RIC most likely have to do first upon reaching the area where the hole is?

   A. Send a member of the RIC down through the hole to search for the fire fighter.

   B. Try to establish communication prior to engaging any activities.

   C. Call for a ladder crew and attempt to place a ground ladder down into the hole.

   D. Call for more help before committing members of the RIC.

2. How would you have a fire fighter descend through the hole to reach or search for the downed fire fighter?

   A. Lower the fire fighter with a pre-rigged haul system.

   B. Use a ground ladder.

   C. Use the "rescuer hose slide" method.

   D. Try to access the basement level from an exterior way.

3. Upon reaching the floor of the basement, you locate the downed fire fighter. He is breathing but not responsive. What method should be used to bring the fire fighter up through the hole?

   A. Using the rope rescue bag, tie a handcuff knot around the fire fighter and haul him up.

   B. Place the fire fighter on the hose and utilize the Nance drill technique.

   C. Try to use a ground ladder and carry the downed member up the ladder.

   D. Rig a haul system using the above ceiling as an anchor point.

4. When a fire fighter falls through a floor, he or she will always be located directly under the hole.

   A. True

   B. False

5. What method would you use to remove a responsive fire fighter from a lower level via a hole in the floor?

   A. Lower a drag strap and pull the fire fighter up.

   B. Use a 1¾ inch (4 cm) hand line and the Nance drill technique.

   C. A pre-rigged haul system.

   D. If the fire fighter is responsive, take the time to get a ground ladder and have him or her climb to safety.

CHAPTER

12

Fire Fighters Trapped in Attics

Knowledge Objectives

After studying this chapter you will be able to:

- Understand the different types of attic spaces. (p 223–227)
- Understand the differences between finished and unfinished attic spaces as they relate to rescuing a fire fighter. (p 223–227)
- Understand the variables of attics in regards to construction. (p 224–227, 228–230)
- Know the dangers that truss construction pose to fire fighters during a rapid intervention crew (RIC) rescue operation. (NFPA 1407, 7.1(1) , p 224–227)
- Know the types of access into attics and the dangers they pose. (NFPA 1407, 7.1(1), 7.7(3) , p 227–230)
- Understand that an RIC rescue of a fire fighter in an attic, finished or unfinished, poses many dangers and includes numerous variables. (NFPA 1407, 7.7(3), 7.12(7) , p 231–233)
- Understand the rescue exit options for the distressed fire fighter. (NFPA 1407, 7.4(3), 7.12(7) , p 231–233)
- Recognize and understand that there are very limited times a fire fighter should enter an unfinished attic and what precautions should be taken. (NFPA 1407, 7.12(7) , p 233)

Skills Objectives

- There are no skill objectives for Rapid Intervention Crew candidates. NFPA 1407 contains no Rapid Intervention Crew Job Performance Requirements for this chapter.

You Are the Rapid Intervention Crew Member

The RIC company, of which you are a member, is staged and standing by, when all of a sudden a MAYDAY is called that a fire fighter is trapped in an unfinished attic space. You look up and see moderate smoke drifting from the ridge vent.

1. What concerns should you have about this rescue attempt?
2. If you can get to the fire fighter, what are your directional options (up, down, or to the side) for removing him to safety?
3. How would truss construction affect this operation?

Introduction

Because there are so many types of attics and the buildings they are a part of, it makes it very difficult to develop standard rescue techniques that fit all needs. Therefore, for the purposes of this book, specific skills for an attic rescue will not be presented here; rather, the different types of attics (finished and unfinished) will be defined and the dangers they present examined. If there is an activation for a downed fire fighter in an attic, the application of the rescue techniques reviewed previously can, in many cases, then be applied. We will also discuss the few reasons fire fighters should go into attics and what can be done to attempt a rescue if something goes wrong. An important part of any fire fighter survival and rescue program is to learn not to get into trouble. This includes knowing how to conduct a risk assessment and make decisions based upon logic and life safety.

Attics

Attics are mostly found with <u>Type V wood frame construction</u> buildings. In <u>Type III ordinary construction</u>, we will encounter cocklofts, but they are not considered, by definition, an attic. An attic can be defined as the space below the roof of a structure. In the most basic of terms it is a <u>confined space</u>, with limited access, that can be unfinished or finished living space.

■ Finished Attic Space

<u>Finished attics</u> typically function as additional living space. They have stairways leading to them, doors, and windows, and can be a single room, multiple rooms, apartments, or boarding house rooms **FIGURE 12-1**. Finished attic space is found in single and multiple occupancy residences, mixed occupancy buildings, and commercial occupancy buildings. An RIC operation would, in most likelihood, use the same skills and techniques as for any other situation to get to and move a downed fire fighter to safety. However, finished attic space has its own inherent dangers and concerns. Your primary concern with finished attic space is to always remember that it is an attic,

A.

B.

FIGURE 12-1 An example of finished attic space as shown from **A.** the inside and **B.** the outside.

and that the space above you is a modest void below the roof. Finished attic space will usually also have <u>knee walls</u> in the living areas **FIGURE 12-2**.

If you are dealing with <u>balloon construction</u>, you must be aware of any fire getting into the walls, traveling into the knee wall voids, and getting over your head unchecked. Attempting

FIGURE 12-2 A knee wall in a finished attic.

FIGURE 12-3 An attic with open ceiling joists with mechanical systems and blown-in insulation.

a rescue of a fire fighter in an attic with unchecked fire traveling through the walls and overhead is an extremely dangerous situation. Vertical ventilation is a priority in this type of situation, along with aggressive fire control. The RIC will need the support of hoselines (from the Engine Company) and ventilation (from the Truck Company).

Unfinished Attic Space

Unfinished space creates a lot of difficulties for an RIC operation. There are many types of unfinished attic spaces fire fighters can encounter. Almost every residential structure will have an attic, but these attics can greatly vary from a small tight crawl space to a large, open attic that might be full of mechanical systems such as the heating, ventilation, and air conditioning (HVAC) systems. Many attics may have open ceiling joists filled with insulation **FIGURE 12-3**, while others will have floors, typically plywood, that are used for storage. Attics that do not have any floor covering are inherently dangerous for fire fighters to enter. Without a floor, it is next to impossible to drag or slide a fire fighter over the open ceiling joists.

Variables of Attics

Attic variables are based on the construction type, the style, and how the space is used. The primary variables are the construction, which includes the pitch and style of the roof (gables, mansards, and dormers) and the access available to the attic itself.

Construction

Construction will play a significant role with attics, and as previously mentioned, attics are primarily found in Type V, wood frame buildings. For the most part, these buildings are residential and can be single or multi-occupancy. The intent of this section is not to teach building construction, but rather to create awareness for the fire fighter of

what additional knowledge they must have. An excellent source for information and one that is considered to be a "fire service bible" is *Brannigan's Building Construction for the Fire Service*. Having a good understanding of building construction is important because of the variations and differences in the types of buildings you will encounter, including older construction such as post and beam construction, balloon frame construction, traditionally framed platform construction (non-truss), platform construction using engineered lumber, truss joist I-beams and roof trusses, construction with multiple gabled roofs and numerous rooflines using roof trusses, and subdivided unfinished attic space that has a finished area undetectable from the street. All of these construction variables are important to know and understand.

In older buildings (typically up until the middle of the 1900s), you will most likely encounter balloon construction where, if the fire is below the attic space and gains access into the walls and floor voids, it will soon be in the attic. In more modern buildings, you will find truss construction. With many of the roof truss systems, the web members, which make the truss system strong, make it almost impossible to use the attic as a storage area **FIGURE 12-4**. However, HVAC mechanicals are often found in this space, and homeowners will still utilize what they can of this space for storage. This makes for a very dangerous situation for any fire fighter to enter. Considering it is taught that a truss roof construction is dangerous for fire fighters to stand on because it is susceptible to failure under heat and fire conditions, it makes sense that it also cannot be safe to enter that same attic from below.

Safety

If you know that a truss roof engaged in fire is unsafe to be on, then why would it be safe to be under? Understand building construction and beware of truss construction!

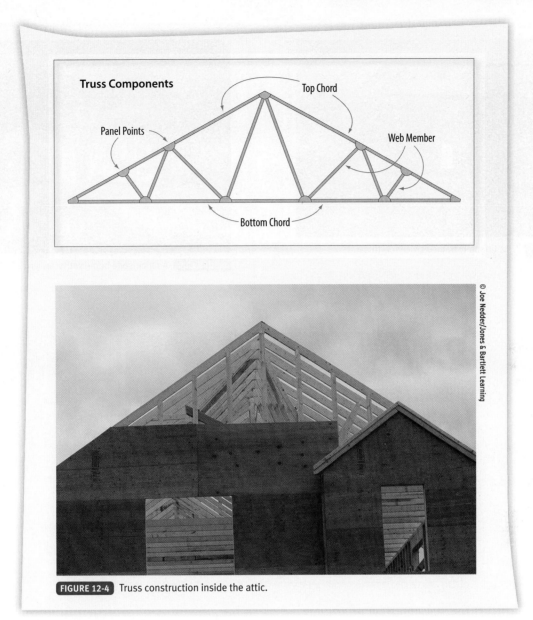

FIGURE 12-4 Truss construction inside the attic.

The most obvious way to identify a truss roof would be from preplanning—gather the information from your building departments, observe a new neighborhood going up and how the homes are constructed, and, during smoke alarm inspections, take the opportunity to look into the attic space. None of this, however, is fool proof, and if on arrival you do not know if the building has a truss roof or not, always err on the side of caution until it is determined exactly what type of construction you are dealing with. The most probable way a truss roof will be discovered will be either when a portion of the roof has burnt through, exposing the attic area; the Truck Company discovers it is a truss roof while venting, or the truss roof is discovered by a company hooking ceilings.

Regardless of how it is discovered, this information needs to be passed onto the Incident Commander (IC) quickly. If information is communicated by radio, all companies on the fireground will hear the transmission, become immediately aware of the dangers, and be able to take the appropriate

precautions. If the roof trusses are on fire, companies operating on the roof should evacuate immediately, as they are in imminent danger. There is no warning sign for a truss failure when they are on fire—it will just happen.

Traditionally framed attics (non-truss) allow the occupants to utilize it as storage space **FIGURE 12-5**. The amount of storage will depend on the style of the house (either ranch, cape, or colonial) and the roof pitch. Typically we will find $^4/_{12}$ to $^6/_{12}$ pitch roofs on ranch style homes and $^9/_{12}$ to $^{10}/_{12}$ pitch roofs on two-story colonial style homes. The less the roof pitch, the less area for storage **FIGURE 12-6**. Conversely, on the colonial-style homes, you will find large open attics with the ridgepole 10 to 12 feet (3 to 4 m) off the attic joist. In many cases, these large open attics will have plywood flooring allowing the occupants to utilize the attic for storage. When you couple these large open areas with HVAC ductwork, air handlers, and then add in storage, an attic like this is a very dangerous situation for a fire fighter

FIGURE 12-5 A traditionally framed attic (non-truss).

© Joe Nedder/Jones & Bartlett Learning

FIGURE 12-6 A ranch-style home with a lowerroof pitch.

© Joe Nedder/Jones & Bartlett Learning

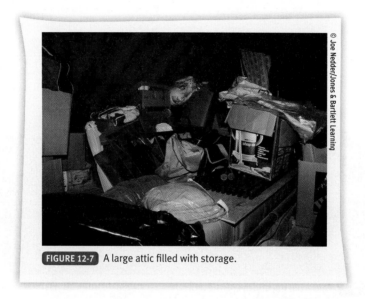

FIGURE 12-7 A large attic filled with storage.

© Joe Nedder/Jones & Bartlett Learning

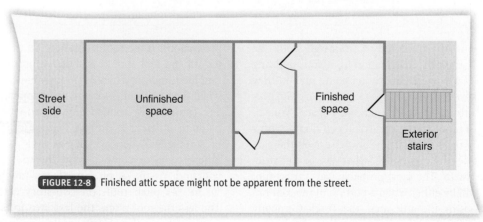

FIGURE 12-8 Finished attic space might not be apparent from the street.

to enter during a building fire FIGURE 12-7 . The fire fighter will be exposed to all types of entanglements, with few or no landmarks to maintain orientation and only one way out— the way he came in.

Sometimes an attic can be very deceiving. You might find a house with what appears from the street to be an unfinished space; however, the homeowner divided the attic in half and has put an apartment or another finished room in the rear of the attic space FIGURE 12-8 , which may be accessible from an interior or exterior stairway. Whenever a separate exterior stairway (typically in the rear of a building, hidden from the street) leading to the attic is found, it is a strong indicator of a separate apartment or boarding room.

Today's variety of home styles can include multiple gables and numerous rooflines. The <u>dormers</u> can be shed or gabled and can vary in size **FIGURE 12-9**. These can also be very deceiving from the outside. In homes with multiple gables and rooflines, expect to find hidden voids, knee walls, and crawl spaces that can be used for storage. Full shed dormers allow the full use of finished space in areas that might have been traditionally an unfinished space. It can also be used as semi-finished space (i.e., an unheated space with a floor but without finished walls) for walk-in storage or a large walk-in closet. During size-up this type of space can be observed by getting a 360-degree view of the structure. The attic space above the full dormer is very limited and without access. Dormers are usually found on Cape Cod-style homes, both on the front and rear roofs, and on attached garages to create additional living space **FIGURE 12-10**. Dormers can also be found, however, in $^{10}/_{12}$ and $^{12}/_{12}$ pitch two-story farmhouses and other styles homes.

Regardless of what type, finished or unfinished, an attic is still a confined space with limited access (one way in and out) just below the roof. When entry must be made, have an additional company with a line charged and at the ready with you. Fire in an attic space is a dangerous situation with an increased potential for something wrong to happen. Always be proactive and get more help.

■ Access

The access to unfinished attic space is limited to a set of walk-up stairs, a set of fold-down stairs, a small crawl-through scuttle hole either in the ceiling or through a knee wall, or no access at all.

Occasionally you will find a full set of finished stairs leading to the attic, in which the original intent is usually to create a finished living space in the future. Fold-down attic stairs are most likely found with an attic that is very large, giving the occupant easy access to the storage area. The stairs, when folded down, create an opening approximately (24 × 48 inches [61 × 122 cm]) and are usually found in hallways **FIGURE 12-11**.

A newer option with fold-down attic stairs is the inclusion of gas springs or struts to raise and lower the stairs effortlessly, replacing the traditional spring mechanisms. These units have been known to explode and become missiles during fires, flying past fire fighters and occasionally embedding themselves into walls. This advancement in technology can complicate matters for fire fighters and members of the RIC. The only way you can determine if the attic stair unit has the traditional spring mechanism or gas struts is to slowly pull the attic stair unit downward while listening for the traditional squeaky spring sound. If you do not hear any sound, assume the unit does not contain springs and proceed with caution.

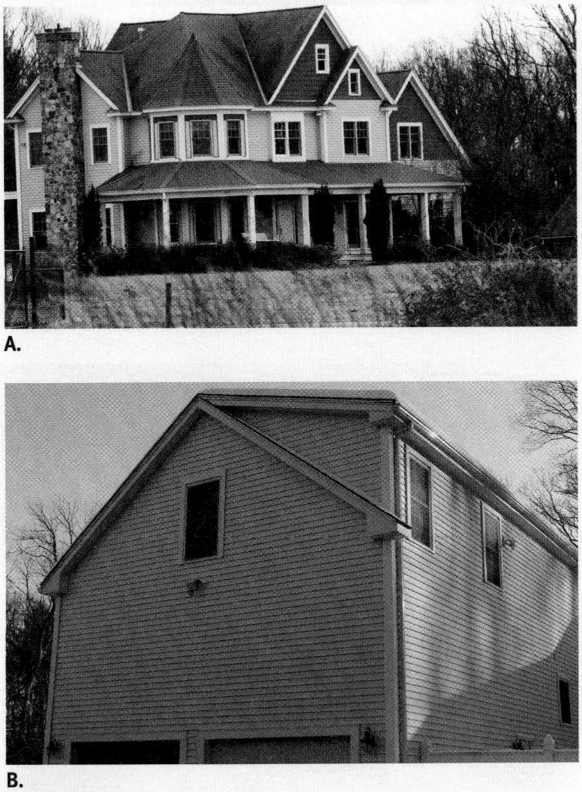

FIGURE 12-9 Types of roof styles include **A.** numerous gables and **B.** shed dormers.

FIGURE 12-10 A Cape Cod-style home **A.** with a rear dormer **B.**

A.

B.

FIGURE 12-11 A fold-down attic ladder **A.** closed and **B.** open.

Older homes will typically have a scuttle hole through the ceiling. These holes are small (18 to 24 square inches [46 to 61 sq. cm]) and will have a wood hatch cover that drops in place from the top FIGURE 12-12 . To access the attic, the cover up must be pushed up and slid to the side FIGURE 12-13 . Scuttle holes were never designed for fire fighters to crawl through to gain attic access, and are often found in closets or hallways. To gain access to the attic through a scuttle hole, a folding or pencil ladder is most often used. Finally, there are attic spaces with no access, which are found in areas where the attic space is very small, such as in a shed dormer with the voids behind knee walls.

■ Other Construction Considerations

Modular Construction

For the purposes of this book, modular construction will refer to prefabricated homes that are wood frame structures built off-site and then delivered to the site as a "box" and assembled. Modular construction is not referring to

FIGURE 12-12 An attic scuttle access hole.

FIGURE 12-13 A scuttle door being pushed to the side.

manufactured homes that are typically factory-built trailers or mobile homes. Modular homes present some different construction techniques that fire fighters and members of the RIC need to be aware of. First, the ceiling wallboard or drywall is not fastened by traditional mechanical means, such as with screws. The drywall is typically glued to the frame using a two-part urethane adhesive **FIGURE 12-14**. The challenge that this method presents to fire fighters is that the adhesive will begin to lose its mechanical strength at 482°F (250°C). Second, the construction method for a two-story home involves each box or floor being stacked one upon the other, which can present issues to fire fighters. Depending on how the boxes are designed and assembled on site, wherever the first and second floors meet, there will exist a large void, which at times can be as great as 20 to 22 inches (51 to 56 cm).

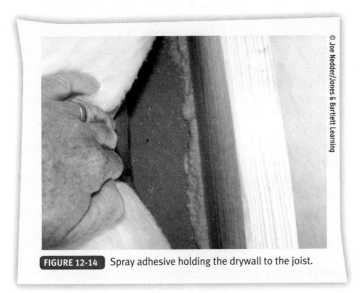

FIGURE 12-14 Spray adhesive holding the drywall to the joist.

© Joe Nedder/Jones & Bartlett Learning

If the fire gets into the void between the first and second floor, it will engage the glue, which is flammable, and spread the fire into this void quickly. For fire fighters, it means if you are on the second floor, there is hidden fire below you, and if you are on the first floor, the only thing holding the wallboard onto the ceiling (i.e., the glue) is losing its mechanical strength and could collapse at any moment. This large void space in-between the first and second floors has been described as balloon construction on the horizontal plane.

If you are in an attic and the drywall below you fails, it could allow the fire to rapidly spread into the attic. If the RIC is trying to get to a downed fire fighter in an attic and the decision is made to pull the ceiling to execute the rescue, it could mean that as you pull the ceiling in the traditional manner, large portions of drywall could drop on you unexpectedly.

Fires in these type structures have been known to advance rapidly once the adhesive is burning and engages the structure faster than traditional platform construction. When you look at a home that was constructed in a modular manner, it looks no different than a traditionally framed home. However, if it is discovered, either by preplanning or fireground operations, that the house is of modular construction, extra caution should always be taken.

Green Energy Generation

With a national focus on conserving and generating energy, more people are installing energy-generating systems in their homes, two of which can affect RIC attic operations; namely photovoltaic (PV) roof panels and solar water heating systems. PV roof panels incorporate arrays or individual panels or the use of PV shingles. Both these systems are roof-mounted and create hazards and problems for fire fighters and the RIC, including electric shock, the dead load of the system on the roof as the fire burns below, and the batteries used to store the generated electricity. With both systems, you cannot break the panels to vent the roof or create a rescue hole for a fire fighter trapped in an attic **FIGURE 12-15**. Because these panels are energized, great caution should be taken not to place any rescue ladders on or within proximity to them. Solar water heating systems use the sun's energy to heat a fluid (typically antifreeze in cold climates and water

Courtesy of NIST

A.

Courtesy of Glenn Corbett

B.

FIGURE 12-15 A PV roof shingle system **A.** and PV roof panels after a roof fire **B.**

in warm climates). The primary concern with these systems is the weight and dead load they add to the roof, because an attic involved in fire with this additional weight might fail earlier than expected **FIGURE 12-16** .

Safety

Fire fighters should *never* break PV roof panels in an attempt to gain access below the roof.

■ Reasons for Entering Attic Space

Attics that are finished and used for living and working space will have traditional access (stairs) and can be treated like any other living space within the structure, including for rapid intervention. However, they will have significant dangers that must be identified, considered, and addressed in the Incident Action Plan, because even though the attic is finished living space, the fire could be moving aggressively within the walls, crawl spaces, or overhead. Unchecked, the fire could breakout and overrun the RIC and other companies operating.

Unfinished attics are, and must be treated as, a totally different situation based upon the variables presented previously. The reasons for fire fighters to enter an unfinished attic space include:

- Fire extinguishment and overhaul.
- To check for fire and or fire extension.
- Utility control (HVAC units and air handlers are in the attic).
- To rescue a fire fighter who has fallen through the roof into the attic.

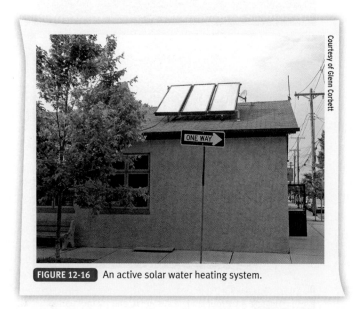

Courtesy of Glenn Corbett

FIGURE 12-16   An active solar water heating system.

In any of these situations, a good size-up and risk assessment of the given situation needs to be done *before* anyone enters the attic space.

Attic fires fall into two categories: (1) the fire has originated in the attic or (2) the fire has extended into the attic. With a fire that has *extended* into the attic, you are there already fighting a fire that has originated elsewhere and extended. A fire that *originates* in the attic might be heavy fire showing on arrival or small and smoldering when the fire department arrives on scene, and many times starts off as an investigation.

How many times has there been a call reporting either light smoke in the attic, the smell of smoke, or a ceiling light fixture that has blown the circuit breaker and is hot or glowing? Many times, these types of calls happen after a severe thunderstorm with a possible lightning strike. Your initial reaction may be to get right into the attic, but perhaps you should utilize other tools you possess, such as a thermal imaging camera (TIC) to seek out the "hot spots" and make a decision from the information gathered. Is the "hot spot" isolated or spread out? If you open the access panel to the attic, is there a light smoke condition? Is there high heat? Have you scanned the attic with the TIC? Is the attic large enough to allow you to stand up or is it a low, confining space? The answers to these questions should help your risk assessment survey and to determine if it is truly safe to place a fire fighter into such a space.

When fire is in an attic, access is needed quickly. For an attic that is heavily engaged, it is dangerous and risky to suggest putting a fire fighter through a small scuttle hole that will require him to either reposition or remove his self-contained breathing apparatus (SCBA) to fit through. The most traditional way to fight a fire in an attic space is to either use the scuttle opening to place and direct a stream of water into the attic space or to aggressively pull the ceiling below the attic and fight the fire through the openings you have created. In many cases, penetrating nozzles and Bresnan nozzles have also been used successfully. A few obstacles that might be encountered with large open attics used for storage include:

- The attic ceiling joists have been covered with boards or plywood to make a floor for storage **FIGURE 12-17** .
- In cases where there is no floor installed, but items have been stored on the open joists, as you pull the ceiling and insulation down, some of these items might fall through onto the fire fighters **FIGURE 12-18** .
- The attic contains HVAC mechanicals and ductwork **FIGURE 12-19** .

In commercial structures that are Type V construction, it is possible to find the building's HVAC units and air handlers inside the attic. If utility control is required, question whether you should enter the attic space to do so, or if there is another means to gain utility control without entering the confined, limited-access attic space.

FIGURE 12-17 Attic ceiling joists covered with plywood.

FIGURE 12-18 Items being stored in the attic on the open joists.

FIGURE 12-19 An attic with HVAC mechanicals and ductwork.

Rescuing Fire Fighters From an Attic

Many of the specific skills previously reviewed can be applied and used in an attic rescue. Each situation is different, and you need to apply the skill that will best serve the problem presented. Unfinished attics, however, can and will present different challenges and levels of difficulty. Here the different available avenues of egress and the variables and issues that may potentially increase the level of difficulty associated with each avenue of egress will be reviewed. Keep in mind that, while addressing the concerns of a particular means of egress for an attic rescue, the overall concerns for the RIC operation are still:

- Locating the downed fire fighter.
- Monitoring the downed fire fighter's air supply status and providing air.
- Removing the downed fire fighter rapidly. Because the attic is a confined, limited-access space, consider which skills and techniques can be used to remove the downed fire fighter.
- Controlling the fire; it is the best way to prevent it spreading into the attic space.
- Vertical ventilation for life, if it will not draw the fire onto the downed fire fighter.
- Attempting efforts to redirect the smoke and heat (ventilation).
- For an unfinished attic space, if there is sufficient manpower on scene, multiple access points should be attempted simultaneously to reach the downed fire fighter.

The goals of the RIC are always to find the downed fire fighter, get him air, and get him out quickly.

■ Finished Attic Space

A fire fighter who has to be rescued from a finished attic space has gotten there by either going up the staircase or via a window from a ladder, perhaps for a vent, enter, search (VES). Regardless of how he got there, if a MAYDAY arises, the RIC will need to be ever-vigilant and cautious. The IC will have to commit additional resources for fire control, protection of the downed fire fighter, and ventilation for life, and the RIC Operations Chief will need additional resources for the rescue operation. The issues and variables of removing a fire fighter from *finished* attic space include:

- Where is the downed fire fighter? Are you in a single finished room above a garage or are in a multiple-room boarding house?
- Fire moving through the walls and behind the knee walls into the open void above the ceiling is a critical concern; are there hoselines in place for confinement or suppression? What about opening the roof for venting? You will need to lift the heat and smoke off of the fire fighter while searching for him.

- Is there sufficient personnel standing by ready to assist in removing the fire fighter out of the building once he is located?
- What is the current heat, smoke, and fire situation in the finished attic space? What was it when the MAYDAY was sounded and where is it now? Be proactive in on-going size-up. The last thing needed is for fire to break out of the walls or ceiling and drop down on the downed fighter or the RIC.

Unfinished Attic Space

A fire fighter who has to be rescued from an unfinished attic has typically gotten there by either climbing or crawling in through the access opening in the ceiling or he has fallen through the roof into the attic. In either case, this can become one of the most difficult and demanding rescues you will ever be involved in. If the RIC does not have safe access to the attic, how can they accomplish the rescue? The RIC's job is to stay proactive by thinking and planning at least one step ahead. Consider all the issues and variables of removing a fire fighter from an *unfinished* attic space, including the four means of possible egress:

1. Down (ceiling)
2. Up (roof)
3. To the side (gable)
4. Out a window (gable or dormer)

The issues and variables of removing a fire fighter down through a ceiling include:

- Where is the downed fire fighter? Can you safely cut through the ceiling without knowing his location?
- Is this a truss roof, and if so will you be operating below it and is it engaged in fire?
- To open a ceiling with hand tools is very manpower-intensive—do you have the available personnel and tools?
- The use of gas power saws might not be practical, as a smoky environment will deny oxygen to the gas combustion motor and prevent or inhibit the saw from running.
- Do you know what is above the ceiling? Has the attic floor been covered over with wood boards or plywood, and do you know what is being stored there?
- If the floor is covered with plywood or boards, do you have the manpower and tools to flip over your pike poles and drive them up, hammering into the floor above in an attempt to lift it?
- Does the attic have HVAC and mechanicals in it? Depending on where the air handlers and ductwork are located, they might inhibit or prevent access through the ceiling.
- Know what the current heat, smoke, and fire conditions are in the attic as well. Has the fire gained and overrun the attic space? Is there any fire suppression in place trying to control the fire?
- Is the fire elsewhere in the structure? Is it under control or is the fire still burning uncontrolled?

Along with these issues and variables of removing the fire fighter, what about the downed fire fighter?

- Is he entangled?
- Is he responsive or unresponsive?
- Has he run out of air in his SCBA?
- Once you locate him, you will need to lower him down.
- Will you have to cut away some ceiling joists?
- In order to lower him by ropes, you will need to put fire fighters into the attic; is it safe to do so?
- If it is not safe to lower him by ropes, how will you lower him down? Can you carry him down to safety by ground ladders?

The issues and variables of removing the downed fire fighter up through a roof include:

- First and foremost, if the fire fighter fell through the roof into the attic, is the roof safe for a rescue operation?
- What is the roof construction? Is it truss construction and weakened by fire, heat, or vent holes?
- What are the heat, smoke, and fire conditions currently in the attic? Has the fire gained and overrun the attic space? Is there any fire suppression in place trying to control the fire?
- If the fire fighter has crawled into the attic and got trapped, do we know where to release the heat (vertical venting) without pulling the fire up and out, overrunning the downed fire fighter?
- How steep is the roof, is it a $^{10}/_{12}$ pitch or a $^5/_{12}$ pitch?
- Is the roof covered with snow and or ice? Is there an imminent danger of additional collapse? Are trees or wires blocking access with aerial devices? What aerial equipment is available that would support the rescue? Remember, if the downed fire fighter fell through the roof, you cannot put additional personnel on a roof that has already collapsed or is seriously weakened!
- What if the roof is covered with solar panels? Do you have enough knowledge to deal with this?
- Can you rapidly and safely get rescuers into the space below the attic to pull ceilings and remove the downed fire fighter?

Once you have dealt with these issues, you will still be faced with the basic RIC issues of exactly where the downed fire fighter is located, what his air supply situation is, and what techniques you can use to remove him.

The issues and variables of removing a fire fighter down through a gable end include:

- Typically gables are not structural, meaning that fire fighters can cut a large opening into the gable without fear that the roof above it will collapse.
- What manpower efforts would be needed to open a large hole in an attic gable from the outside?
- An open gable will also function as a vent hole, drawing the fire out.
- If the attic is unfinished with no floorboards, once the hole is open, how will you drag the downed fire fighter over the open joists?

- What is the current heat, smoke, and fire conditions in the attic? Has the fire gained and overrun the attic space? Is there any fire suppression in place trying to control the fire?
- Do you know what gable the fire fighter is located closest to?
- Do you have the personnel to raise ground ladders if needed, to cut open the gable, and to remove the downed fire fighter?

Again, the issues outlined are structural and personnel issues; the basic RIC issues of where the downed fire fighter is located, what his air supply situation is, and what techniques to use to remove him still need to be dealt with.

If Fire Fighters Need to Enter an Unfinished Attic

There will be limited times when fire fighters need to enter an unfinished, restrictive space attic. In those situations you need to be prepared and proactive in planning survival skills, which include:

- If you have to crawl into an attic that is charged with smoke, tie off a rope to the pencil or folding ladder being used to access the attic space before stepping off and making entry. This is not a search rope application but rather the use of a life safety rope. If you become disoriented or become trapped, the rope will assist in leading you back to the ladder and safety or lead rescuers in to you. Have your partner with a flashlight positioned at the top of the ladder; if you get disoriented, he can call out to give you a direction.
- Be cautious with electrical wiring—not all electrical devices, outlets, and wiring that is found will meet electrical codes, which presents the danger of electrocution.
- Vent for life if possible. You need to release the heat and smoke.
- Have at least one if not two hoselines in place: one for suppression and one for protection of the downed fire fighter.
- Go only as deep as required to accomplish your objective and get out.
- Most importantly, make sure that the skills training and capabilities of the person entering the attic make him suitable for the assignment.

Crew Notes

There will be limited times when fire fighters need to enter an unfinished, restrictive space attic. In those situations, you need to be prepared and proactive in planning survival skills.

■ RIC Attic Rescue Considerations

Many of the basic skills presented in this book can be utilized for an attic rescue. If you, as a fire fighter, become lost or disoriented in an unfinished attic, there are a few basic skills to assist in your self-rescue. First, call a MAYDAY right away! The RIC will be trying to determine where you are. If fire conditions below you allow, use your tool or foot to break through the ceiling below. If the attic is covered with plywood, try to break through the plywood and then the ceiling below. Use your radio to communicate with the RIC informing them of what you are doing. If you are unable to do this, try to determine where you are in the attic and provide this information when you call your MAYDAY. Are you on the B side of the structure or the D side? Are you near the bathroom vent pipes or a chimney? Any clues you can provide the RIC will assist them in trying to locate you. If you are unable to communicate, bang your tool—it can help lead the RIC to your location from below.

As a member of an RIC looking for a downed fire fighter in an unfinished attic, you can utilize your thermal imaging camera (TIC) to scan the attic to try to locate him. If the downed fire fighter's approximate location is known, open up ceilings below and use the TIC and look and listen. If the downed fire fighter is found and you need to cross open joists to get him to safety, consider, if the scuttle hole is large enough or if there are fold-down stairs, popping a door from a room and using it like a backboard to place the downed fire fighter on and drag him to the opening. If you have to open up from below and bring the downed fire fighter through the ceiling joists, if he is conscious and able to assist in his rescue, he can utilize his SCBA low-profiling techniques to pass through the space between the joists. If the downed fire fighter is unconscious, you will have to open up the space between the joists—utilize a battery-powered sawzall-type tool, as gas saws might not operate in a smoke-filled environment.

Conclusion

The best way to avoid having to rescue a fire fighter trapped in an attic is to avoid or strictly control the entry into these spaces. If the need to enter exists, due consideration to the construction, fire condition, and overall goal needs to be evaluated before any fire fighters are placed in such a dangerous situation. Conditions, available personnel, and time will dictate what can be done. Attic areas are a confined space, with one way in and out; therefore, prior to entering, a risk assessment must be conducted and the findings honored by all.

Wrap-Up

Chief Concepts

- Attics are mostly associated and found with Type V wood frame construction.
- Type III ordinary construction may be encountered in cocklofts, but they are not considered, by definition, an attic.
- RIC operations would use the same skills and techniques for any other situation to get to and move the downed fire fighter to safety.
- Your primary concern with finished attic space is to remember it is an attic, and the space above you is a modest void below the roof.
- Beware of truss roofs! If trusses are on fire and it is discovered from "below," all companies operating under the truss must evacuate immediately as they are in imminent danger.
- There is no warning sign for a truss failure when they are on fire—it will just happen.
- Good size-up and risk assessment needs to be done *before* anyone enters an attic space.

Hot Terms

<u>Balloon construction</u> An older type of wood frame construction in which the wall studs extend vertically from the basement of the structure to the roof without any fire stops.

<u>Confined space</u> A space that is limited or restricted and is not meant for continuous occupancy.

<u>Crawl spaces</u> Small spaces in which access is gained to certain areas of a structure.

<u>Dormers</u> Structural element of a building that protrudes from the plane of a sloping roof surface. Dormers are used, either in original construction or as later additions, to create usable space in the roof of a building by adding headroom and usually also by enabling addition of windows.

<u>Finished attics</u> Typically an additional living space. They have stairways leading to them, doors, and windows and can be a single room, multiple rooms, an apartment, or boarding house rooms.

<u>Knee walls</u> Short walls supporting rafters at some intermediate position along their length. The term is derived from the association with the vertical location of the human knee. Knee walls are common in old houses that are typically not a full two stories in height, in which the ceiling on the second floor (in the "attic" area) slopes down on one or more sides.

<u>Type III ordinary construction</u> Buildings with the exterior walls made of noncombustible or limited-combustible materials, but interior floors and walls made of combustible materials.

<u>Type V wood frame construction</u> Buildings with exterior walls, interior walls, floors, and roof structures made of wood.

<u>Unfinished attic space</u> Attic space in which there is no living space.

References

Corbett, G. P. and F. L. Brannigan. *Brannigan's Building Construction for the Fire Service, Fifth Edition.* Burlington, MA: Jones & Bartlett Learning; 2015.

National Fire Protection Agency (NFPA) 1407, *Standard for Fire Service Rapid Intervention Crews.* 2015. http://www.nfpa.org/codes-and-standards/document-information-pages?mode=code&code=1407. Accessed February 11, 2014.

Attics come with many different variables that pose significant challenges to RIC and firefighting operations. Extreme caution should be used if access to an attic is needed, although these spaces should be avoided at all cost when possible. As a member of the RIC, you are sent to an unfinished attic space for a MAYDAY call at a residential home that was most likely built before World War II. There is heavy fire on the first floor of the two-story structure.

1. Typically, what is the major difference between a finished attic space and an unfinished attic space?
   A. There is no major difference.
   B. A finished attic space typically has a stairway, doors, and windows, whereas unfinished attic space does not.
   C. The difference is based on the construction type.
   D. Finished and unfinished attics spaces are determined by the building permit issued at the time of construction.

2. Why does the fact the building is over 50 years old concern you?
   A. Fire tends to burn hotter in older buildings.
   B. There could be asbestos in buildings over 50 years old.
   C. It does not concern you.
   D. More than likely it is a balloon frame construction.

3. What possible reasons would a fire fighter enter an attic space even though it is extremely dangerous to do so?
   A. Fire extinguishment and overhaul.
   B. Utility control.
   C. To rescue a fire fighter who has fallen through the roof into the attic.
   D. All of the above.

4. What are the dangers in cutting the ceiling below the attic space to gain access to a fire fighter trapped in an attic?
   A. If the fire fighter's exact location is not known, it is unsafe to cut with a power saw, because you could injure him or her.
   B. The danger does not equal the urgency in getting the fire fighter to safety.
   C. You could accidently cut through electric wires.
   D. You could cut into a space in were an HVAC is located.

5. Typically, what are the ways in which to remove a fire fighter from an attic?
   A. Down, through the ceiling.
   B. Up, through the roof.
   C. To the side, through a gable end, or out a window of a gable or dormer.
   D. All of the above.

The Multi-Purpose Prop: Purpose and Construction

Here you will find directions for constructing the Multi-Purpose Rapid Intervention Training Prop, which is designed to support the following drills and techniques taught in this text:

- Denver drill: this prop is based upon an actual problem and line-of-duty death (LODD). The interior width of 28 inches (71 cm) and the windowsill height of 42 inches (1.1 m) are key **FIGURE 1**
- Wire entanglement
- Wall breeching
- Self-contained breathing apparatus (SCBA) low profiling
- Window lift drill

Using props such as this will greatly enhance your training and will create a better learning environment. The Multi-Purpose Prop is designed to be:

- Modular in design and easily reconfigured for different drills
- Stored flat in an area approximately 96 inches × 48 inches × 30 inches (2.4 m × 1.2 m × 76 cm) with all components

- Assembled in 10 minutes
- Cost-effective; inexpensive and easy to build
- Able to serve multiple rapid intervention crew (RIC) and survival training needs
- Easy to repair; being modular, if a part is damaged or breaks, it can easily be replaced

In addition to requiring very little space when stored, the advantages of the using the Multi-Purpose Prop in your training include that it can be moved from station to station easily and, most importantly, that it will aid in simulating the conditions and obstacles the various drills and techniques were designed for.

Constructing and Assembling the Multi-Purpose Prop

The basic unit of the multi-purpose prop is the Denver drill configuration **FIGURE 2**, which consists of the following sections **FIGURE 3**:

- Base
- Two sides
- Large end panel for the Denver drill window
- Small end filler panel for the Denver drill window
- Back exit brace

■ The Basic Denver Prop

To construct each section of the Denver prop, follow the instructions for the materials and dimensions shown in **FIGURE 4**.

FIGURE 5 illustrates the assembly required for building the Denver prop. The steps for assembly are:

1. Layout the base and two sides. Place one wall on the base and hold in place by dropping in four ³/₈ inch × 4 inch (9.5 mm × 10 cm) bolts with washers **FIGURE 6**. There is no need to use nuts to fasten the bolts, as the weight of the panel will hold it in place. Repeat this with the second wall.
2. Install the large front panel for the Denver prop window and bolt in as per Figure 5.

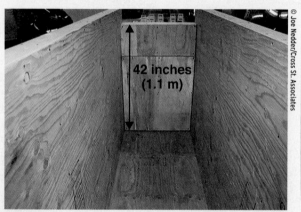

FIGURE 1 The distance from the floor to the top of the windowsill [42 inches (1.1 m)] is important with the Denver drill.

A.

B.

C.

FIGURE 2 The Denver prop: **A.** side-view; **B.** window end; **C.** end opposite the window.

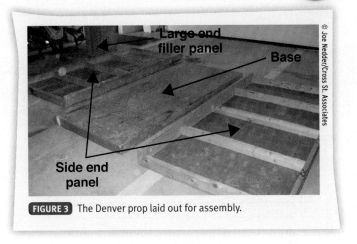

Large end filler panel

Base

Side end panel

FIGURE 3 The Denver prop laid out for assembly.

3. Install the small front panel on top of the large front panel to give you the 42-inch (1.1-m) window lift required from the inside of the prop. Bolt in securely as shown.

4. Install the Rear Brace as shown in Figure 5.

FIGURE 7 shows the completed Denver prop with the students inside preparing to remove the downed fire fighter.

■ Converting the Prop for a Window Lift Scenario

The Multi-Purpose Prop can be used to practice lifting a downed fire fighters or civilians out a window by simply remove the small front panel. This provides a 36-inch (91-cm) lift from the outside of the prop to the windowsill **FIGURE 8** .

Crew Notes

If you desire a shorter height for the window lift, the prop can be modified by making the large front panel lower and making the small panel higher, provided that the overall height is maintained at 42 inches (1.1 m) for the Denver drill.

■ Converting the Prop for Wall Breaching

For wall breaching, construct the wall breach panel as shown in **FIGURE 9** . The 14½-inch (37-cm) pocket must be maintained to simulate standard wall framing spacing. Once constructed, simply remove the two front panels and install the wall breach panel, making sure it is bolted securely **FIGURE 10** . The prop is then ready for fire fighters to practice wall breaching **FIGURE 11** .

■ Converting the Prop for SCBA Low Profiling

The Multi-Purpose Prop can also be used in low profiling skill drills. Once the breach is performed the fire fighters, their SCBA low-profiling skills to "pass through" the opening they created can be practiced. If the prop is to be used strictly for low profiling skill drills, the breach prop panel can be used. To add challenge to the drill and enhance fire fighters' skills, items

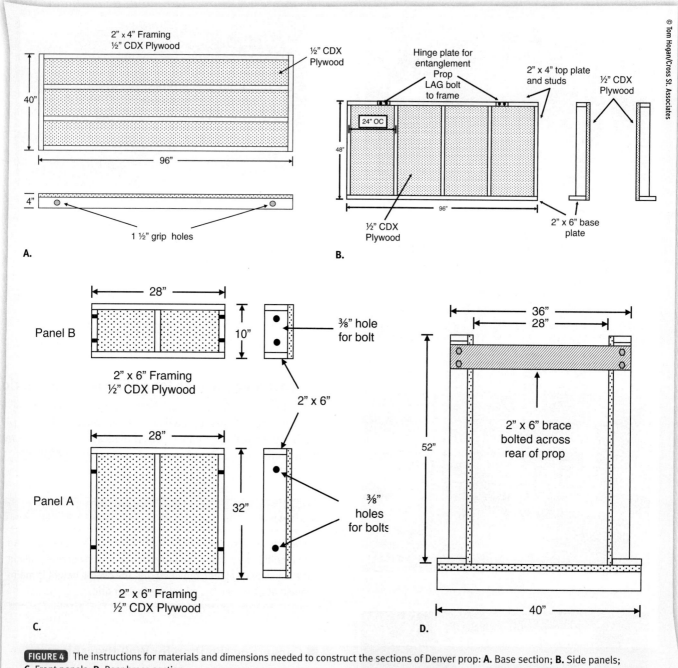

FIGURE 4 The instructions for materials and dimensions needed to construct the sections of Denver prop: **A.** Base section; **B.** Side panels; **C.** Front panels; **D.** Rear brace section.

such as MC cable can also be added **FIGURE 12**. Cable or wires can be added to the prop by simply drilling a series of holes in each of the two side panels and then weaving cable or wire through the holes to make a web the fire fighter must negotiate.

■ Converting the Prop for Wire Entanglement

The conversion of this prop for wire entanglement drills is easy to do and offers several options **FIGURE 13**. Using **FIGURE 14**, construct another frame. This frame is attached to the side wall

with a hinge **FIGURE 15**, which allows the frame with the wires to either be raised while the fire fighter enters the prop and then dropped down on him, or the frame with the wires can be left in place and the fire fighter needs to crawl in and through the entanglement **FIGURE 16**. When constructing this frame, different kinds of wire should be used **FIGURE 17**, and they should be woven around the top of the frame and let them loop down onto the floor of the prop **FIGURE 18**. The goal for this drill is to have a loop or loops entrap the fire fighter as he passes through the prop **FIGURE 19**.

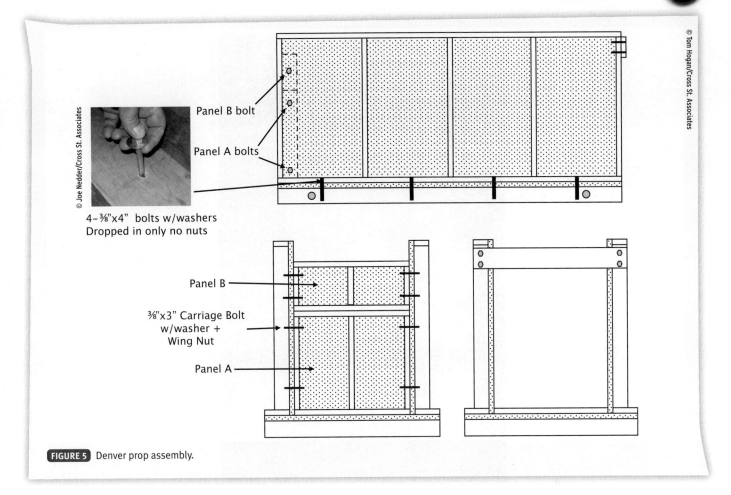

© Tom Hogan/Cross St. Associates

© Joe Nedder/Cross St. Associates

Panel B bolt

Panel A bolts

4–⅜"x4" bolts w/washers
Dropped in only no nuts

Panel B

⅜"x3" Carriage Bolt
w/washer +
Wing Nut

Panel A

FIGURE 5 Denver prop assembly.

© Joe Nedder/Cross St. Associates

FIGURE 6 A bolt with a washer used to hold the wall in place.

© Joe Nedder/Cross St. Associates

FIGURE 7 Students inside the completed Denver prop preparing to remove the downed fire fighter.

FIGURE 8 Exterior view of the Multi-Purpose Prop configured for a window lift.

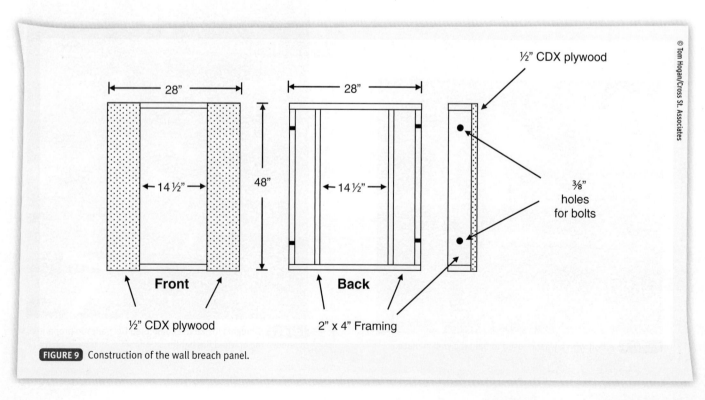

FIGURE 9 Construction of the wall breach panel.

FIGURE 10 The wall breach panel bolted securely into place on the prop.

A.

B.

FIGURE 11 Fire fighters **A.** ready to breach the wall on the prop, and **B.** the interior of prop after the wall is breached.

FIGURE 12 MC cable being used for a low profile drill.

FIGURE 13 The Multi-Purpose Prop assembled for wire entanglement.

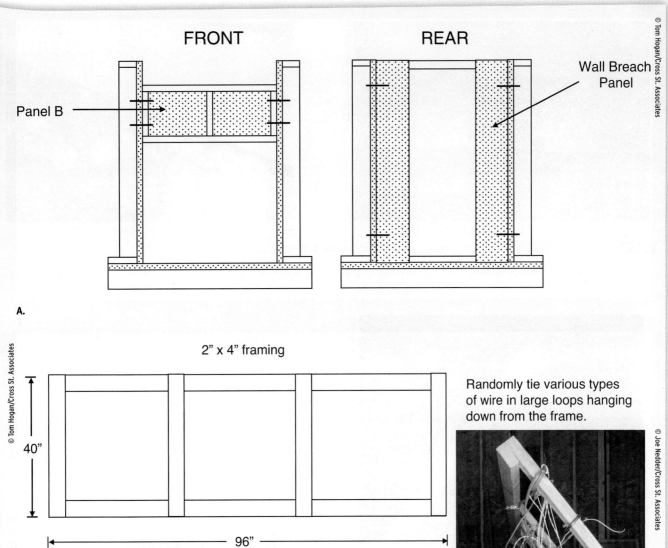

FRONT

REAR

Panel B

Wall Breach Panel

A.

2" x 4" framing

40"

96"

Hinge plates for attachment
to main prop

Randomly tie various types
of wire in large loops hanging
down from the frame.

B.

FIGURE 14   The wire entanglement frame: **A.** Attached to the side of the prop **B.** With a hinge and hinge pin.

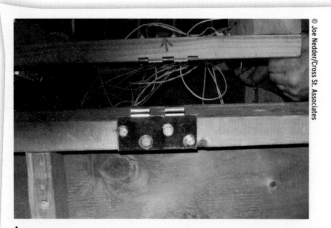

A.

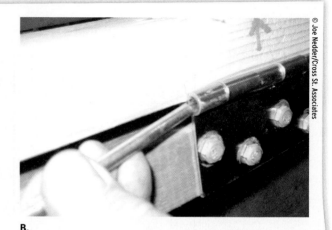

B.

FIGURE 15   The wire entanglement frame: **A.** Attached to the side of the prop **B.** With a hinge and hinge pin.

FIGURE 16   The wire frame can be raised and then lowered down on the fire fighter to entangle him.

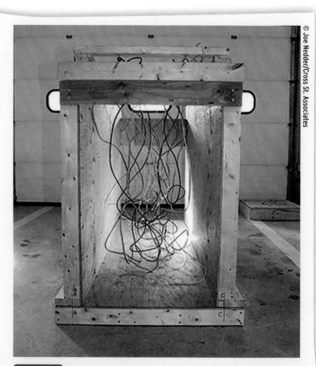

FIGURE 17   Various types of wire tied randomly in large loops hanging down from the frame.

FIGURE 18 Wires woven around the top of the frame and then looped down onto the floor of the prop.

FIGURE 19 A fire fighter in the entanglement prop.

Crew Notes

When constructing the side walls, place the receiving end of the hinge first; then to install the wire frame, all that is necessary is to align the frame and wall, and then insert the hinge pin.

APPENDIX B

NFPA 1407 Correlation Guide

NFPA 1407, *STANDARD FOR FIRE SERVICE RAPID INTERVENTION CREWS*, 2015 EDITION	CORRESPONDING CHAPTERS	CORRESPONDING PAGES
4.1.1	1	4, 7, 8
4.2.4	1, 2	4, 7, 8, 12
4.3.1	2	12
5.1	2	12–14
5.1.1	2	12–14
5.1.2	2	12–14
5.1.3	2	12–14
5.2	2	12–14
5.2.1	2	12–14
5.2.2	2	12–14
5.2.3	2	12–14
5.2.4	2	12–14
6.1	2	12
6.1.1	2	12
6.2.1	2	12–13
6.2.2	2	12–13
6.3	2	12–13
7.1(1)	4, 12	52–53, 224–230
7.1(2)	4	47–48
7.1(3)	3, 4, 8	41, 52, 135–138, 144
7.1(4)	4	48, 49
7.1(5)	4	47–48
7.1(6)	4	48
7.1(7)	4	48, 49
7.1(8)	4	48, 49
7.1(9)	4	47–48
7.2	4	48–52
7.3	2	12–14
7.4(1)	4, 5	46–47, 60–63
7.4(2)	6	97–99
7.4(3)	8, 9, 10, 11, 12	132, 143, 150, 155, 169–170, 210–218, 231–233
7.4(4)	6	109
7.4(5)	6	99
7.4(6)	4, 7	50, 117
7.4(7)	8, 9	132, 143, 150, 155
7.4(8)	5	60–63, 77–82
7.5(1)	4, 5	46–47, 60–63

NFPA 1407, *STANDARD FOR FIRE SERVICE RAPID INTERVENTION CREWS*, 2015 EDITION	CORRESPONDING CHAPTERS	CORRESPONDING PAGES
7.5(2)	4, 5	46–47, 49, 60–63
7.6(1)	6	97–99, 100–113
7.6(2)	6	97–99, 100–113
7.6(3)	6	94–99, 100–113
7.7(1)	5	64–72
7.7(2)	5	64–72
7.7(3)	5, 12	64–72, 74–75, 227–230, 231–233
7.7(4)	5	74–76
7.7(5)	5	73–75
7.8(1)	7	117–127
7.8(2)(a)	7	117–127
7.8(2)(b)	7	117–127
7.8(2)(c)	7	117–127
7.8(3)(a)	7	117–127
7.8(3)(b)	7	117–127
7.8(3)(c)	7	117–127
7.8(3)(d)	7	117–127
7.8(3)(e)	7	117–127
7.8(3)(f)	7	117–127
7.9	3	39–40
7.10(1)	2	25
7.10(2)	2	27
7.10(3)	2	15–16, 27–28
7.10.1	2, 10	13, 185
7.10.2(1)	2, 10	13, 185
7.10.2(2)	2, 5	13, 77–89
7.11(1)	7	117–127
7.11(2)	7, 11	117–127, 210, 213–216
7.11(3)	4	49
7.11(4)	7	117–127
7.12(1)	8	132–142, 143–150
7.12(2)	9	153–161, 162–163, 164–166
7.12(3)	10	185–187
7.12(4)	10	169, 170–176, 188–189
7.12(5)	10	169–176, 178–185, 190–193, 194–197
7.12(6)	11	210–218
7.12(7)	12	231–233
7.12(8)	9, 11	155–161, 162–163, 210–218
7.12(9)	10	198–203
7.12.1	6, 8	104–105, 132, 143, 150
7.13.1(1)	5	60–63, 72–73, 75–77
7.13.1(2)	5	74–76
7.13.1(3)	5, 6	75–89, 108
7.13.1(4)	6	109
7.13.2	5	77–89
7.14	2	14–26
7.14.1	2	14–26
7.14.2	2	14–26
8.1.1	3	33–41

Glossary

360-RECON A 360-degree view of a structure for the purposes of gathering information such as fire conditions, building construction, access and egress, and hazards found around the building. Commonly conducted by an RIC after arriving on the scene.

Accountability system A method of accounting for all personnel at an emergency incident and ensuring that only personnel with specific assignments are permitted to work within various zones.

Airman RIC member #4 is the airman, responsible for maintaining the RIC emergency air system.

Anchor hook Hook used in conjunction with a personal escape system that is placed on a window sill and used as an anchor prior to bailing out of a window.

Anchor point A single secure connection point at which a rope can be secured.

Back scanning Using the TIC to look behind you to monitor heat and fire conditions while maintaining a visual contact with the crew.

Balloon construction An older type of wood frame construction in which the wall studs extend vertically from the basement of the structure to the roof without any fire stops.

Baluster An upright support piece used to hold the railing of an open stairway.

Basic grab points Various points on a fire fighter's SCBA and turnout gear that are easily grabbed under adverse conditions in order to move the downed fire fighter.

Belay To protect against falling by managing an uploaded rope (the belay line) in a way that secures one or more individuals in case the main line rope or support fails, or to protect against falling while practicing dangerous skills at a height greater than head level.

Buddy breathing Sharing one's air with another via a system on some SCBAs.

Carabiners Metal snap links used connect elements of rope rescue equipment.

Chicago ladder carry A technique used to carry either a fire fighter without a SCBA or a large fire fighter down a ladder. It can also be used for civilian rescue.

Chicago lift A technique for lifting a fire fighter out of a window.

Confined space A space that is limited or restricted and is not meant for continuous occupancy.

Crawl spaces Small spaces in which access is gained to certain areas of a structure.

Crew integrity The ability to maintain control of an organized group of fire fighters under the leadership of a company officer, crew leader, or other designated official.

Critical incident stress debriefing (CISD) Counseling designed to minimize the effects of psychological/emotional trauma on those at a fire or rescue incidents who were directly involved with victims suffering from particularly gruesome or horrific injuries.

Denver drill Term used to describe the removal of a fire fighter from a restricted space.

Descent control device (DCD) A device used in conjunction with a personal escape system that creates friction by means of a rope running through it.

Door chocks Small wedge-like devices used to hold a door open, usually made from wood, plastic, or metal.

Dormers Structural element of a building that protrudes from the plane of a sloping roof surface. Dormers are used, either in original construction or as later additions, to create usable space in the roof of a building by adding headroom and usually also by enabling addition of windows.

Drag rescue device (DRD) A device integrated into the turnout coat that can be utilized to drag a fire fighter.

Drywall Also known as plasterboard, wallboard, or gypsum board; drywall is a panel made of gypsum plaster pressed between two thick sheets of paper. It is used to make interior walls and ceilings.

Emergency traffic Used on the fireground when you need to get a message out; indicating an imminent hazard or a situation that is of immediate danger.

Engineered lumber Larger spans of lumber used for floor or ceiling construction; typically glued together under pressure.

Escape DCD An emergency rope escape system that can be standalone, attached to the escape harness, or part of a system integrated into the SCBA.

Feet-first window lift A technique for removing a fire fighter out of a window feet first.

Finished attics Typically an additional living space. They have stairways leading to them, doors, and windows and can be a single room, multiple rooms, an apartment, or boarding house rooms.

Fire and smoke behavior The science of understanding fire, it stages and how it grows and spreads along with understanding what he smokes volume, velocity, density and color indicate.

Flashlight A hand-held battery-powered light source. Usually the light source is a small incandescent light bulb or a light-emitting diode (LED).

Forcible entry tools Tools used by fire fighters to gain access to areas that are blocked or locked.

Handcuff knot A knot used to secure the arms/wrists of a victim that is typically utilized when hauling an unconscious fire fighter from a lower level.

Harness Class I or class II: a device used to secure the body to an anchor point, found either as part of the PPE or it can be separate.

Hazardous materials (HazMat) Any materials or substances that pose an unreasonable risk of damage or injury to persons, property, or the environment if not properly controlled during handling, storage, manufacture, processing, packaging, use and disposal, or transportation.

Head-first window lift A technique for removing a fire fighter out of a window head first.

Heavy debris Debris that cannot be removed by hand or with the use of hand tools, but requires the use of technical expertise in collapse rescue and extensive manpower and the use of equipment such as hydraulic jacks, air bags, saws, struts, and shoring to remove the fire fighter from entrapment.

HVAC Heating, ventilation, and air conditioning; a system designed to heat and cool a building.

Hydraulic cutters A powered rescue tool consisting of at least one moveable blade used to cut, shear, or sever material.

Hydro-force style tool (rabbit tool) Hydraulically operated hand tool used to force doors.

Immediately dangerous to life and health (IDLH) Used to describe an environment that could cause harm and/or death if not immediately escaped.

Incident Command System (ICS) The combination of facilities, equipment, personnel, procedures, and communications operating within a common organizational structure that has responsibility for the management of assigned resources to effectively accomplish stated objectives pertaining to an incident or training exercise.

Incident Commander (IC) The person in charge of the incident site and responsible for all decisions relating to the management of the incident.

Infrared technology Used in thermal imaging, identifies the different temperatures viewed by displaying light and dark images on the camera screen.

K-12 style saw A rotary cut off saw used for cutting through wood, metal, or concrete. The saw blade can be changed to a specific blade to cut the desired material.

Knee walls Short walls supporting rafters at some intermediate position along their length. The term is derived from the association with the vertical location of the human knee. Knee walls are common in old houses that are typically not a full two stories in height, in which the ceiling on the second floor (in the "attic" area) slopes down on one or more sides.

Knife A cutting tool with an exposed cutting edge or blade, hand-held or otherwise, with or without a handle.

Knot A fastening made by tying together lengths of rope or webbing in a prescribed way, used for a variety of purposes.

Landing An area where one stairway ends and another begins.

Life safety rope Static kernmantle rope in lengths of 75 to 100 feet (23 to 30 m) and typically $1/2$ or $7/16$ inch (13 or 11 mm) diameter, kept in a rope bag for easy deployment, and used for hauling or lowering a victim.

Light debris Debris that, although trapping the fire fighter, can be removed by other fire fighters by hand or with the use of the hand tools they have with them.

Line-of-duty death (LODD) Death of a fire fighter while on duty or operating in his or her official capacity.

LIP Location, identification, problem; an acronym used when calling a MAYDAY to advise the IC of the situation being encountered by the fire fighter in peril.

Long lug out Technique used for self-rescue by which a fire fighter identifies the longer lug on the female hose coupling, which will lead out of the structure to the water source.

LUNAR Location, unit number, name, assignment, resources needed; an acronym used when calling a MAYDAY to advise the IC of the situation being encountered by the fire fighter in peril.

Mask-mounted regulator (MMR) Low-pressure regulator that attaches to the mask section of an SCBA.

Masonry block wall The building of structures from individual units laid in and bound together by mortar; the term *masonry* can also refer to the units themselves. The common materials of masonry construction are brick, stone, marble, granite, travertine, limestone, cast stone, concrete block, glass block, stucco, and tile.

Mechanical advantage system A system usually consisting of ropes and pulleys used to assist in the hauling of a heavy object. Systems can be pre-rigged for fire service use as in the use in RIC applications. Mechanical advantage is the ratio of the output force produced by a machine to the applied input.

Mutual aid An agreement among emergency responders to lend assistance across jurisdictional boundaries. This may occur because an emergency response exceeds local resources, such as a disaster or a multiple-alarm fire, or it may be a formal standing agreement for cooperative emergency management on a continuing basis, ensuring that resources are dispatched from the nearest fire station regardless of which side of the jurisdictional boundary the incident is on.

Nance drill A series of skills and techniques that are used to rescue a fire fighter who has fallen through a hole and who has no other means of escape except through the hole.

One-rescuer drag The technique of one fire fighter moving another fire fighter who is downed.

Oriented search Searching an area while keeping your basic search skills in mind, such as keeping a wall to your left or right as a landmark, maintaining contact with the rest of the search crew, and utilizing tools to search wide areas.

Personal alert safety system (PASS) A device worn by fire fighters that sounds an alarm if the fire fighter is motionless for a period of time.

Personal basic tools Tools that every fire fighter should carry on his/her person. The tools should be limited to the following: webbing, flashlight, knife, door chocks, wire cutters, and a personal rope with carabiner or a personal escape rope system.

Personal escape rope system An NFPA-rated escape system kept on the fire fighter that has 50 feet (15 m) of rope, an anchor device (hook or carabiner), and a mechanical device to control the escape and decent, designed for easy deployment to assist a fire fighter in escaping a situation in which his or her life is in imminent danger. The entire system must be NFPA-compliant.

Personal protective equipment (PPE) Gear worn by fire fighters that includes the helmet, gloves, hood, coat, pants, SCBA, and boots. PPE provides a thermal barrier that protects fire fighters against intense heat.

Personnel accountability report (PAR) An accountability report or roll call taken in place.

Proactive Within the rapid intervention discipline, being proactive refers to an officer or company that tends to prepare for an occurrence *rather* than react to an event after the fact. It is a position of anticipation!

Radio A battery-operated hand-held transceiver used to communicate information.

Rapid intervention A team of fire fighters who are trained, prepared, and standing by at the scene of a fire to rescue a fellow fire fighter who is in peril.

Rapid intervention crew (RIC) A crew or company that is assigned to stand by at the incident scene, fully dressed, equipped for action, and ready to deploy immediately to rescue lost or trapped fire fighters when assigned to do so by the Incident Commander.

Rapid room search Left- or right-hand primary search, off the main line, that is done very quickly.

RASP Control Officer Officer in charge of monitoring the RASP search. He or she needs to be close to where the RIC has entered to conduct the search and have direct communication with the RIC Operations Chief and/or Incident Command.

Rescue Those activities directed at locating endangered persons (fire fighters) at an emergency incident, removing those persons from danger, treating injured victims (fire fighters), and providing for transport to an appropriate healthcare facility.

Rescue rope bag A dedicated bag for life safety rope, which is 75–100 feet (23–30 m) of ½-inch (13-mm) or ⁷/₁₆-inch (11-mm) static kernmantle rope; is used for hauling and lowering; and is equipped with extra-large carabiners, one which is attached at one end of the rope with a figure eight on a bight knot. The bag may also contain a pulley and extra carabiners to rig a 2-to-1 mechanical assist. This bag is used exclusively for the extraction and rescue of fire fighters.

RIC company basic tools Tools used by the RIC for the purposes of finding, assisting, and removing a downed fire fighter. The minimum tools needed include a portable radio, 200 feet (61 m) of search rope, a thermal imaging camera, forcible entry tools, and an air supply for the downed fire fighter.

RIC Operations Group Supervisor This position is best filled with a Chief Officer who is RIC trained.

RIC rescue air unit A fully functioning SCBA in a protective carrying device that is able to deliver air to a downed fire fighter. The device can simply be a SCBA or a specific manufacturer's device designed for RIC operations.

Riser The near-vertical element in a set of stairs, forming the space between steps.

Risk management A tool used in both preplanning and on scene to evaluate and reduce fire fighters' exposure to injury, loss, or death.

Rope-assisted search procedure (RASP) bag A 200-foot (61-m) search rope designed to enable the RIC to move quickly and maintain an awareness of where they are (distance in) on the line.

Rope-assisted search procedure (RASP) search Rope techniques typically used in open areas.

Scan, target, and release technique A technique used while advancing forward with a TIC: looking at the area, looking at the target area checking for fire and heat conditions, and then proceeding.

SCBA harness conversion A technique for turning the straps of a fire fighter's SCBA into a harness for use in dragging the downed fire fighter.

Search rope A guide rope used by fire fighters while conducting searches in a structure that allows them to maintain contact with a fixed point for easy exit or to allow easy access for others.

Self-contained breathing apparatus (SCBA) A respirator with an independent air supply used by fire fighters to enter toxic or otherwise dangerous atmospheres.

Self-rescue A fire fighter's use of techniques and tools to remove him or herself from a hazardous situation.

Situation-specific tools Tools or equipment needed to address specific situations, such as a tool needed for a specific task determined by the type, size, and construction of a building.

Stringer The part of the stairway used to support the steps and risers.

Survival skills Learned skills and techniques used by fire fighters to increase chances of surviving an unexpected event while performing fire suppression/rescue operations.

Thermal imaging camera (TIC) Electronic devices that detect differences in temperature based on infrared energy and then generate images based on that data. These devices are commonly used in obscured environments to locate victims.

TIC sweep Slow, deliberate side-to-side view of the conditions in front of you at the floor, mid-level, and ceiling level.

Tool man RIC member #2 or #3, responsible for carrying the irons and/or specialty tools. These two positions do the majority of searching.

Tread plate The part of a stairway that is walked on.

Tunnel vision To focus and concentrate on just one thing.

Two-rescuer drag The technique of two fire fighters moving another fire fighter who is downed.

Two-rescuer drag/push The technique of two fire fighters moving a downed fire fighter—one is dragging and while the other is pushing.

Two-rescuer side-by-side drag A technique used to move a downed fire fighter by two fire fighters dragging side-by-side.

Type III ordinary construction Buildings with the exterior walls made of noncombustible or limited-combustible materials, but interior floors and walls made of combustible materials.

Type V wood frame construction Buildings with exterior walls, interior walls, floors, and roof structures made of wood.

Unfinished attic space Attic space in which there is no living space.

Universal air connection (UAC) A device installed on all SCBA and a requirement per NFPA, it allows a direct air fill to a SCBA cylinder from another SCBA/RIC emergency air system.

Wall breaching Survival skill technique where the use of mechanical force is applied to open or penetrate a wall for the purpose of escape from and IDLH atmosphere.

Water knot A knot commonly used to tie webbing. This knot is used to create a webbing drag strap that is part of the fire fighter's personal basic tools. It is constructed by tying an overhand knot in one end of the webbing and then following through the knot with the other end of the webbing.

Webbing High-strength nylon or polyester material woven in the same fashion as rope, but made to be flat. Can be constructed in a single layer (flat) or in a tube shape (tubular). When tied with a water knot, can create a drag strap.

Webbing cinch harness A technique used to rapidly make a harness around a downed fire fighter utilizing personal webbing (caution should be used as a cinch harness will squeeze the chest when tensioned is applied).

Webbing shoulder harness A technique used to make a harness from personal webbing around a downed fire fighter that does not compress the chest.

Wire cutter Hand-held device used to cut many types and sizes of wire.

Wood lathe and plaster A building process used mainly for interior walls until the late 1950s. Typically made of strips of wood covered by plaster-type cement.

Index

Note: Page numbers followed by *f*, or *t* indicate material in figures, or tables respectively.